devised, researched,
compiled, and designed by

Herbert Lindinger
Egon Chemaitis
Michael Erlhoff
Sibille Riemann
Helmut Staubach

# Ulm Design

# Ulm Design

The Morality of Objects

Hochschule für Gestaltung Ulm 1953–1968

edited by Herbert Lindinger
translated by David Britt

The MIT Press
Cambridge, Massachusetts

Herbert Lindinger

# Ulm: Legend and Living Idea

Design has become a central concern in every industrialized country, not only in its economic but also in its social and cultural guise. This has not been the case for long. It came about only after more than a hundred years of debate and conflict. The underlying impulse for this development undoubtedly came from designers, architects, and industrialists in the Benelux countries, in Germany, in Britain, in Italy, in Scandinavia, and in the USA.

Among the most influential factors in the process were two schools in Germany: the Bauhaus, in Weimar, Dessau, and Berlin in the 1920s; and after World War II the Hochschule für Gestaltung (HfG) in Ulm, which has since – like the Bauhaus – acquired an aura of myth. The "legendary Ulm school" is spoken of not only in Europe but also among initiates in America, India, and Japan: and those "initiates" include not only architects, designers, motion picture makers, and graphic designers, but sometimes painters, musicians, and poets.

It remains surprising that this extraordinarily small, almost monastic college should still be so well known, two decades after its closure. It was in full operation for just twelve years, if we leave out of account the makeshift courses of 1953 and 1954 and the last years, 1967 and 1968, when most of the effort was directed against the College's adversaries and toward a desperate struggle for survival. It can only be the extraordinary ideas and personalities involved, or the unusual nature of the whole enterprise, that keeps the memory and the presence of Ulm alive.

## International Aspirations

One thing that was definitely unusual about the HfG was its international scope. By comparison with a present-day average of around 10 percent of foreign students at German universities, the 40-50 percent at Ulm constituted a notable fact. Although most of these were from neighboring European countries, from the United States, and from Japan, the fact that the students came from 49 different countries points to a noteworthy diffusion of Ulm ideas as early as the 1960s, and to the timely and stimulating quality that enabled those ideas to transcend regional boundaries. Ulm ideas were ambassadors. Students were, after all, being drawn to a country that was not exactly well liked, eight years after World War II.

Another unusual thing was the explicit antifascist intention of the founders. The school was supported by a private foundation that bore the names of a brother and sister, Hans and Sophie Scholl, who were executed by the Nazis: names that won the school many new friends, but also served to mobilize a number of covert foes. The HfG's experimental attitudes and critical approach to society seemed to demand that it be independent of the public educational bureaucracy; and the Scholl foundation, set up as a result of a gift of money from the USA and Norway, obtained through the good offices of the US High Commissioner, John J. McCloy, was there to guarantee that independence. The College was marked by hope and unclouded optimism, long sustained by the belief that it was involved in what Max Bill called "the making of a new culture."

## The New Objectives

The Hochschule für Gestaltung viewed itself as an international center for teaching, development, and research in the field of the design of industrial products. This category included, on the one hand, those objects that are intended for domestic, manufacturing, office, and scientific use, and for the construction industry, and on the other hand the visual and linguistic vehicles of information disseminated by the modern mass media.

The HfG was divided into four departments: Product Design, Visual Communication, Building, and Information. From 1961 a motion picture department, the Institut für Filmgestaltung, was associated with these. The course of study lasted four years — one year of the Basic Course, three years in a specialized department — and could end with the diploma of the HfG. The teaching was made up approximately one-half of practical design work and one-half of lectures and seminars. The students were taught the new scientific knowledge and methods that seemed best suited to raise the general level of industrial design in the future. Some classes were for students from all departments; most were departmental. Then there were numerous guest lectures by distinguished scientists and designers. The aim was not only to produce a highly qualified designer but to foster a critical social and cultural awareness.

Alongside this teaching activity there were institutes in which research and development projects were undertaken on behalf of industry.

## Dynamics and Conflict

The HfG was planned as a place for experiment, an institution open to new hypotheses, theories, and developments. In itself, the enormous numerical preponderance of guest instructors (around 200) as opposed to permanent faculty members (20) led to a sustained dynamic, a constant state of mental unrest. The list of those guest instructors, then still young and largely unknown, now looks like a Who's Who of science, literature, and art.

Among the guest instructors were Josef Albers, Frei Otto, Karl Gerstner, Etienne Grandjean, Ralf Dahrendorf, Hans Magnus Enzensberger, Walter Jens, Josef Müller-Brockmann, Martin Walser, and Kokei Sugiura; and visiting lecturers included Walter Gropius, Ludwig Mies van der Rohe, Charles Eames, Norbert Wiener, R. Buckminster Fuller, and Alexander Mitscherlich.

At the same time, the HfG inevitably became embroiled in conflicts and contradictions. With the benefit of hindsight, we can see that conflict was part and parcel of the institution in a way that has no parallel elsewhere. Its utopias resisted the indignity of translation into reality. The experimental ethos clashed with the need to provide some kind of warrantable syllabus. The partisans of various theses wanted them not only proclaimed but also tested and carried through in practice. In so small a campus — a monastery, as some people called it — diversity of opinion inevitably came up against its limits.

## Ulm Design Philosophies

At least six very different phases define the image of the HfG in those fifteen years. Essentially, they reflect an endeavor to find a new identity for the designer who gives visible form to industrial society, but also to all conceivable life in an industrial world of commodities and media.

## New Educational Ideals

The first Ulm phase, the founding phase, covers approximately the period 1947-53. That was when Inge Scholl, Otl Aicher, Max Bill from Switzerland, and Walter Zeischegg from Austria, together with many like-minded individuals – such as the writers of *Gruppe 47* – wrestled with basic concepts, strove to raise money, and looked for suitable institutional structures and people to occupy them.

The original intention was to found a *Hochschule* — a specialized university-level institute — for sociopolitical questions, as a contribution to a new, democratic education; new educational models were to be explored on a purpose-built campus on a hill at the edge of the small city of Ulm. In the end it was decided to concentrate on the design problems of the industrial society of the future. A generous donation from the Americans made the launch possible.

## A New Bauhaus

The second phase, from 1953 through 1956, was dominated by the new Rector, Max Bill, an alumnus of the Bauhaus. As the HfG building, designed by Bill himself, took shape, it gave tangible and convincing form to the new program. Bill's professionalism, and his high reputation, lent the College international status and respect. His program, "To participate in the making of a new culture, from spoon to city" — exerted a great fascination on students who faced the spectacle of a devastated Europe. This, and his preference for the rational in the design process, left a durable mark on the institution.

The link with the Dessau Bauhaus was part of the program and was detectable in every department of theory and practice. The presence of former Bauhaus instructors — Albers, Itten, Peterhans, and Klee's pupil, Nonné-Schmidt — decisively influenced the form of the Basic Course from 1953 through 1956. Their predominantly sense-based teaching soon met with resistance from the students. Even Bill, although himself a product of the Bauhaus, saw this as a transitional solution, intended to last only until instructors with a new conception could be found. He eventually found them in the persons of the Argentine painter Tomàs Maldonado, the Dutch architect Hans Gugelot, and a former member of the *De Stijl* group, Friedrich Vordemberge-Gildewart. Maldonado and Aicher, in particular, pressed for a radical shift away from the fundamentally craft-based Bauhaus tradition, and for a reorientation toward science and modern mass production technologies. A successful initial collaboration with industry (Braun) strengthened their argument.

In their eyes, however, such a reorientation was obstructed by Bill's own philosophy of the predominance of art in design. The conflict was irresoluble. And since Bill, as Rector, had no desire to rely on gaining a majority in the planned Rectorial collective, he left the College. Under protest! That was the first major row that found its way into the press.

## Design and Science

The third phase, 1956–58, was dominated by the teaching of Otl Aicher, Maldonado, Gugelot, Zeischegg, and Vordemberge-Gildewart. These instructors tried to build a new and markedly closer relationship between design, science, and technology. This was the first manifestation of the *Ulmer Modell,* the Ulm model, which has still lost none of its relevance. The HfG evolved a model of training that aimed at giving designers a new, and rather more modest and cautious, understanding of their own role. As design was now to concern itself with more complex things than chairs and lamps, the designer could no longer regard himself, within the industrial and aesthetic process in which he operated, as an artist, a superior being. He must now aim to work as part of a team, involving scientists, research departments, sales people, and technicians, in order to realize his own vision of a socially responsible shaping — *Gestaltung* — of the environment. Under Maldonado, a new Basic Course came into being, which broke away more and more clearly from Bauhaus concepts and absorbed the lessons of perceptual theory and semiotics.

Industry began to entrust the HfG with large-scale projects, such as the Hamburg Metro; and the various departments evolved, for the first time, a methodology of design work. Under the influence of Conrad Wachsmann and Herbert Ohl, the architecture department transformed itself into a Department of Industrialized Building.

At the same time, it began to become clear that the idea of a private school was ultimately impossible to realize. The HfG's dependence on governmental sources of finance increased, and so did the influence of those sources. At the same time, the birth of the Ulm model was implicitly also the first step toward the next crisis: the hegemony of science over design.

## Planning Mania

The fourth phase covered the approximate period 1958–62. The crucial moment was when people recognized the necessity of integrating the human and social sciences, ergonomics, operational research, planning methodology, and industrial technology more completely into the curriculum. This tendency was reinforced by the fact that the designers suddenly found themselves outnumbered by the scientists in the management of the College. Among the scientists who now came to the fore were the mathematician and planning theorist Horst Rittel and the industrial sociologist Hanno Kesting. The result was a valuable deepening and scientific underpinning of the design process. It was at this time, for instance, that the British guest instructor Bruce Archer formulated the design methodology developed and practiced in Ulm from 1958 onward. At the same time, from 1958 onward, there was a growth of scientific positivism at Ulm: universal ideas *à la* Bill, such as "from spoon to city," or "building a new culture," gave way to a new scientific caution, and in place of manifestos there were now "working hypotheses."

At the same time, planning methodology took such a hold that some students made it almost a religion. It seemed only a matter of time before scientific precision, system, and the computer — already a distant presence on the horizon — would free design of all its irksome, irrational weaknesses.

The analytical preliminary studies that the students were making for their project assignments, and the idea of the "value free" approach, took such a hold that questions of morality, aesthetics, and normative values came to be regarded as positively indecent. In the end, the designers and architects at the HfG rebelled.

The teachers of design, Aicher, Gugelot, Zeischegg, Maldonado, and Ohl, had long since been driven into a sort of inner emigration, and they eventually saw that the only way to get the situation under control again was by forcing through a keenly contested change in the College constitution. The instructors who suffered a loss of power as a result of this constitutional change, and their student supporters, appealed to the public. Some sections of the press, led by *Der Spiegel,* took their side and denounced the HfG.

## The Ulm Model

The fifth phase, under Aicher and Maldonado (1962–66), was marked by an attempt at a new balance between theory and practice, between science and design. The designer's self-image was redefined. The scientific knowledge necessary for design underwent a change of emphasis toward practical utility, without any reduction in the amount of time devoted to theoretical subjects in the schedule. Additional members of the permanent faculty were Gui Bonsiepe, Claude Schnaidt, Herbert Lindinger, Herbert Kapitzki, and the French philosopher and specialist in communication theory, Abraham Moles.

The so-called Ulm model of training received its definitive formulation in this phase. The proportion of theory in the curriculum tended to increase. The theoretical part of the diploma projects was progressively shifted away from speculative and toward experimental studies. The field of operation was extended to cover the areas of mass transportation, personal transportation, and electronics. In order to reintegrate the departments, some interdisciplinary project themes were worked on.

The first ecological themes were starting to appear, and the conception of the Basic Course underwent an enormous change. In the realm of theory, the first steps were taken toward design analysis. There was a new profession to be defined, and the instructors in the Product Design department exerted considerable influence by formulating job definitions and outlining a framework for professional training.

On the other hand, there was no escaping from the fact that the financial structure of the College was heading for collapse. The first and second crises had made waves; the College's adversaries had taken heart; and by 1963 the pressure from the local Baden-Württemberg parliament, the Landtag, was taking the form of ultimatums. The older faculty members were beginning to see the catastrophe as inevitable, and they began to move their research and development activities elsewhere. The younger ones still refused to acknowledge what was happening.

## Death Throes and Collapse

The sixth and last phase of the HfG's existence (1967–68) bore the marks of impending dissolution. There were efforts to resist the inevitable, and a succession of programs and rescue plans were drawn up. A postgraduate program was worked out, and there was talk of converting the HfG into an institute of environmental planning.

The class schedule was cut back for lack of money, and the students had no way of knowing whether they would ever finish their course. In this highly charged situation the Landtag, now the main source of funding, ordered the HfG to make sure that any future proposals for its continued existence were worked out with full student participation — and this in 1968, when student representatives in Ulm were naturally seeing the future in terms of the radical demands made by students in Berlin and Paris. The result was that students and faculty simply wore each other down.

As a result, on December 5, 1968, when the Landtag voted to close the College down, its adversaries were able to represent it as having destroyed itself.

## The Continuing Relevance of Ulm Ideas

In its time, Ulm succeeded in convincing society at large that cultural aspirations were relevant not only to home life but also to everyday activities in schools, offices, factories, hospitals, and transportation. A number of designs were both pioneering and lasting achievements, such as the design concept for the Braun company, the Hamburg subway, TC 100 hotelware, the first Braun hifi system, the Kodak slide carousel (still in use), or the corporate image for Deutsche Lufthansa.

Many Ulm ideas have remained surprisingly full of life to this day, not in Ulm but elsewhere, in many countries. At over fifty universities and academies in Europe, North and South America, India, and Japan, former "Ulmers" are teaching. The curriculum that was so painfully elaborated in Ulm has become the basis of design training in many countries. Many ideas and designs have become common property and everyday reality. They were conceived and designed — as can now be seen — for the future, so that design in general has profited by the example of Ulm, often indirectly and unconsciously.

Much, of course, has become dated and now seems light-years away. We now think more critically about the limits of planning and the limits of rationalism. We give vastly more emphasis to emotion, imagination, play, and individualism, especially in private areas of life. But for the functional departments of everyday existence the Ulm approach has proved to be prophetic and, in parts, entirely relevant to the present day.

The history and the fate of the HfG might possibly have lingered in obscurity for another thirty years, and valuable sources of information might have been lost forever, had not two patrons made its cause their own.

The Stiftung Volkswagenwerk (Volkswagen foundation), famous for its extensive support of research, granted a generous subsidy to finance production of the original edition of this book.

At the same time, the Olivetti company undertook, as part of its program of exhibitions and other cultural commitments, to present an international HfG touring exhibition, in conjunction with the Institut für Industrial Design at Hanover University. It is no coincidence that this particular Italian company has involved itself with the presentation of Ulm. To all architects, designers, and typographers — to anyone involved with design — Olivetti, together with a few other resonant names, is in a sense part of the family. So it is no surprise to find that Olivetti was one of the first firms to be attracted to Ulm, and among the first to invite people from Ulm to visit its centers in Ivrea and Milan.

The editor and authors of this book are grateful to these two patrons; we are sure that the specialist public will not be alone in recognizing the value of their contribution.

It would have been impossible to document the life of the HfG without the active assistance of many former faculty members and students. There was no central archive, and the material and sources were scattered over almost every continent; some of them lay under thirty years of dust.

Finally, we are grateful for the support of the Bauhausarchiv, Berlin; the Museum of Modern Art, New York; the Centre national d'Art et de Culture Georges Pompidou, Paris; and Musashino Art University, Tokyo.

Hanover 1990

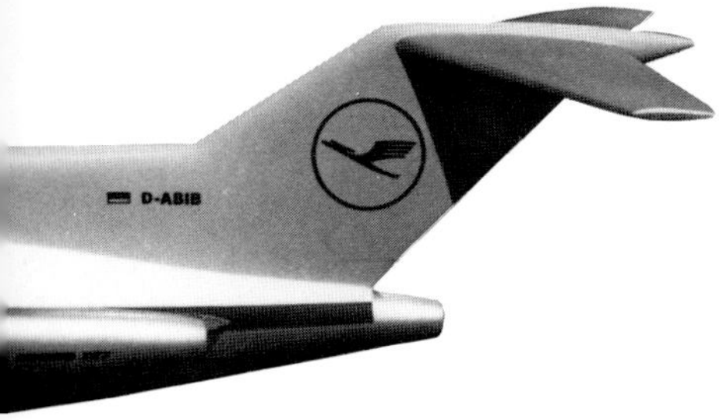


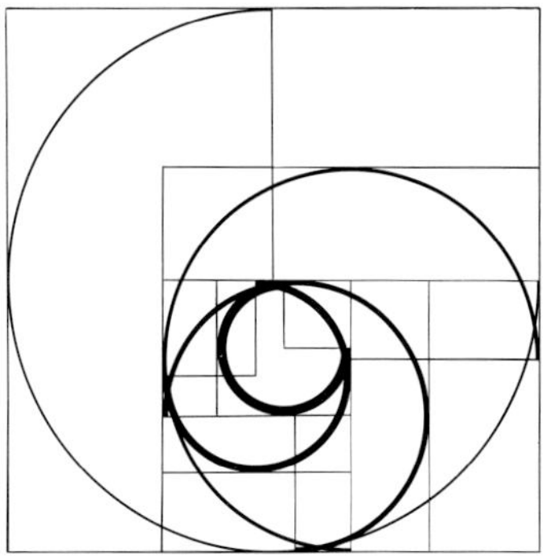

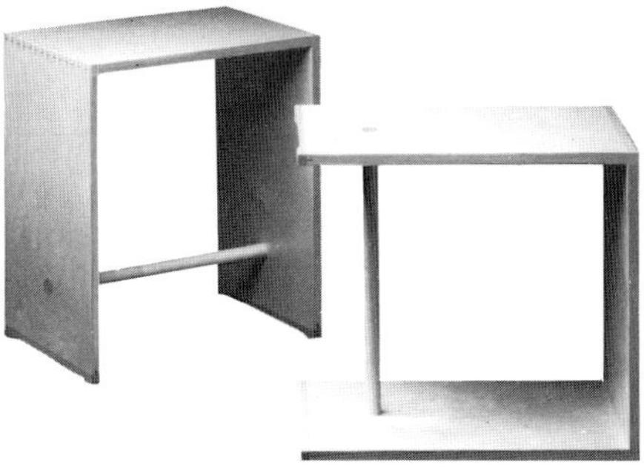

when we say grotesque we refer to a
specific form of alphabet without serifs

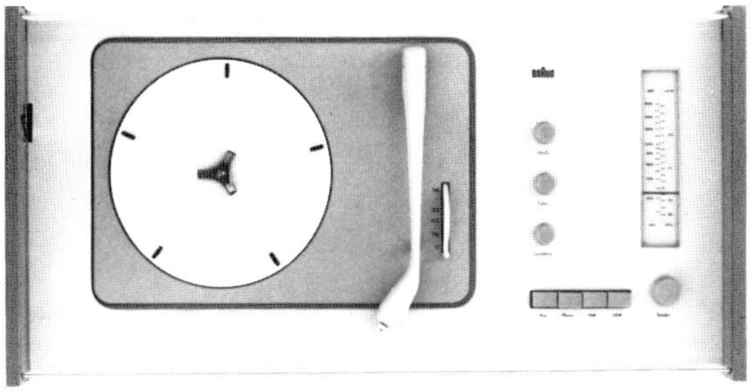

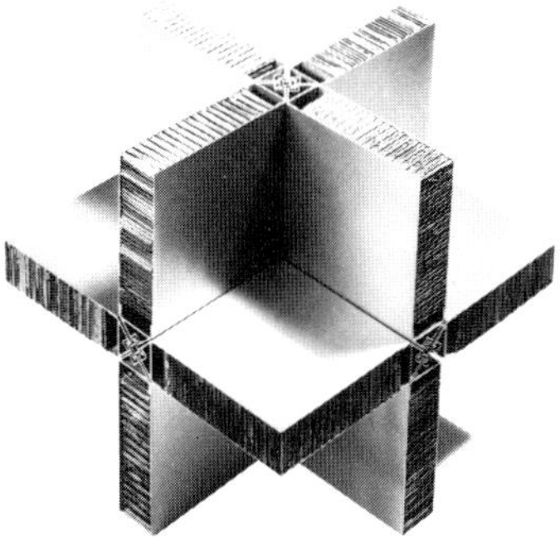

brass in modern jazz
miles davis
philips longplay
bl 286150

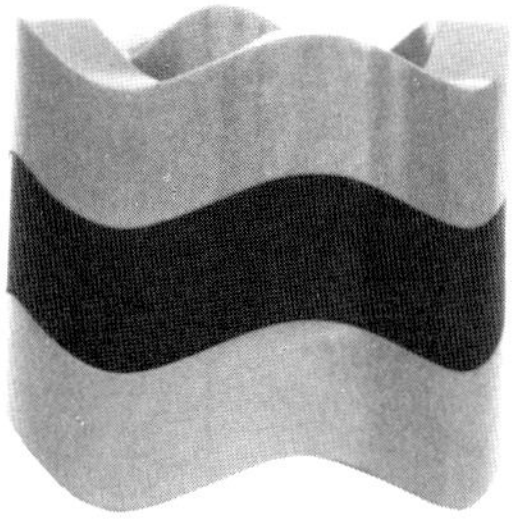

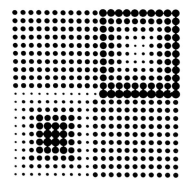

# Chronology

## The Roots of Ulm

Philosophy of the Enlightenment,
rationality, positivism,
primacy of reason

Educational ideas: Pestalozzi, Montessori,
"active learning,"
Kerschenstein: the "work college"

The culture of modernism,
Constructivism, De Stijl,
Bauhaus

Institutional concepts:
campus universities, the Pratt Institute

Wassily Kandinsky, Nina Kandinsky, Georg Muche, Paul Klee, Walter Gropius, 1926

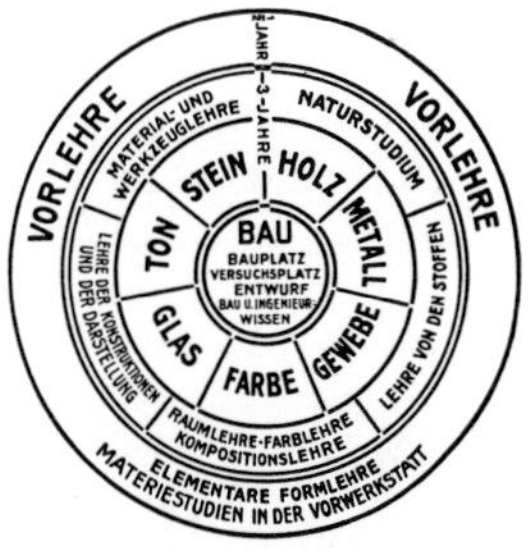

## 1943

Hans and Sophie Scholl
Members of the White Rose resistance group
Executed by the Nazis on a charge of high treason, February 22, 1943

**"Somebody must make a start somewhere. What we did and wrote was what so many people are thinking. It's just that they daren't say it."**
Sophie Scholl, before the People's Court, in Munich, February 22, 1943

## 1946

Inge Scholl founds the Volkshochschule, for adult education, in Ulm

vh

Otl Aicher and Inge Scholl are in contact with
Martin Buber, Albrecht Goes,
Marie-Luise Kaschnitz,
Franz Schnabel,
Carl Zuckmayer,
Werner Heisenberg,
Konrad Lorenz,
Carl Orff, Walter Jens,
Hans Magnus Enzensberger,
Hans Werner Richter

## The Young Seek New Paths

Max Bill's book
*Reconstruction*

Otl Aicher designs posters for the Volkshochschule

## 1947

Scholl and Aicher in contact with Max Bill in Zurich

First contacts with former members of Bauhaus (Winter, Helene Nonné-Schmidt). Bauhaus ideas are taken up

Scholl and Aicher make plans for a college that will educate its students for citizenship and democracy. Curriculum: "social and political sciences."

## A New School for a New Humanity?

First contacts with a source of finance: Norwegian Fund for European Assistance

## 1948

## Why Not a New Bauhaus?

First discussions with Bill on the curriculum of the proposed Scholl college (Geschwister-Scholl-Hochschule)

Hans Werner Richter speaks at the Volkshochschule

Herbert Wiegandt, on the beginnings of the Volkshochschule in Ulm: "In those days a light bulb was a valuable item, and if one was left unguarded it soon disappeared. And so it happened that the instructors who gave classes in the Wagner School used to fetch their light bulbs from the office beforehand, screw them in, and so become bearers of illumination in more senses than one; the purely physical light source was handed back on the following day."

## 1949

End of 1949: contacts with Shepard Stone (Information Section, United States High Commission in Germany)

Inge Scholl presents the Hochschule project to the United States High Commissioner, John J. McCloy

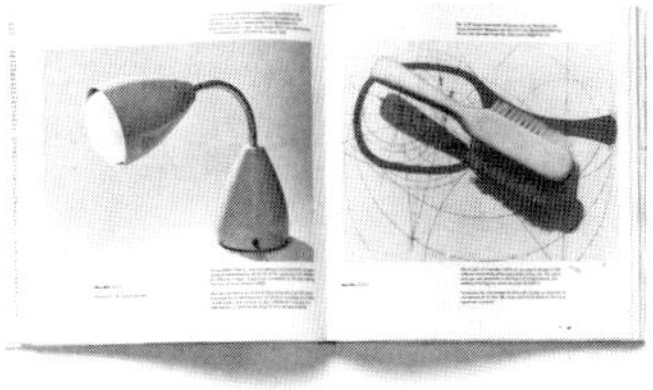

Max Bill's touring design exhibition, *Die gute Form*, shown in Basel (subsequently also in Ulm)

# 1950

"The young intellectuals of West Germany, who feel themselves responsible for the age in which they live and want to assume that responsibility — a numerically small group, but one that is important for the whole — will have a decisive influence for good or ill on the spiritual, economic, and political future.

"Here the College sees a great task to be accomplished. It intends to be a point of crystallization for a younger generation of thinking people who now lack a precise goal, or who can see no way to achieve the goals they have, and to assume their responsibility in practical terms."

(From Inge Scholl's briefing papers for her interview with McCloy)

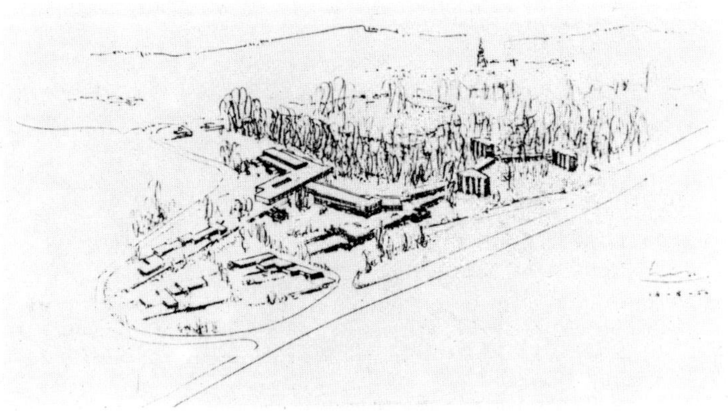

Bill pushes the curriculum in the direction of environmental design and the Bauhaus inheritance

# 1951

## "From Spoon to City?"

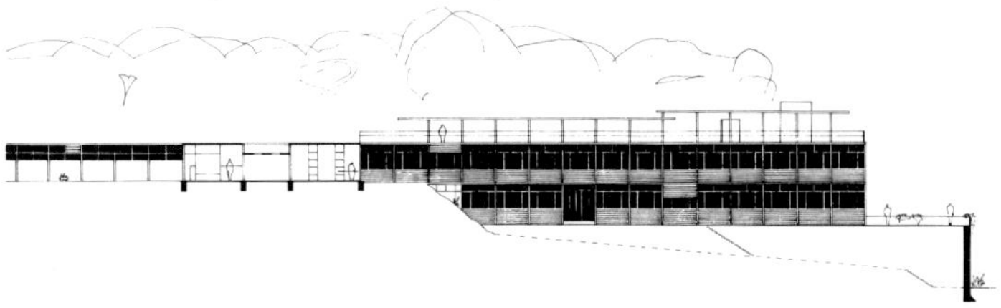

First building plans from Max Bill for a steel structure (a company offers to donate steel)

Walter Zeischegg arrives to draw up plans for a research institute

Walter Zeischegg: exhibition of handle design, *Hand und Griffe,* Vienna

May 1951
Educational policy:
College system
Student involvement in administration
Campus idea
Work in small groups
Learning by doing
Students trained to argue and justify their actions
Interdisciplinary rather than specialized training

## Not Art: Gestaltung

Hans Werner Richter withdraws

January 26, 1950
McCloy presents the plans for the "Geschwister-Scholl-Hochschule" to the United Council of World Affairs in Boston: there are to be departments of Politics, Journalism, Broadcasting, Photography, Advertising, Industrial Design, Urban Planning

Max Bill
Grand Prix of the Triennale, Milan

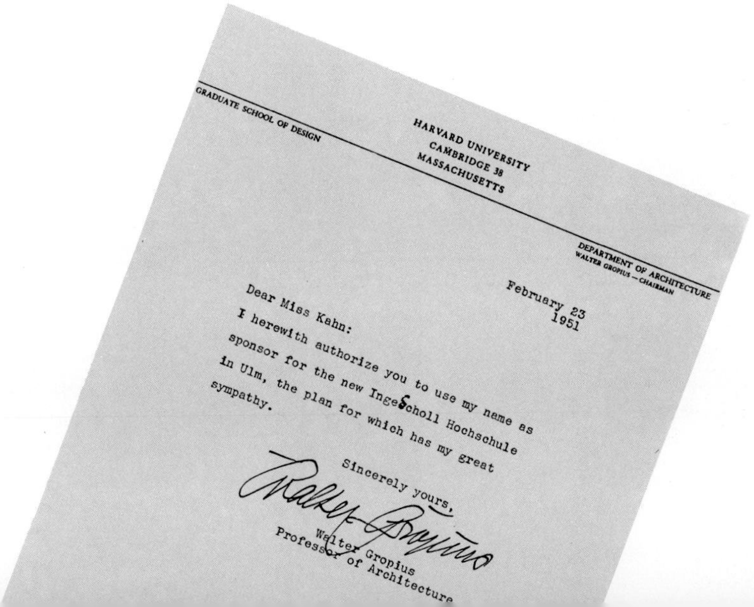

GRADUATE SCHOOL OF DESIGN

HARVARD UNIVERSITY
CAMBRIDGE 38
MASSACHUSETTS

DEPARTMENT OF ARCHITECTURE
WALTER GROPIUS — CHAIRMAN

February 23
1951

Dear Miss Kahn:

I herewith authorize you to use my name as sponsor for the new Inge Scholl Hochschule in Ulm, the plan for which has my great sympathy.

Sincerely yours,

Walter Gropius
Professor of Architecture

## 1952

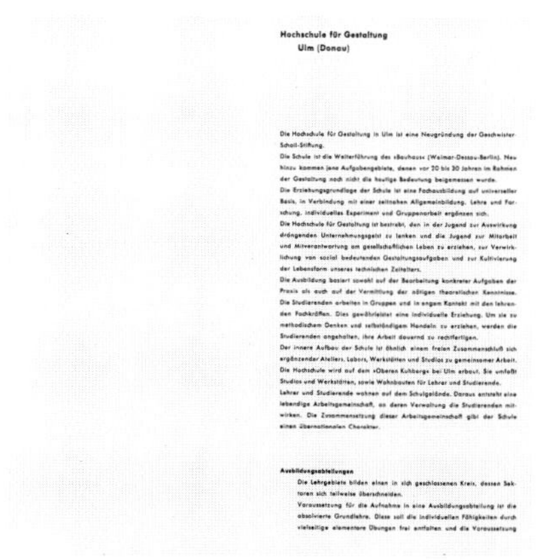

Hochschule für Gestaltung
Ulm (Donau)

Die Hochschule für Gestaltung in Ulm ist eine Neugründung der Geschwister-Scholl-Stiftung.
Die Schule ist die Weiterführung des »Bauhaus« (Weimar-Dessau-Berlin). Neu hinzu kommen jene Aufgabengebiete, denen vor 20 bis 30 Jahren im Rahmen der Gestaltung noch nicht die heutige Bedeutung beigemessen wurde.
Die Erziehungsgrundlage der Schule ist eine Fachausbildung auf universeller Basis, in Verbindung mit einer zeitnahen Allgemeinbildung. Lehre und Forschung, individuelles Experiment und Gruppenarbeit ergänzen sich.
Die Hochschule für Gestaltung ist bestrebt, den in der Jugend zur Auswirkung drängenden Unternehmungsgeist zu lenken und die Jugend zur Mitarbeit und Mitverantwortung am gesellschaftlichen Leben zu erziehen, zur Verwirklichung von sozial bedeutenden Gestaltungsaufgaben und zur Kultivierung der Lebensform unseres technischen Zeitalters.
Die Ausbildung basiert sowohl auf der Bearbeitung konkreter Aufgaben der Praxis als auch auf der Vermittlung der nötigen theoretischen Kenntnisse. Die Studierenden arbeiten in Gruppen und in engem Kontakt mit den lehrenden Fachkräften. Dies gewährleistet eine individuelle Erziehung. Um sie zu methodischem Denken und selbständigem Handeln zu erziehen, werden die Studierenden angehalten, ihre Arbeit dauernd zu rechtfertigen.
Der innere Aufbau der Schule ist ähnlich einem freien Zusammenschluß sich ergänzender Ateliers, Labors, Werkstätten und Studios zu gemeinsamer Arbeit.
Die Hochschule wird auf dem »Oberen Kuhberg« bei Ulm erbaut. Sie umfaßt Studios und Werkstätten, sowie Wohnbauten für Lehrer und Studierende.
Lehrer und Studierende wohnen auf dem Schulgelände. Daraus entsteht eine lebendige Arbeitsgemeinschaft, an deren Verwaltung die Studierenden mitwirken. Die Zusammensetzung dieser Arbeitsgemeinschaft gibt der Schule einen übernationalen Charakter.

Ausbildungsabteilungen
Die Lehrgebiete bilden einen in sich geschlossenen Kreis, dessen Sektoren sich teilweise überschneiden.
Voraussetzung für die Aufnahme in eine Ausbildungsabteilung ist die absolvierte Grundlehre. Diese soll die individuellen Fähigkeiten durch vielseitige elementare Übungen frei entfalten und die Voraussetzung

From the prospectus:
"The Hochschule für Gestaltung in Ulm is a new foundation, set up by the Scholl foundation. The College is a continuation of the 'Bauhaus' (Weimar – Dessau – Berlin), with the addition of those areas of work that, twenty or thirty years ago, were not considered so vital to the discipline of design as they are today. The basis of the teaching in the College is a professional training on a comprehensive basis, in conjunction with an up-to-the-minute general education. Training and research, individual experimentation and group work, are mutually complementary."

## "No Cash for Communists!"

After an anonymous denunciation of the Scholl family, the promised donations, including the steel for the building, are canceled

## Success!

The founder members of the HfG succeed in obtaining donations of timber and concrete

From the deed of gift:
"The United States High Commissioner hereby makes a donation to the Scholl foundation of DM 1,000,000.00 in pursuance of the shared objectives of the United States and the Federal Republic of Germany. This donation is made for the purpose of supporting the Scholl foundation in the building of a college of design, an institution that shall be conducted on an interdenominational and non-political-party basis and shall in particular aim to give a coordinated program of instruction in the areas of civic responsibility, cultural productivity, and technical skills to the end of improving the quality, form, and usefulness of the consumer goods manufactured in Germany."

## 1953

September 8, 1953
Building work starts on the Kuhberg, a hill overlooking Ulm

August 1953
The first classes are held in the Volkshochschule, on Marktplatz, Ulm, and in the Langemühle building.
Helene Nonné-Schmidt, Walter Peterhans, Josef Albers, and Johannes Itten teach 19 students

## "Life as a Work of Art"

Max Bill, 1953:
"We regard art as the highest form of expression in life, and it is our endeavor to organize life as a work of art. As Henry van de Velde once proclaimed, we want to fight against the ugly with the help of the beautiful, the good, and the practical. The Bauhaus, as the successor to the art institution founded in Weimar by Van de Velde, had the same objective. We in Ulm are taking matters further, by attaching even more value to the design of objects, by extending the scope of urban planning, by bringing the Department of Visual Design up to date, and finally by adding a Department of Information. All this arises from the natural needs of our time."

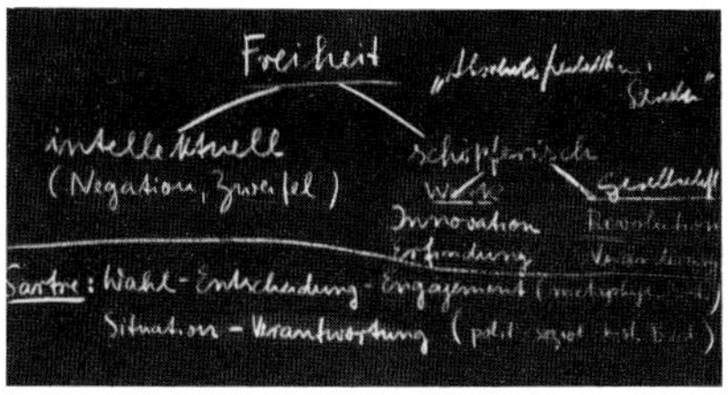


August 1953
Ludwig Mies van der Rohe and Hugo Häring visit Ulm

# Opening

July 5, 1954
Topping-out ceremony for the first section of the building

## 1954

Max Bill becomes the first Rector of the HfG

## 1955

October 2, 1955
Official opening of the HfG

From Inge Aicher-Scholl's speech:
"And, you see, the fact that this college, and its mission, have brought together so many utterly different people and challenged each of them to produce an entirely personal initiative: this marks its uniqueness and represents the profoundest justification of its existence. This building is like a victory over the weariness and resignation, the pessimism and skepticism, of our time. And this is not by chance, but because it has a task to perform on behalf of the present and of the future."

From Max Bill's speech:
"It is not every day, not every year, not every ten years, but much more rarely, that someone somewhere sets up a new institution of higher education. When something new appears, it is because that new thing corresponds to a need. And so it is here. This college answers a need, that of placing young people in the best position to further the evolution of those objects that are used in everyday life."

## On the Kuhberg: Moving In . . .

Otl Aicher's posters for the Volkshochschule Ulm win the Grand Prix at the Triennale, Milan

At work on the interior

From Walter Gropius's inaugural speech:
"A broad education must point the right way to the right kind of future collaboration between the artist, the scientist, and the businessman. Only together can they evolve a standard of quality in production that takes the human being as its measure: that is to say, one that takes the imponderables of our existence just as seriously as our physical needs. I believe in the growing importance of teamwork for the raising of the mental level of life in the democracies."

## 1956

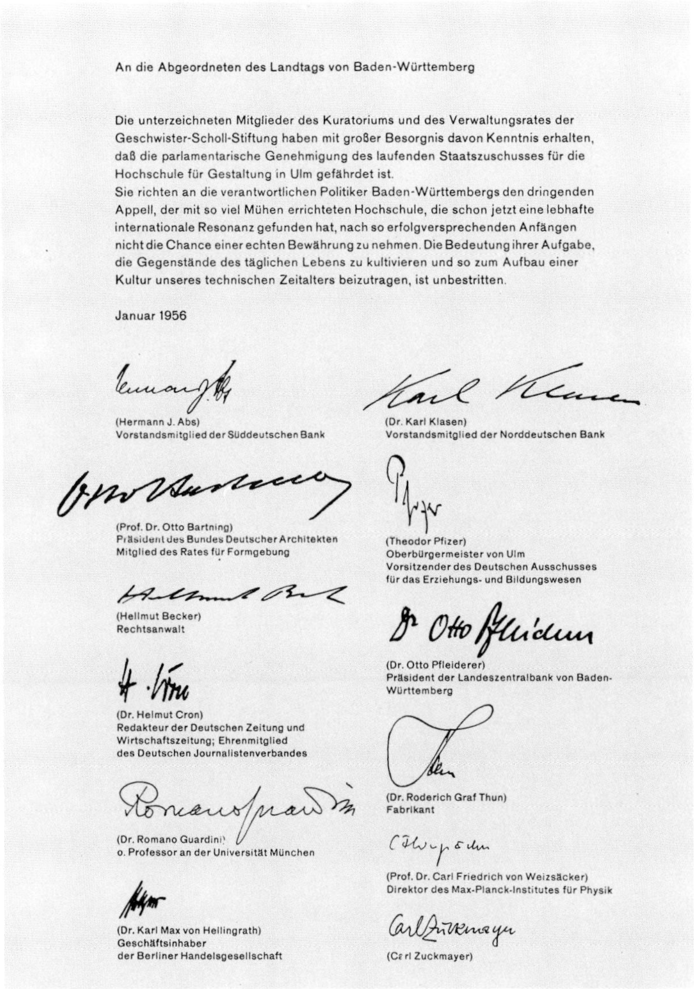

An die Abgeordneten des Landtags von Baden-Württemberg

Die unterzeichneten Mitglieder des Kuratoriums und des Verwaltungsrates der Geschwister-Scholl-Stiftung haben mit großer Besorgnis davon Kenntnis erhalten, daß die parlamentarische Genehmigung des laufenden Staatszuschusses für die Hochschule für Gestaltung in Ulm gefährdet ist.
Sie richten an die verantwortlichen Politiker Baden-Württembergs den dringenden Appell, der mit so viel Mühen errichteten Hochschule, die schon jetzt eine lebhafte internationale Resonanz gefunden hat, nach so erfolgversprechenden Anfängen nicht die Chance einer echten Bewährung zu nehmen. Die Bedeutung ihrer Aufgabe, die Gegenstände des täglichen Lebens zu kultivieren und so zum Aufbau einer Kultur unseres technischen Zeitalters beizutragen, ist unbestritten.

Januar 1956

(Hermann J. Abs)
Vorstandsmitglied der Süddeutschen Bank

(Dr. Karl Klasen)
Vorstandsmitglied der Norddeutschen Bank

(Prof. Dr. Otto Bartning)
Präsident des Bundes Deutscher Architekten
Mitglied des Rates für Formgebung

(Theodor Pfizer)
Oberbürgermeister von Ulm
Vorsitzender des Deutschen Ausschusses für das Erziehungs- und Bildungswesen

(Hellmut Becker)
Rechtsanwalt

(Dr. Otto Pfleiderer)
Präsident der Landeszentralbank von Baden-Württemberg

(Dr. Helmut Cron)
Redakteur der Deutschen Zeitung und Wirtschaftszeitung; Ehrenmitglied des Deutschen Journalistenverbandes

(Dr. Roderich Graf Thun)
Fabrikant

(Dr. Romano Guardini)
o. Professor an der Universität München

(Prof. Dr. Carl Friedrich von Weizsäcker)
Direktor des Max-Planck-Institutes für Physik

(Dr. Karl Max von Hellingrath)
Geschäftsinhaber
der Berliner Handelsgesellschaft

(Carl Zuckmayer)

Petition signed by prominent architects, bankers, academicians, writers, and industrialists, to the members of the Landtag (legislature) of Baden-Württemberg, appealing to them not to cut off the subsidy to the newly founded HfG

Five co-Rectors: Otl Aicher, Max Bill, Hans Gugelot, Tomàs Maldonado, Friedrich Vordemberge-Gildewart

Tomàs Maldonado and Georges Vantongerloo

Herbert Lindinger (student in the Department of Visual Communication) wins first prize in Paris international photographic competition

## 1957

May 1957
Inge Aicher-Scholl on goodwill tour of the United States

The firm of Max Braun AG, Frankfurt, wins the Grand Prix of the Milan Triennale for products designed mainly by faculty members of the HfG (Aicher, Gugelot)

... and Moving Out

Max Bill leaves the HfG after differences of opinion on structure of teaching and syllabus

Disagreements among the students over Bill's departure

The press is divided on the issue

## 1958

Three co-Rectors: Otl Aicher, Hans Gugelot, Tomàs Maldonado

Draft constitution:
Training curriculum more specific; increased emphasis on theoretical specialties

### Science and Design – Will It Work?

From Maldonado's address at the beginning of the academic year 1957–58:
"To set out to continue the work of the Bauhaus, in any literal sense, would mean trying to restore the past. The best of the former Bauhäusler will certainly agree that to continue the Bauhaus's work means going, in a sense, against the Bauhaus. We adopt only its progressive, anticonventional attitude, the effort to contribute to society in the specific historical situation in which we find ourselves. In this sense, and in this sense only, we are continuing the work of the Bauhaus."

ulm 1

October 1958
Charles Eames and R. Buckminster Fuller visit the College

October 1958
First issue of the magazine *Ulm* published

The President of the Federal Republic, Theodor Heuss, pays an unofficial visit to the HfG

## 1959

Three co-Rectors:
Otl Aicher, Hanno Kesting, Tomàs Maldonado

Inauguration of the Optical Perception Research Institute

New objectives of basic curriculum defined by Maldonado

### Methodology: 1 + 1 = 2?

March 1959
A provisional constitution is drawn up

Herbert Bayer and Friedrich Vordemberge-Gildewart

Mia Seeger and Professor Dr. Hirzel, of the Design Council (Rat für Formgebung), visit the HfG

HfG exhibition in the hall and canteen of the College

Max Bill exhibition in the Ulm city museum

## 1960

Three co-Rectors: Tomàs Maldonado, Herbert Ohl, Horst W. J. Rittel
From April: Gerd Kalow, Horst W. J. Rittel, Friedrich Vordemberge-Gildewart

Proportion of foreign students: approximately 40 percent
Proportion of female students: approximately 12 percent

The filmmaking sector is built up and developed within the Visual Communication Department

### Science versus Design

Conference of *Gruppe 47*.
Theme: radio drama

July 1960
HfG exhibition, Triennale, Milan

April 1960
HfG representatives attend World Congress of Design in Tokyo

Pfaff-Gritzner-Keyser AG, Karlsruhe, wins the Silver Medal of the Milan Triennale for a sewing machine (design development by Hans Gugelot, Herbert Lindinger, Helmut Müller-Kühn)

## 1961

Two co-Rectors: Gerd Kalow, Hans Gugelot
From November: Horst W. J. Rittel, Friedrich Vordemberge-Gildewart

July 1961
Faculty Conference
(entrance qualifications changed)

### A Slight Change of Course: Specialization in the First Year

Fall 1961
The Basic or Foundation Course *(Grundlehre)* is transformed into the first year of the specialized courses

"The 1960–61 Basic Course was the last in which all the students worked together . . . . Some of the so-called 'theoretical' instruction remains common to all first-year students, but what used to be called the Basic Course is now divided into four specialist areas in which students work separately both from students in other specialties and from those in the higher years."

First issue of *Output,* HfG student magazine

Hans Roericht wins the Culture Prize of the Federation of German Industry for his diploma work, "Design for Hotel Tableware"

Five students of the Department of Building win second prize in the international competition mounted by the oil company Avia International for the design of a system-built filling station

## 1962

Two co-Rectors: Herbert Ohl, Tomàs Maldonado
From October 1, 1962: Rudolf Doernach, Otl Aicher, Christian Staub

Disagreements on the degree of emphasis to be given to planning subjects

December 15, 1962
Consultative Council issues new constitution for the HfG: co-Rectors to be replaced by a single Rector: Otl Aicher

Expansion of the motion picture and television sector (Oberhausen Group)

"The motion picture is not a discipline that concerns itself esoterically with its own problems but an instrument for analyzing and changing the world around us. It consequently requires that the filmmaker make a study of aspects of that world on his own initiative. For him, the film is the medium that gives expression to his own individual free will. These requirements presuppose a student on whose education, and on whose character, great demands can be made. The student's decision to become a creative filmmaker must be a passionate one."

## "Design Is More Than Analytical Methodology"

December 19, 1962
Death of Friedrich Vordemberge-Gildewart (at the HfG since 1954, a co-Rector in 1955–57 and 1960–62, and an instructor in typography and graphic design for the Department of Visual Communication)

Max Braun AG wins the Compasso d'Oro for its house design style, to which the HfG made a major contribution

Gerhard Mayer and Heinz Wäger win third prize in an international ideas competition for a sanitary installation unit, and their design is purchased

Reinhart Butter wins the cultural prize of the Federation of German Industry for the design of a precision weighing instrument

Michael Conrad, Pio Manzú, and Hans Werner win first prize in an international competition to design the bodywork of a Gran Turismo automobile

## 1963

Rector: Otl Aicher

KULTUR

HOCHSCHULEN

ULM

Auf dem Kuhberg

Ein Blatt des Deutschen Gewerkschaftsbundes feierte sie als „Hochschule einer neuen Gesellschaftsordnung". Carl Zuckmayer nannte die Gründung „ein Ereignis, das über Deutschland und Europa hinaus ein Signum neuen Geistes ist". Der ähnlich enthusiasmierte John McCloy verhalf ihr zu einer Million Mark amerikanischer Subventionen, und Baden-Württembergs ehemaliger Kulturminister Schenkel frohlockte: „Das neue Bauhaus!"

Keine andere Akademie-Neugründung im Nachkriegsdeutschland wurde mit mehr Vorschußlorbeeren bekränzt als die der „Hochschule für Gestaltung" (HfG) in Ulm am 2. Oktober 1955.

Keine Gründung hat aber auch so viel mehr versprochen als gehalten, keine andere akademisch-künstlerische Lehr- und Forschungsanstalt wurde bis auf den heutigen Tag von so vielen Krisen und Querelen heimgesucht, von Mitgründern und Mitarbeitern so scharf kritisiert wie das von der „Geschwister-Scholl-Stiftung" getragene Unternehmen auf dem Ulmer Oberen Kuhberg.

Nach der Meinung des kleinschreibenden Züricher Malers, Architekten und Designers Max Bill — Erbauer und erster Rektor der Kuhberg-Akademie — haben „[illegible] fanta[illegible] und tüch-

HfG-Gründerin **Inge Aicher-Scholl**
Hinter kahlem Beton ...

Hochschule kaum mehr für sich in Anspruch nehmen".

Perrine, dessen von der Ford-Stiftung finanzierte „Forschungsstelle für optische Wahrnehmung" an der Ulmer Schule 1961 einging, schrieb im vergangenen Jahr an den Verwaltungsrat der Geschwi[illegible]

Designer-Akademie sei neben ähnlichen Instituten in Chicago und London eine der drei führenden Schulen ihrer Art, hauptsächlich auf das neue Image, das die Frankfurter Radio- und Elektrogeräte-Firma Braun in den Jahren 1955 bis 1956 von den Ulmer Gestaltern, vornehmlich von dem Designer Hans Gugelot, bezogen hatte.

Ruhm und Erfolg dieser kompromißlos modernen, auf internationalen Ausstellungen mehrfach preisgekrönten Formgebung trug den Ulmer Entwerfern weitere Aufträge ein: so von der Badischen Anilin- und Soda-Fabrik (BASF), von Olivetti, Krupp, den Farbwerken Hoechst, der Hamburger Hochbahn und der Lufthansa.

Für diese Form-Maßschneiderei wurden in Ulm besondere Entwicklungsgruppen installiert, deren Mehrheit heute in einem von Gestalter Hans Gugelot geleiteten „Institut für Produktentwicklung und Design e. V." in Neu-Ulm zusammengefaßt ist. Dem eigentlichen Hochschulbetrieb indes, dem Gugelot nur noch als Gastdozent verbunden ist, war eine ähnlich kontinuierlich-erfolgreiche Entwicklung nicht beschieden. Krisen-Signale im vergangenen Jahr:

▷ drei Rektoratswechsel, die Einführung einer neuen Schulverfassung und damit verbundene Mißhelligkeiten im Lehrkörper;

▷ zwei Protestaktionen von HfG-[illegible] denten, bei denen [illegible] „undemokr[illegible] polemisie[illegible]

Attacks on the HfG in the news magazine *Der Spiegel,* contrasting the success of Hans Gugelot's semiautonomous design consultancy with the discord among faculty and students within the College

## Another Change of Course

Touring exhibition of the HfG in Stuttgart, Ulm, and Munich

The financing of the HfG is endangered

The Baden-Württemberg legislature (the Landtag) presents the College with a ten-part ultimatum, including the abolition of student participation in decision making; the ultimatum is accepted and new statutes are drawn up

The Motion Picture Evaluation Center (Filmbewertungsstelle) in Wiesbaden gives a "valuable" rating to a 10-minute documentary film, *Thema Fotografie*

Declaration by faculty and students in response to the hostile press campaign:

"At the instigation of a number of members of the HfG, a press campaign has been mounted against this institution, which has had or may still have the following consequences:

"1. A reduction in the budget and consequently a reduction in the scope for development of the HfG (extensions to the building and to the curriculum).

"2. A devaluation of the diploma and of the social recognition accorded to graduates of the HfG.

"3. A reduction of public and private scholarship funds.

"4. Damage to the reputation of the HfG in business and industrial circles.

"5. A weakening of the whole approach to design promoted by the HfG and by other institutions and movements inside and outside Germany.

"We therefore protest the unwarranted defamation of this college and those who work there. We consider that the image of the HfG presented by a section of the press, based on the actions, statements, and complaints of a small group of HfG members, has no basis in reality.

"We declare emphatically that the HfG still stands for a vital concern within our society.

"It is our view that it has proved beyond any possible doubt that it deserves public support. The HfG has sufficiently demonstrated that it is making a vital contribution to the solution of the problems with which it concerns itself, not only in Germany but far beyond.

"We are firmly convinced that its significance will continue to grow in future years.

"Ulm, May 27, 1963."

In a report prepared at the request of the Baden-Württemberg government in 1963, Professors Hellmut Backer, Theodor Eschenburg, and Alexander Mitscherlich warn against a state takeover of the HfG, which they consider inadvisable in view of its experimental character and the consequent exceptional values and priorities involved

## 1964

Rector: Tomàs Maldonado

First and second prizes of the Federation of German Industry for the design of an automobile dashboard

The film *Portrait einer Bewährung,* by Alexander Kluge, wins one of the four main prizes in the documentary film category at the 11th West German Short Subject Festival at Oberhausen

The Motion Picture Evaluation Center (Filmbewertungsstelle) in Wiesbaden gives the rating "of exceptional quality" to *Unendliche Fahrt – eher begrenzt,* by Alexander Kluge and Edgar Reitz

October 1964
Dr. Mia Seeger, of the Design Council (Rat für Formgebung), visits the HfG

September 9, 1965
Death of Hans Gugelot

## 1965

The board chairman of the Scholl foundation, Thorwald Risler, resigns. He is succeeded by Dr. Friedrich Rau

May 1965
HfG exhibition at the Stedelijk Museum, Amsterdam

April 1965
Deutscher Werkbund, architecture and design convention

July 1965
HfG criticized over fund-raising activities for Vietnam

The United States Ambassador, George McGhee, visits the College

ICSID (International Council of Societies of Industrial Design) Board Meeting at the HfG

## 1966

Rector: Herbert Ohl

Otl Aicher moves his design consultancy to Munich to work on the corporate image for the 1972 Olympic Games

## Financial Position Critical

The Scholl foundation in dire financial straits; faculty posts cut back, teaching program curtailed

January 1966
Deutscher Werkbund, Baden-Württemberg section, architecture and design convention at the HfG

April 1966
Building Department hosts convention of British teachers of architecture

## 1967

June 1967
Tomàs Maldonado leaves the HfG after 13 years

June 1967
HfG students demonstrate to protest the shooting of Benno Ohnesorg

The Landtag of Baden-Württemberg demands that the HfG be amalgamated with the Ingenieurschule (college of engineering). Federal subsidies withdrawn. Financial situation untenable.

October 1967
Inaugural speech by Rector Herbert Ohl: "If the state is to take over, then autonomy must be retained."

## External Pressures Grow

October 1967
The Film Department becomes an independent institution, the Institut für Filmgestaltung

Proportion of foreign students: 45 percent
Proportion of female students: approximately 17 percent

# The End

## 1968

October 1968
The faculty members refuse to start their courses on the grounds that available resources and personal financial guarantees are inadequate; Rector Ohl announces that materials and plans for a properly organized opening to the session do not exist

November 1968
The Landtag votes to close the HfG

December 1968
The Premier of Baden-Württemberg, Hans Filbinger, declares: **"We want to make something new, and for this we need to liquidate the old."**

## The Design World and the Press Support the HfG: In Vain

Gui Bonsiepe:
"What was heroic was not the end of the HfG but the hopes that there were at its beginning. The HfG is not to be measured by what it achieved but by what it was prevented from achieving."

# The Scholl Foundation

The Scholl foundation (Geschwister-Scholl-Stiftung) was the body legally and financially responsible for the HfG. It was set up in 1950 on the initiative of Inge Scholl, in memory of her brother and sister, Hans and Sophie Scholl, executed by the Nazis in 1943. Its purpose was to set up and maintain a college of design in association with a research institute for product design.

The constitution of the foundation was the work of Hellmut Becker (later director of the Max Planck Institute for Educational Research, Berlin). He was among the earliest advisers to the founders and remained a member of the governing board until the College's closure.

The directors of the foundation were as follows: Inge Aicher-Scholl, Otl Aicher, and Max Bill (1950–53); Inge Aicher-Scholl (1954–58); Inge Aicher-Scholl, Professor Max Guther, and Thorwald Risler (1958–63); Thorwald Risler, Professor Walter Erbe, and Hans Zumsteg (1963–65); Dr. Friedrich Rau (1965–68).

The main decision-making and executive body of the foundation was its governing council, whose 14 members included representatives of the Federal government, the state of Baden-Württemberg, the city of Ulm, and public institutions. A board of trustees, consisting of well-known and influential personalities, supported the aims of the foundation in its dealings with the outside world. The administration of the HfG was in the hands of an administrative director, H. Schlensag, who was succeeded by G. Schweigkofler.

Governing Council:
Hellmut Becker, Kressbronn
Brigitte Bermann-Fischer, Frankfurt
Helmut Cron, Stuttgart
Karl Max von Hellingrath, Munich
Karl Klasen, Hamburg
Otto Pfleiderer, Stuttgart
Werner Plappert, Heidenheim
Roderich Graf Thun, Jettingen

Board of Trustees:
Hermann J. Abs, Frankfurt
Otto Bartning, Darmstadt
Walter Gropius, Cambridge, Massachusetts
Romano Guardini, Munich
Odd Nansen, Oslo
Herbert Read, Stonegrave, England
Ignazio Silone, Rome
Henry van de Velde, Ober-Ägeri, Switzerland
Carl Friedrich von Weizsäcker, Göttingen
Carl Zuckmayer, Chardonne, Switzerland

The history of the Scholl foundation and of the HfG were marked by a constant struggle to find funds for instruction, study facilities, and research.

In spite of a certain income from work undertaken with and for commercial concerns in Germany and elsewhere, there remained a constant reliance on subsidies from Federal, state and city governments. Even the public advocacy of the Trustees was ultimately unable to stabilize the foundation's precarious financial situation.

Comments on the Founding of the HfG

This Institute will be of greater importance, on a European scale, than the College of Arts and Crafts that I founded in Weimar in 1906, and the Bauhaus that Gropius developed from it.
Henry van de Velde, architect and designer

Since my first visit to war-torn Germany, I have been in constant touch with those associated with the project and have been privileged to witness the growth and work of a circle of lively, open-minded talents. Its influence may well be a significant and productive one for Germany as a whole, if it proves possible to build on the plans that have emerged from work done so far, and translate them into reality.
Carl Zuckmayer, writer

As a consequence of the total exhaustion caused by the war years, the younger generation has regrettably produced far fewer new initiatives than was the case after World War I. I am all the more pleased to hail your experiment. I am convinced that from this plan a healthy nucleus of intellectual life can evolve in Germany.
Professor Werner Heisenberg, physicist

| Yearly budgets (DM) | 1957–58 | 1958–59 | 1959–60 | 1960–61 | Apr.–Dec. 1961 |
|---|---|---|---|---|---|
| Baden-Württemberg Education Ministry | 180,000 | 180,000 | 180,000 | 180,000 | 250,000 |
| Baden-Württemberg Commerce Ministry | | | 100,000 | 125,000 | 130,000 |
| Federal Republic Interior Ministry | 75,000 | 90,000 | 170,000 | 150,000 | 152,000 |
| City of Ulm | 75,000 | 64,000 | 61,000 | 62,000 | 61,000 |
| Own income | 312,000 | 405,000 | 516,000 | 671,000 | 437,000 |
| Total budget | 642,000 | 739,000 | 1,027,000 | 1,188,000 | 1,030,000 |

| Yearly budgets (DM) | 1962 | 1963 | 1964 | 1965 | 1966 | 1967 |
|---|---|---|---|---|---|---|
| Baden-Württemberg Education Ministry | 500,000 | 600,000 | 600,000 | 600,000 | 900,000 | 900,000 |
| Federal Republic Interior Ministry | 300,000 | 269,550 | 246,000 | 200,000 | 200,000 | |
| City of Ulm | 95,000 | 150,000 | 200,000 | 200,000 | 200,000 | 200,000 |
| Own income | 602,000 | 950,000 | 636,240 | 339,654 | 649,778 | unknown |
| Total budget | 1,497,000 | 1,970,252 | 1,682,240 | 1,339,654 | 1,949,778 | unknown |

The German has a tendency to fly off into another world, and not to keep his cultural and educational world in harmony with his daily work. I therefore think that what you are proposing and planning is exactly what we need.
Hermann J. Abs, President of the Reconstruction Bank

The objectives of the Scholl Foundation seem to me to be based on ideas that are valuable and worthy of encouragement.
Dr. Ludwig Erhard, Federal Trade Minister

When we see how the inadequacy of our colleges has its origin in organizational deficiencies – in a complete failure of communication between faculties and departments – it is possible to see the intended program as the first, brave attempt to attain universally valid results on the basis of present-day priorities.
Professor Will Grohmann, Hochschule der Künste, Berlin

"The Hochschule für Gestaltung is an international center for training, development, and research in the field of industrial product design.
"The term 'industrial products' is taken to cover objects designed for everyday use and for use in production, in administration, and in science, as well as in the building industry, and also the visual and linguistic vehicles of information disseminated through the modern mass media."

The HfG was divided into four departments: Product Design, Visual Communication, Building, and Information, to which Filmmaking was added as an affiliated institute (Institut für Filmgestaltung).

The course of study lasted for four years — a one-year Basic Course and three years in a specialized department — at the end of which an HfG diploma might be awarded. The teaching consisted of practical design work, alongside classes and seminars.

The students were made aware of the scientific knowledge and methods that play a part in design work. Some classes were held in common for students from all departments; others were specific. This teaching was supplemented by guest lectures given by distinguished academicians and designers.

The training in the profession of designer was intended to be simultaneously an education in social and cultural responsibility. The teaching policy of the HfG was based on the acquisition of a solid professional competence allied with a critical faculty.

Alongside the teaching departments there were institutes within the HfG in which research and development work in product design and communications was done for various industries and organizations.

# Basic Course

All HfG students were enrolled in the Basic Course for their first year: it was a precondition of entry to any of the four departments. Its goal was to impart general basic concepts of design together with theoretical and scientific knowledge. It offered an introduction to design work, including model making and techniques of visual rendering, enhancement of visual sensitivity, and experimentation with the basic materials of formal creation. The curriculum of the Basic Course — as with the four specialized departments — contained virtually equal proportions of practical and theoretical subjects. As time went on, it was modified several times.

Initially influenced by the Bauhaus, the Basic Course moved around 1955 in the direction of establishing exact mathematical and geometrical principles that were to lead to a visual methodology.

Around 1960, questions of abstract methodology came increasingly into prominence.

In 1962, the shared Basic Course was replaced by first-year courses specific to the individual departments. In the Product Design department, for example, students continued to work on themes formerly associated with the Basic Course throughout their time in the College.

Instructors: departmental faculty members, plus Josef Albers, Johannes Itten, William S. Huff, Helene Nonné-Schmidt, Walter Peterhans, Horst W. J. Rittel

Basic Course
In the Basic Course the student is trained to approach every task with an open mind and to carry it out in accordance with the demands of its function, its social consequences, and its cultural significance. One of the major foundations of the Basic Course is mutual criticism, leading students to justify their own work. Ingrained views and preconceived opinions need to be shaken up in order to make it possible to work independently.

The Basic Course consists of:
Elementary design theory; work with color, form, and light.
Elementary expressive exercises; work with different materials and tools.
Confrontation with the political, social, cultural, and scientific issues of the day; criticism and debate, participation in general discussions and expressions of opinion.

From *Programm der Geschwister-Scholl-Stiftung,* 1952

Disrupted order
Mary Vieira, 1954 – 55
Instructor: Josef Albers

## This Book as an Object

Michael Erlhoff

Every book is an object: within its form and content it stores, and thus objectifies, the concepts and thoughts that are uttered within it. And so, like other objects, the book presents itself first of all as a unity, and then it opens to reveal specific ideas, differing functions, different sequences, and a process specific to itself, without thereby disintegrating into a mere miscellany, and without giving up its paradoxical singleness. If the book is an intelligent one, that is.

But at the same time a book is only a derivative object: a meta-object, a tool, an adjunct. It is not self-sufficient, but tells of something else; it encapsulates a large range of objects and sets itself above them. The book as a meta-object conserves (in the present case the HfG is being recorded and held fast); it consumes (the HfG becomes tangible, edifying, manageable, known, objectified); and it construes (the HfG becomes public, accessible to experience, criticizable, exposed to being known in new ways). And so this book, if it is an intelligent one, does not describe some specific objective reality: it sets forth an attackable objective reality of its own.

But this objectivity in itself is purely a dependent one; this book is not a novel (with the HfG as its ambivalent hero). It is an attempt to present the historical reality and contemporary significance of the HfG: its genesis, its structure, its theories, its influences; for the people whose names once meant "Ulm" must now get used to being a part of history.

Josef Albers teaching
1954 – 55

Paper-folding studies
Christoph Naske finds an unforeseen use for his classmates' work
Instructor: Josef Albers

Creation of textures
Ermanno Delugan, 1953 – 54
Instructor: Josef Albers

Tension/pressure constructions
Ermanno Delugan, 1953 – 54
Instructor: Josef Albers

The only conceivable way to show all this is through a contradictory coupling of disparate materials: documents from the HfG itself, contemporary asides, ahistorical as well as historical photographs, comments made now by those involved, plans and categorizing texts written then.

For this reason the present book is the result of historical, archival, and design work; it owes its shape to anecdotes and events, just as much as to objects and ideas. The result is a form of order, which imposes itself on the events and priorities as they appeared at the time, while attempting to clarify or eliminate all possible connections between actions and objects or theories and history.

In the upper part of the page will be found, in various fonts of type, documents by those involved in the HfG: texts from the magazine *Ulm* and other in-house publications of the time; extracts from interviews held recently with former Ulm instructors; and new texts by the same *dramatis personae,* written specially for this book; in addition there are illustrations of Ulm objects and sidelights on Ulm life.

The lower part of the page contains documentary material from external publications and theoretical, discursive, and expository essays written from today's vantage point on such topics as the political and intellectual situation in Germany after 1945, the theory of the object, the Basic Course, systematization and industry, the cultural context, the aftermath. No attempt has been made to achieve balance in these

Color transparency
Eva-Maria Koch, 1954 – 55
Instructor: Josef Albers

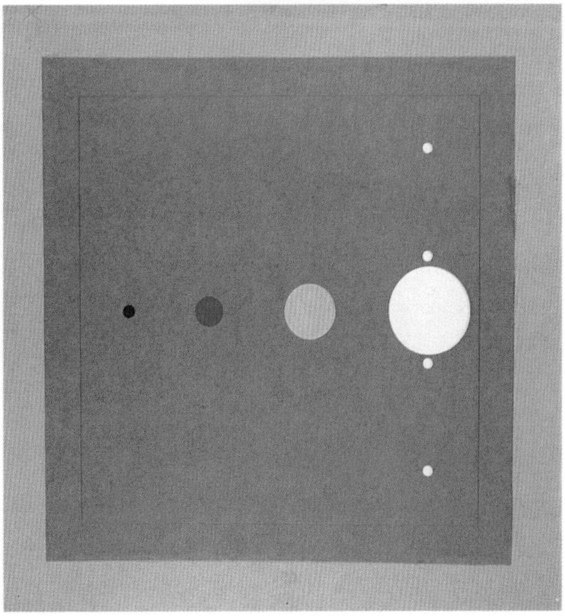

*right*
Changing the effect of color by quantitative means
Maurice Goldring
Instructor: Johannes Itten

*left*
Arrangement of elements according to rule
Almir Mavignier, 1953 – 54
Instructor: Max Bill

essays, which view the HfG both as a historic institution and as a present-day issue. They are therefore different in character from the documents, but neither more nor less subjective.

So the reading of the book can linger in the upper part, as it were, or it can dig into the lower part; or it can skip back and forth, without expecting either part to be a direct explanation of the other. The book takes shape in the head.

As this book is both an object and a meta-object, it comes to fruition only when handled by its readers; and they will use it as a handle to gain an understanding of its subject, the Hochschule für Gestaltung Ulm.

In the Basic Course the students in all the areas taught by the College acquire a common practical and intellectual foundation: a capacity to work creatively and make independent judgments irrespective of tradition, and an understanding of the cultural issues that face our age.

Work in the Basic Course is divided as follows: visual introduction, means of presentation, and practical work; to this is added cultural integration, which is also part of the work done later in the departments.

The visual introduction consists essentially of pieces of work and experiments that the students carry out in accordance with a clearly defined brief, and that they then have to justify in discussion. These exercises are intended to make the student conscious of the different visual resources (color, space, form) and lead on to a methodical, controlled process of work. Linked with this there are systematic analyses based on current scientific principles (theory of perception, morphology, topology, semantics). Through exercises, and through the exploration of methods of presentation, such as photography, writing, and drawing, both freehand and technical, the students learn to present their work. In the workshops elementary practical tasks are set, which are solved with the aid of various materials (wood, metal, plaster) and techniques, in order to develop an understanding of materials and manual skills.

From *HfG-Info,* 1955 – 56

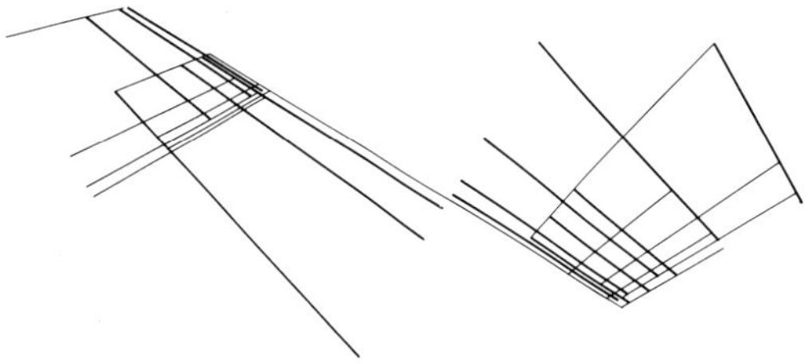

Geometric rules
Almir Mavignier, 1954 – 55
Instructor: Hermann von Baravalle

Finding new kinds of texture
Richard Rau, 1953 – 54
Instructor: Walter Peterhans

Exhibition at the Ulm museum: works by Hans Conrad
On the bench: Helene Nonné-Schmidt

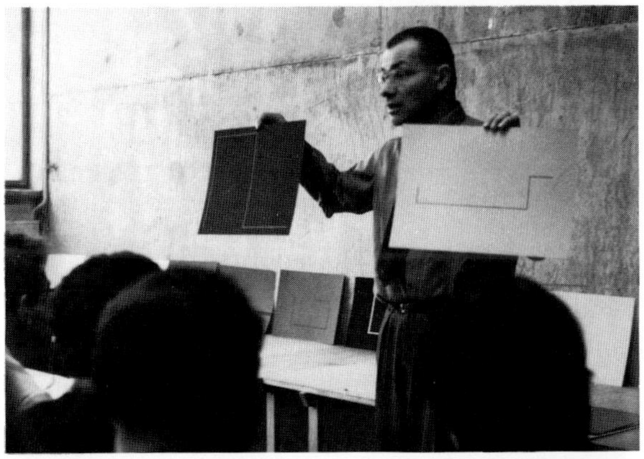

Max Bill teaching

**A curve that divides a space, but not for the curve's own sake (as ornament), that interrupts itself to let the space flow back together, or that simply vibrates like a membrane in space, working in space, binding it together and densifying it.**
Project assignment, Walter Peterhans

Walter Peterhans

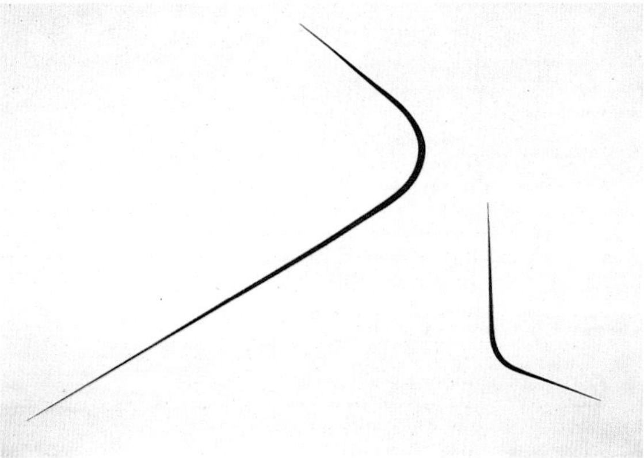

Line and space
Peter Gautel, 1953 – 54
Instructor: Walter Peterhans

# Between Utopia and Reaction

Michael Erlhoff

A Situation Report on the Postwar Years in Germany

As late as the spring of 1945, a printed flyer in Germany carried the menacing message: "We shall never capitulate! We shall come again! Long live the Führer!"

But the Führer killed himself at the end of April 1945, when much of Berlin was already occupied by the Red Army. On May 2 Berlin capitulated, and on May 9 the total, unconditional surrender was signed. Germany was liberated under Allied occupation. The rest of the dates are known: the Potsdam Conference, the Four-Power Agreement in London, the setting up of an Allied Control Commission, and more conferences; then, in 1947, the failure of the Moscow Conference and open conflict between the Western Allies and the Soviet Union (confirmed while the conference was still in progress by the so-called Truman Doctrine, and later known as the Cold War). One consequence of this conflict in mid-1947 was the Marshall Plan, which combated public, and hence potentially state, ownership by translating the desire for a "free enterprise economy" into economic directives (which the American Military Governor, General Lucius Du Bignon Clay, then enshrined in the constitutions of the Länder, the states that made up the Federal Republic). Then came the currency reform, and the founding of the two German republics.

The attitude of the majority of the German population is also well known. Liberation

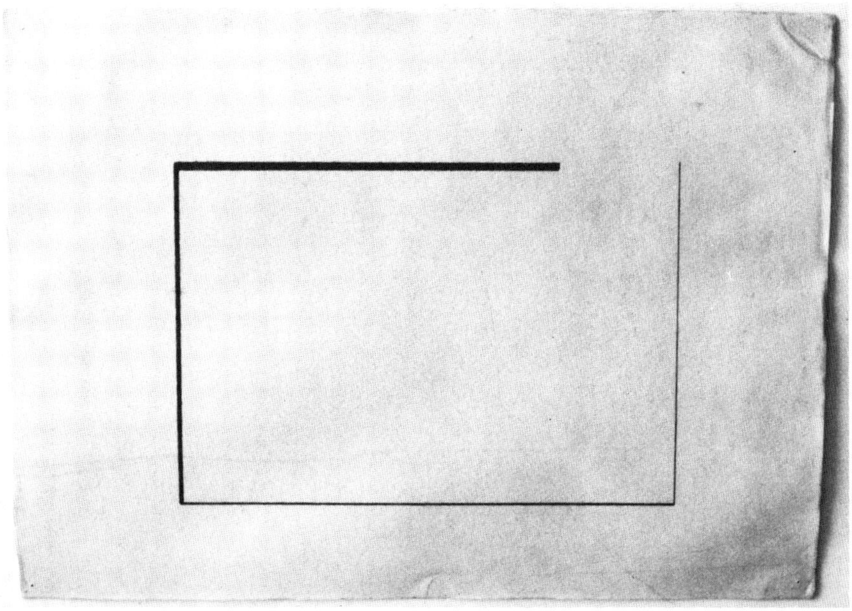

Distribution and differentiation of three lines on a surface
Ermanno and Ello Delugan, 1953 – 54
Instructor: Max Bill

Arrangement of six elements according to rule
Herbert Lindinger, 1954 – 55
Instructor: Max Bill

was regarded as a defeat (just as people still talk to this day of the "end of the war" rather than the "end of Nazism"); there is some logic in this, in the sense that the Germans had not liberated themselves but had been liberated by others. They came to terms with the Allies with the same astonishing equanimity with which they had endured German Fascism. The rubble was diligently cleared away, and reconstruction was embarked on (a few years later an advertising slogan ran: "*Freudig schaffen, froh geniessen*": "Joyfully make, happily enjoy"). Heads were plunged deep in the sand. "After the rain the sun comes out, after the tears it's time to smile," prophesied a hit song of the day, and in 1949, with great success and amazing innocence, Jupp Schmitz sang: "*Wer soll das bezahlen, wer hat das bestellt*": "Who has to pay, who ordered that?"

The majority of a nation repressed the past, as their own deep-seated principles demanded that they should. And so, on October 21, 1945 — during the Nuremberg trials — a returned exile, Stefan Heym, wrote publicly: "To close one's mind to unpleasant facts is a defense mechanism of the human psyche." Those very trials, and the "denazification" process that was inevitably no more than a formality, initiated a tendency to personalize German history between 1933 and 1945 and lent credence to obscurantist theories whereby the nation had been "seduced" into "unreason."

This is a process that has

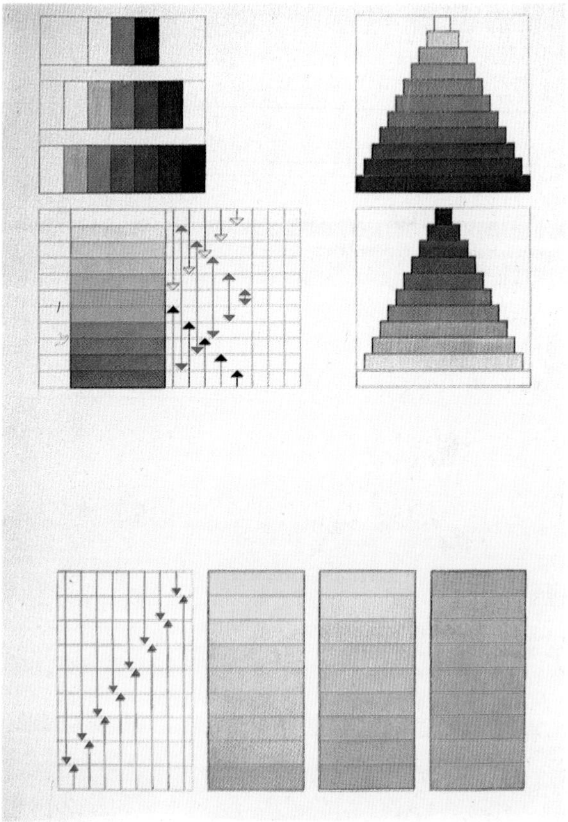

Color Theory, after Paul Klee (watercolor technique)
Instructor: Helene Nonné-Schmidt

*top*
Bernd Meurer, 1956 – 57

*above left*
Klaus Franck, 1955 – 56

*right*
Hans von Klier, 1955 – 56

never been fully understood. But how could it have been otherwise? Any radical analysis of the historical facts and of the circumstances behind them would have implicated the Allies themselves. And how could there have been a change in consciousness after 1945 without such a radical analysis and criticism? New ideological concepts were available, but the language remained the same: in 1948, as Reinhard Lettau shows, the daily press was still blandly using phrases like "pitched battle of debate" and "clenched style," demanding "sound attitudes," and "grilling" people. The language was simply a reflection.
This too should be known: countless people were murdered in Germany between 1933 and 1945 — political opponents, critics, Jews, Gypsies, intellectuals. Very few people resisted, openly or secretly: Hans and Sophie Scholl, a few pastors, some nameless individuals. Quite a few simply kept their heads down without involving themselves: the conformism represented by the fictional character of Serenus Zeitblom, Ph.D. And many had left the country, to live in some foreign environment where they were only reluctantly tolerated; in 1945 they faced the question of a possible return to Germany. Many did return, hesitantly, but with the hope that they could build a new, democratic Germany. Together with the few who had stayed in Germany and had not been corrupted by the Third Reich, they embarked, not on a recon-

Helene Nonné-Schmidt and student William S. Huff

Johannes Itten discusses the results of his 1954 – 55 color course

the point of the basic course is to overcome the opposition between pure knowledge and habitual action. on a basis of practical exercises and allied systematic investigations the theoretical bases of new methods of design will be laid.

the basic course embraces four areas of work. visual introduction. training and experimentation on the phenomena of visual perception (color, form, space). means of presentation. practice and analysis of the elementary methods of presentation (photography, writing, freehand and technical drawing) practical work. practical introduction to the manual techniques (wood, metal, plaster) and analysis of design media. cultural integration. lectures and seminars on contemporary history, present-day art, philosophy, cultural anthropology, morphology, psychology, sociology, economics, and political sciences.

Principles of the Basic Course, formulated by Tomàs Maldonado, 1955

The breathing and meditation exercises that were compulsory at the beginning of every Itten class, 1955

struction but on a new construction, founding magazines and educational institutions: among them were Inge Scholl, Hans Werner Richter, Stephan Hermlin, Hans Mayer, and Alexander Mitscherlich.

In retrospect, one is surprised and impressed by the confidence that then existed in the possibilities of social change, and the enthusiasm with which people were thinking through the process of democratizing Germany: this was a democratization from the top down, seen in terms of education and enlightenment. A symptom of this is the number of magazines that were founded in Germany after 1945, and their titles: *Building, The Other Germany, Change, Sowing, The Ferry, The Golden Gate, The Portal*: optimism, utopianism, a new-start mentality, an attempt to overcome the crippling burden of history. People believed in the potential of the here and now; some sort of a social catharsis seemed possible. With grand gestures, gates were thrown wide open, paths were pointed to, freedoms were dreamed up – soaring beyond the immediate material squalor and the perennial difficulty of finding a point of reference, and hence a usable vocabulary, anywhere in the German past.

These ideas and activities may have borne an uncanny resemblance – in form and style, if not in content – to the initiatives that were being taken by the Allies; they may, indeed have been controlled by them; but they did have a progressive force. They created nuclei of ideas and hard facts

The "Ulmers 1955".
from left:
Gugelot, Vordemberge,
Aicher, Inge Aicher,
Maldonado, Bill, Wachsmann.
Instructors,
Workshopleaders,
Secretaries and Students.

that pointed beyond the increasingly reactionary policies of the Allies themselves and the astounding diligence of the German population and of its later Federal administration. Here and there they had successes: Inge Scholl's achievements bear witness to this, as does the existence of *Gruppe 47*; and so do many ideas that came up in the 1960s as revivals of initiatives taken in these early years.

It would be wrong to take too rosy a view of that time, even so. The Allies imposed their own principles of democracy, and of economics above all, which overrode any utopian idea of a free society: issue 17 of the magazine *Der Ruf* (i.e., *The Call*), edited by Alfred Andersch and Hans Werner Richter, was banned by the Americans, and both editors had to leave, whereupon Richter failed to get a license for a new satirical magazine and assembled *Der Ruf*'s dissident contributors at a meeting in September 1947; this was the beginning of *Gruppe 47*. The bankers, teachers, professors, factory owners, judges, and others who had been in responsible posts under Fascism mostly went back to them; not so very much had changed in real terms. Not only that, but people went on talking as if they still had a German language to use; they went on acting as if everything were normal, and as if survival were the only problem; they pursued security and comfort, as if they thought only of getting settled.

What is more, there evolved an explicit political and intellec-

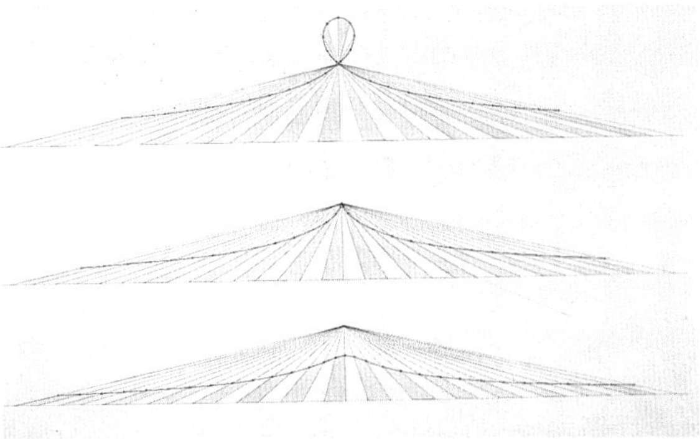

Hermann von Baravalle teaching, 1955 – 56

Geometric studies (conchoid curves) Dominique Gilliard, 1955 – 56 Instructor: Hermann von Baravalle

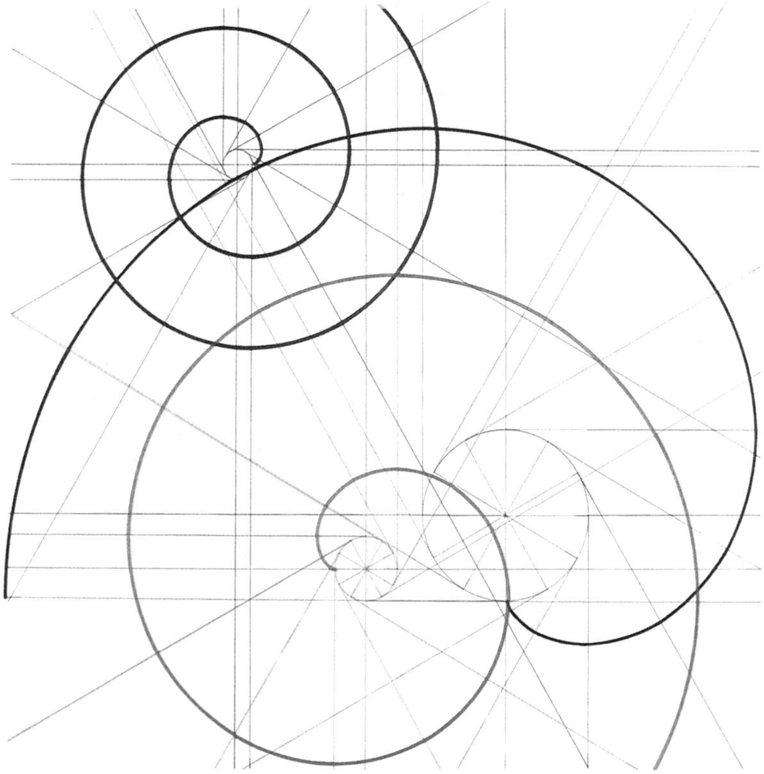

Constructional geometry Eva-Maria Koch, 1954 – 55 Instructor: Hermann von Baravalle

tual climate in which values were once more for sale to the highest bidder, and society's attempt to repress historical fact could be dressed up in ideological guise. Ideals of purity were set up, as if history had not babbled enough of hygiene in the past; "language as art," the "little verse school," and "heightened existence" were praised, for all the world as if there were really something there to praise. What mattered now was absorbing skills, rapid asset growth, neat solutions, and nonstop expansion. With ponderous pragmatism, official thinking cut itself off, ideologically, from what it called "ideology" — by which it meant the irrational. Culture became confetti, showered on the complacent head of reconstruction. The confrontation with German history was neatly packaged — we are still seeing this today — and traded in for anti-Communism; it became, at best, a matter for academics.

The 1950s began, and brought with them the collapse of ingenious projects, magazines, and ideas on all sides: an intrepid few went over to East Germany — where they were soon, and very publicly, destroyed — and just a few projects survived, not by any means unscathed.

Remember: the circle that surrounded Inge Scholl and Otl Aicher in Ulm included Hans Werner Richter, who was disliked by the Americans; and he was involved in the foundation phase of the Hochschule in Ulm. But instead of an Ulm College of Democratic Education or Awareness, whatever

Admission to the College is an initiation rite: the students cut each other's hair. The haircut is the first sacrament. A very short haircut. Very functional and rational: the same length all over the head, just as the hair grows the same length all over. A very monastic hair-style. The second step is the renunciation of capital letters. Not on historical or linguistic-political grounds. On functional grounds. Capital letters are a distraction to the hand and the eye. At Ulm they write in lowercase. The third stage: the loss of the family name, loss of one's burden of origin. Everyone has only a given name. At the same time, conventional habits of address are lost: the familiar "thou" [Du] instead of the formal "you" [Sie]. The last stage: a revolution in mental function. Thinking and feeling are stripped down and reassembled. Mainly through the constant pressure to give a reason for everything. But everything.

Bernhard Rübenach, "Der rechte Winkel von Ulm" (The Ulm Right-Angle), radio documentary, 1959

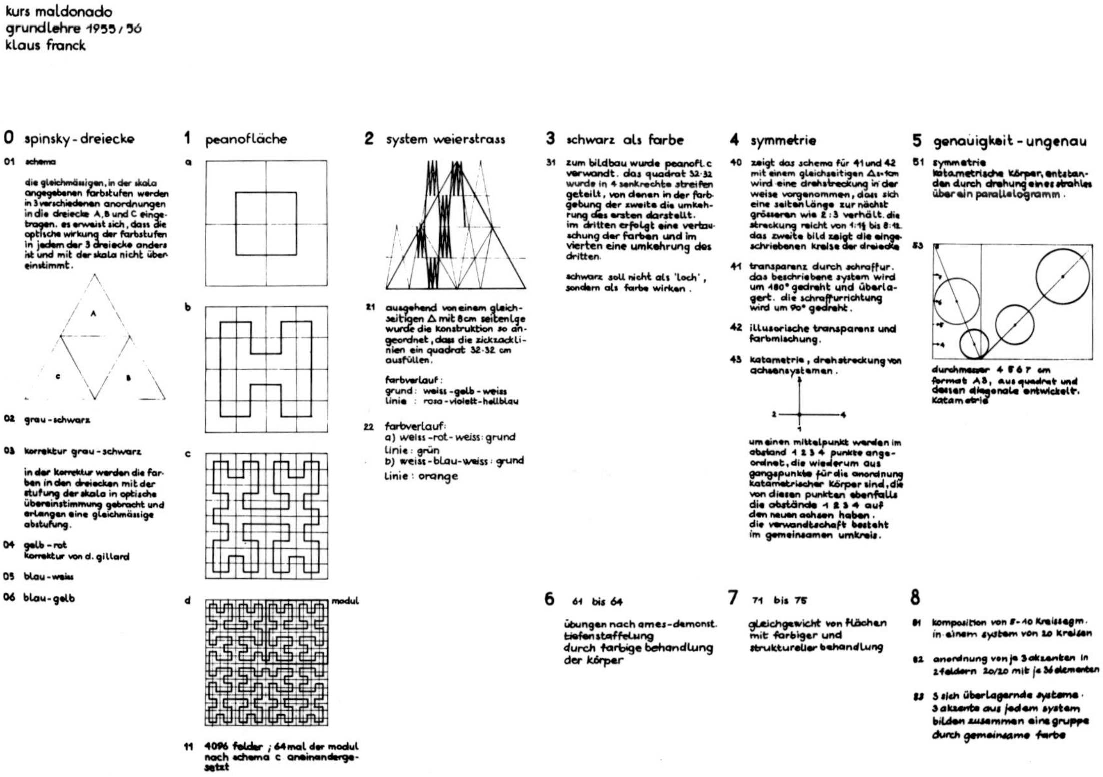


Maldonado course assignment
Klaus Franck, 1955 – 56
Instructor: Tomàs Maldonado

that might have led to, there emerged a College of Design, whatever that led to. Instead of Hans Werner Richter as Rector — which had obviously been a distinct possibility at one time — the neutral Swiss architect and artist, Max Bill, became the founding Rector and the man who built up the HfG.

Or one might extend the catalogue of contradictions, and ask why the American occupation authorities — and those closest to High Commissioner McCloy, in particular — chose to fund the setting-up of the HfG with what at that time was an unimaginable sum of money; or why the same circles were obviously involved in the refounding of the Social Research Institute in Frankfurt.

One thing that is certain is that they did not intend these institutions to be what they then became: centers of intellectual debate and dissent, test beds for new kinds of interaction and intervention.

Two things therefore happened. Patronage ran ahead of itself and afforded — like the proceedings of the Nuremberg Trials, the process of denazification, and the foundation of the Federal Republic itself — not only exoneration and rehabilitation but also the scope, and the urgent need, for resolving conflicts, informing the public, and evolving new patterns of thought.

At the same time, those same institutions were shaped by their own contradictions; they suffered from the dictation, and the inbuilt restrictions, that they received from their American (and also German) the basic course pursues four objectives:

1. it introduces the students to the work of the departments, and above all the methods on which the departmental work is based.

2. it makes the students familiar with the major issues that face our technological civilization and thus sets the context for the practical design work they will have to do.

3. it provides a training in interdisciplinary collaboration and thus prepares the students for teamwork: working in groups of specialists each of whom must understand the issues and perspectives inherent in the work of all the others.

4. it evens out disparities in prior training that arise from the fact that the students are not only from different specialties but from different countries with different educational systems.

From *HfG-Info*, 1958 – 59

Imprecision with precise means
Adolf Zillmann, 1955 – 56
Instructor: Tomàs Maldonado

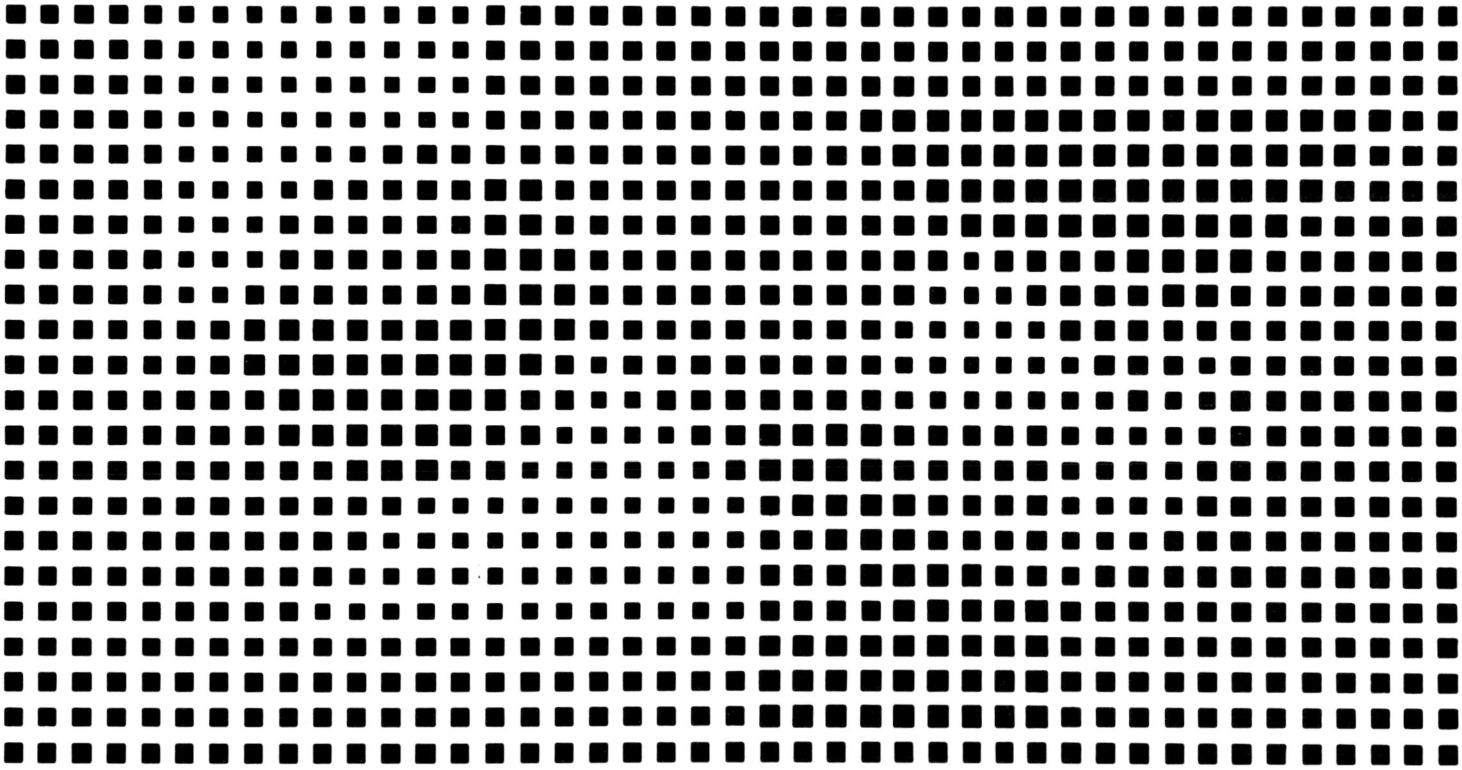

sponsors. Which is why the early phase of the existence of the Federal Republic of Germany was so full of dichotomies.

Dependence of color effects
on ambient color shown on
a Peano surface
Hans von Klier, 1955 – 56
Instructor: Tomàs Maldonado

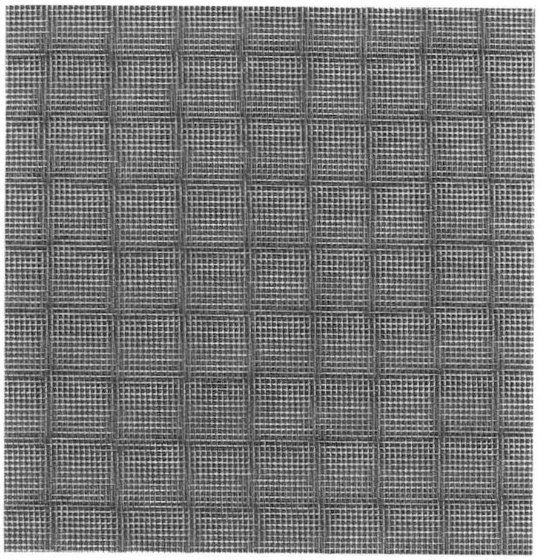

*left*
Black as a color
Combination of contrasts of brightness in which black functions not as a hole but as a color
Urs Beutler, 1956 – 57

*right*
Interference patterns
Superimposition of colored structures
Gunnar Jonsson, 1955 – 56
Instructor: Tomàs Maldonado

*left*
Imprecision produced by exact means
Dominique Gilliard, 1955 – 56
Instructor: Tomàs Maldonado

*right*
Color studies on Spinsky Triangle
Andries van Onck, 1955 – 56
Instructor: Tomàs Maldonado

In considering the basic course exercises that I proposed to students of the Hochschule für Gestaltung during those years, it is above all interesting to observe the use of the curves of G. Peano and W. Sierpinski, which today are a very important part of fractional objects developed by the mathematician Benoit Mandelbrot.
Tomàs Maldonado

Tomàs Maldonado teaching

Otl Aicher teaching

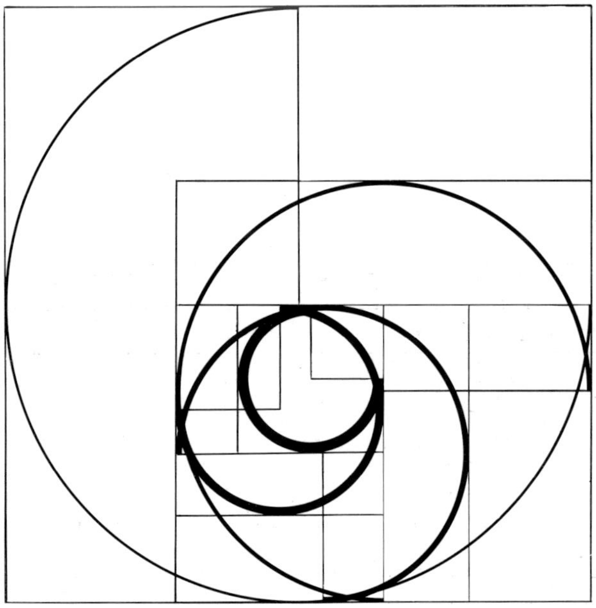

Spiral construction
Marcel Herbst, 1958 – 59
Instructor: Otl Aicher

In 1957 I came to Germany on a DAAD (German Academic Exchange Service) scholarship to work in the electronic music studio at West German Radio, in Cologne. Soon after that I made contact with Tomàs Maldonado — for whose periodical *Nueva visión* I had written in Buenos Aires — and visited him in Ulm. My first impression, in the icy cold of February 1958: a spartan life, marked by a pragmatic kind of idealism and the quest for an aesthetic whose realization must be of relevant practical

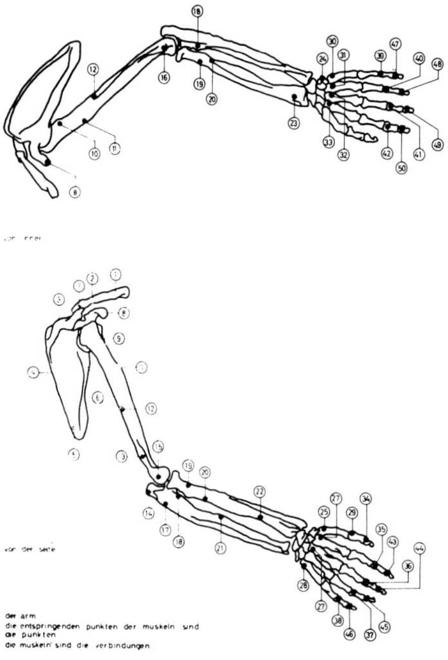


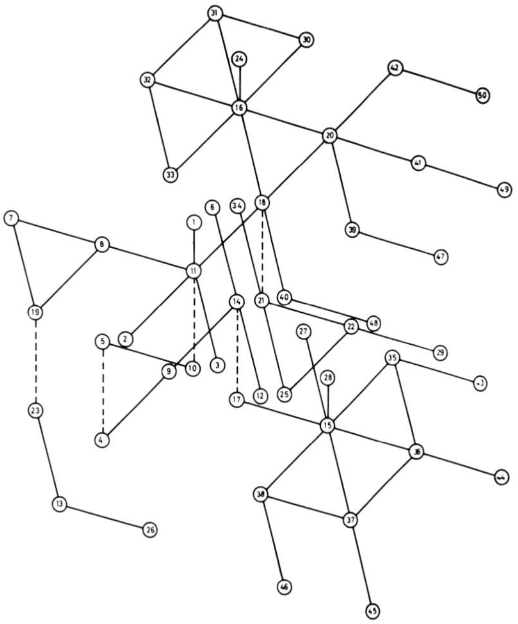

Theory of graphs
Visualization of sequences of movements and functions in graphic representation
John Lottes, 1958–59
Instructor: Anthony Fröshaug

Anthony Fröshaug and students

Seminar on the terrace, 1956

. . . so wherever man becomes fully conscious of his humanity he becomes able, not only to make his environment a technological one, i.e., arrange it so as to supply his basic material needs, but also to give form to his environment. this calls for aesthetic arguments, it calls for the arts, the integration of the arts in daily life . . .

what the ulm program lacks is precisely the decisive components of aesthetic training, namely "training and experimentation in the field of the phenomena of visual perception," "practice and analysis of the elementary methods of presentation." it is these, not an "introduction to departmental work," that are the indispensable foundations for working in a college of design.

Max Bill, "Der Modellfall Ulm," *Form*, 6, 1959

use to society. The instructors and the students pursued the objective of a comprehensive visual education that would one day achieve the right design for every one of the objects that surround us. The ideological basis of this project was open to attack, and there were tensions as a result. But then, is there anything that cannot be called in question? The practical legacy of the Bauhaus to the new college was a reputation for pioneering new teaching methods — a legacy that was not without its burdensome side. The essential point is that the HfG was set up at just the time when its existence was essential for a transformation of the educational landscape in the Federal Republic. All that counts, in the ultimate analysis, is this specific historical role and the consequences that flowed from it.

Mauricio Kagel, 1987

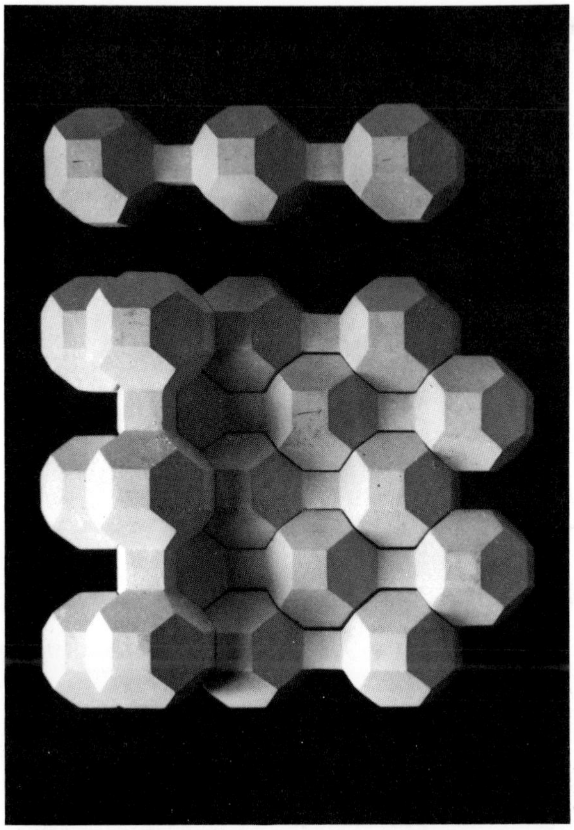

*right*
Stacking forms — completely filling space
Herbert Falk, 1959 – 60
Instructor: Walter Zeischegg

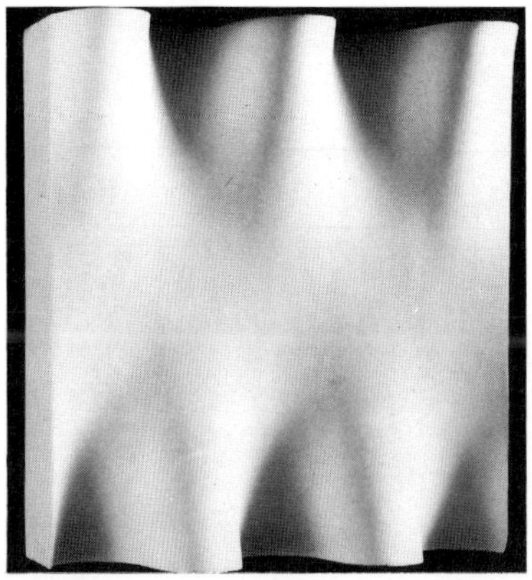

*middle left*
Sine plane
Gerhard Mayer, 1959 – 60
Instructor: Walter Zeischegg

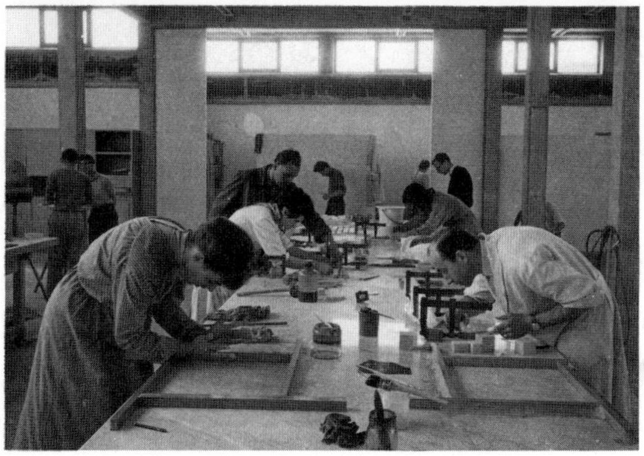

*right*
Plaster workshop
1955 – 56

*left*
Metal workshop
1955 – 56
Workshop leader Cornelius Uittenhout with student Elke Koch-Weser

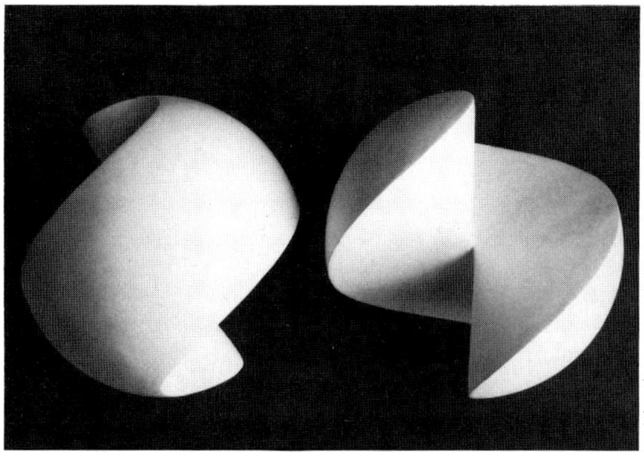

Division of a sphere into equal halves
Eduardo Vargas, 1957 – 58
Instructor: Tomàs Maldonado

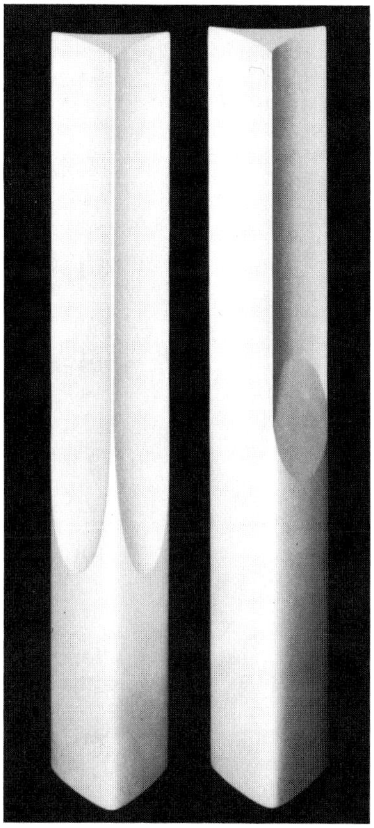

*left*
Studies in transitional form
Reinhart Butter, 1958 – 59
Instructor: Georg Leowald

*right*
Nonorientable surface
This sphere consists in principle of one modified single-sided surface only
Ulrich Burandt, 1958 – 59
Instructor: Tomàs Maldonado

# Ulm as a Model of Modernity

Michael Erlhoff

At a time when there is widespread rejoicing at the supposed demise of the modern, post-Enlightment age and the dawn of a new postmodern age, it is worth defining the Hochschule für Gestaltung in Ulm, not only implicitly — as is done in the other essays in this book — but also explicitly, as a milestone in the modern adventure.

This would make it possible to come closer to an objective definition of the paradox of Ulm, its tendency to oscillate between apologia and critique (although of course it is impossible even to sketch a theory of modernity here; this is a subject that has been discussed at length by writers better qualified to do so: Talcott Parsons, Max Weber, Karl Marx, Jürgen Habermas, and others). The HfG's dilemma is rooted in the inescapable paradox of the modern itself: the modern age possesses a dynamic of its own and at the same time demands commitment because it posits the issue of freedom and unfreedom.

If, at some risk of overcompression, we define the modern age by Max Weber's attractive formula, "the interpenetration of ethics and world," we still must deal with a number of other categories that are inextricably bound up with the concept. These include the following: subject and object; work; the experiential and the conceptual (as entities that define and potentially or ideally mediate both object and subject); morality and function (modernity's secular claims to legitimacy); public awareness

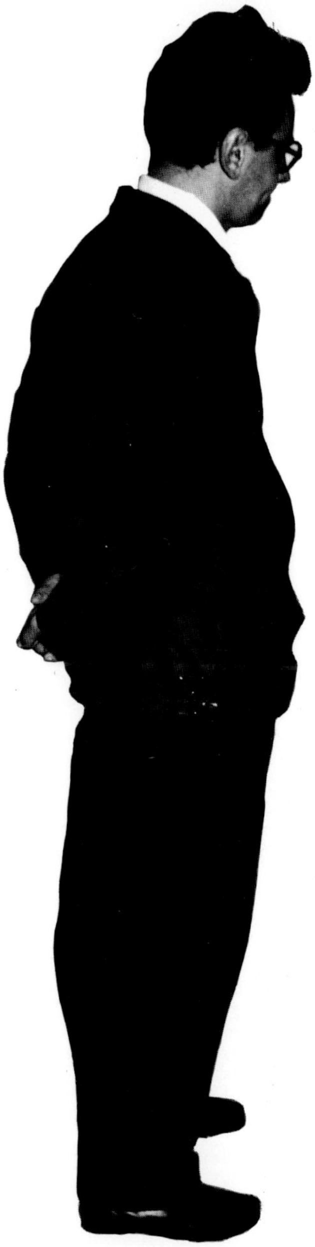

Hans Gugelot

The 1960–61 Basic Course was the last, for the time being, to include all the students. During the course of that academic year its name was changed, on the advice of the Baden-Württemberg education minister, Dr. Storz, to "first-year studies"; and from the beginning of the new academic year the institution itself will be abolished. All new students will be admitted to one or other of the four departments. Some of the so-called theoretical subjects will remain common to all first-year students, but their departmental work, formerly known as Basic Course work, has now been divided into four separate courses, one for each department. The first-year students will now be taught within the departments, apart from the students in other specialties, and also from the students in other years.

In the past the Basic Course students worked with instructors from all the departments, including those who taught theoretical subjects; they will now come into contact only with the small number of instructors from their own departments with whom they will be working for the remaining three years of their course. There is no coordination between departments. The position of head of the Basic Course department, which Mr. Rittel occupied last year, has been abolished. Instead, each department has appointed its own instructor responsible for first-year studies, and determines its own objectives and curriculum autonomously.

Karl-Achim Czemper, in *Output,* 6/7, 1961

The change in question has a long prehistory. I can't give any exact dates. As far as I can remember, it was in 1956 or 1957 that the desire was voiced to step up the specialist instruction within the departments. It was this activation of specialized training that was the main issue in the discussion of the Basic Course. In the next few years there were a number of attempts to give the work of the old Basic Course more specialized content.

Another factor that was taken into account when the change was made was the number of students in the Basic Course. In the last few years the Basic Course, with 40 or 50 students, had become the largest department in the College, far more than one instructor could cope with.

The students rightly complained that there was no time for detailed discussion of individual work. The instructors complained of the unwieldiness of the Basic Course group. The departmental instructors, who spent very little time teaching in the Basic Course, found it impossible to give proper critical attention to Basic Course work alongside their own work in the departments. For instance, last year I had 40 students in the Basic Course and 8 to 10 students in my own department.

The solution that has been found is a logical conclusion of the endeavor to achieve a more individualized training and bring more specialization into the first-year courses. We have found that two years in the departments are not enough for students who have never encountered technological problems before. They have found themselves at the beginning of their second year having to start out to acquire the technical knowledge they need for their work in the department.

Hans Gugelot, interview on the Basic Course, *Output,* 6/7, 1961

and social responsibility; freedom and the possibility of individuation; and lastly — as subcategories, in a sense — potentials for discourse, for self-categorization, for making one discipline transparent to another, for science, for technology, and for industry (with everything that implies by way of honest diligence).

This enumeration (which could of course be expanded) serves to make clear the ambiguity of every one of the categories involved, both grammatically — for all are simultaneously descriptions and aspirations — and, above all, historically: for every last one of them has at some time been perverted and reified. Work has become alienated labor; morality has become bigotry; function has become abstraction; public awareness has become systematic deception; freedom has become the delusive liberty of the consumer; and so on.

This double-edged quality is summed up in the historical fortunes of one idea that is particularly closely bound to that of the modern age: that of rationalism. Every time the call for reason, or for a rational world order, has made itself heard, it has brought in its train the profanation of rationality itself: through specious strategies of dominance, through patterns of rationalization, and through bureaucracy. Furthermore, the post-Enlightenment rationale has remained indissolubly linked to the modern process that has conjured up the capitalist ethos, with its attendant dehumanization, and the primacy of theory

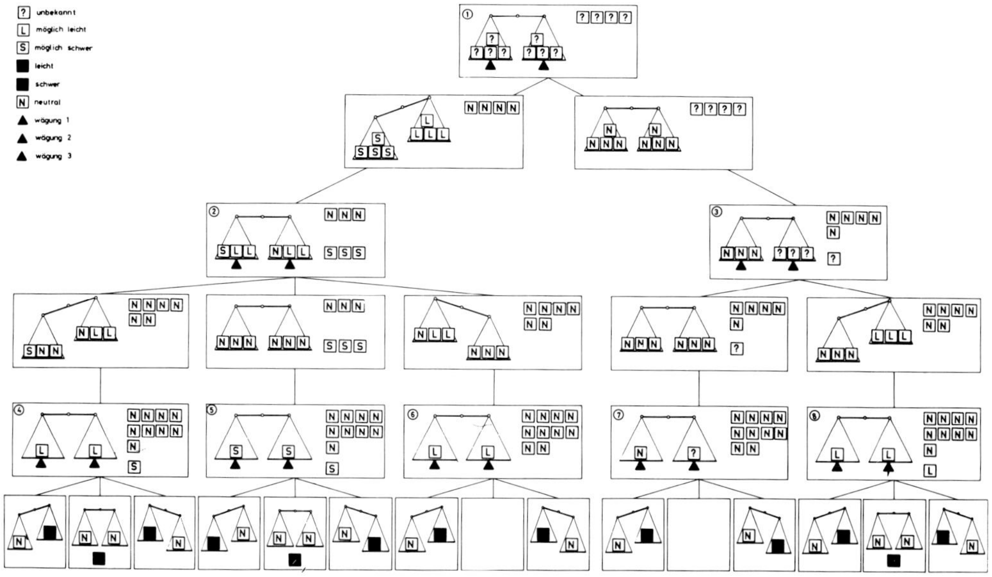


Operational diagram for tracing a fault
Heinz Grüber, 1960–61
Instructor: Bruce Archer

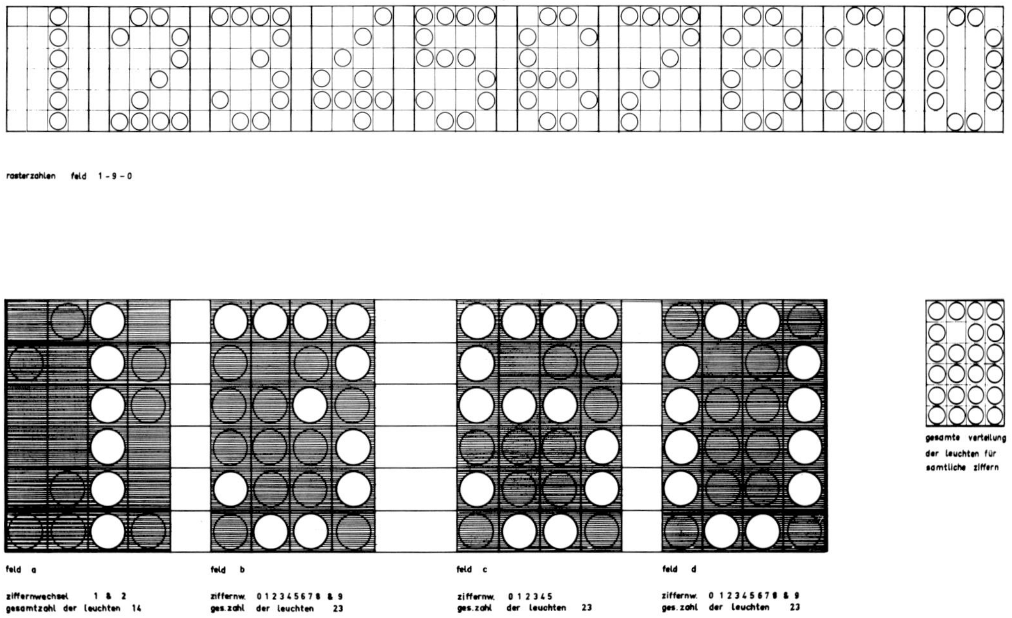


Exercise in optimalization: maximum legibility of numbers with smallest possible quantity of light sources
Robert Graeff, 1959
Instructor: Anthony Fröshaug

over experience. The result has been scientism, dichotomy, and denial of the senses.

The modern age — and this ultimately includes the visual manifestations of modernity — represents a process of abstraction: "The cultural tradition must already have become so fluid that legitimate forms of order can dispense with traditional dogmatic foundations. And persons must be able to act with such a degree of autonomy, within the contingent space of abstractly and universally standardized areas of activity, that without imperiling their own identity they can switch from morally defined contexts of communicative action to legalistically organized areas of activity." (Jürgen Habermas.)

Thus, the modern tendency to create abstractions is the source of the utopian vision of autonomy and liberation that is unique to the modern age. It was precisely this imposition of the abstract onto the real that was resisted, not by mere deluded sectarians, but also by committed modernists themselves. Only "modernism" was truly modern enough for that; but it tended, and still tends, to veil its modernity in the face of historical horrors.

No wonder that those who are commonly credited with having initiated "the modern age" — or who have been praised or vilified as champions of modernity — have been its most persistent critics, in the hope of steering it their way or of putting pressure on it for its own good. The Enlightenment thinkers were unable to accept the foreseeable consequences;

two fields are to be filled with tones from two comparable series of tonal values. the first series consists of a gray scale with ten equal gradations from white to black. the second series also consists of ten squares, of which the first is again white and the last again black. the eight intervening squares are to be filled with different bright hues in such a way that the colors correspond to the steps in the gray scale. for this the following convention is to be observed: a numerical series, 0,1,2,3, . . . 8,9, is applied to the ten degrees of the tone rows. the numbering always begins at white with zero and ends at black with nine. each field to be filled is to consist of 8 by 13 squares of side 10 mm. the colored squares adjoin without any intervening space. the arrangement of the color tones within the fields is determined by the following numerical table:

| | | | | | | | | | | | | |
|---|---|---|---|---|---|---|---|---|---|---|---|---|
| 1 | 6 | 3 | 3 | 8 | 5 | 6 | 3 | 1 | 4 | 0 | 0 | 1 |
| 1 | 1 | 9 | 7 | 1 | 6 | 3 | 5 | 6 | 8 | 0 | 4 | 5 |
| 0 | 3 | 4 | 4 | 2 | 6 | 5 | 3 | 6 | 9 | 3 | 0 | 8 |
| 6 | 9 | 2 | 2 | 6 | 3 | 6 | 6 | 5 | 3 | 9 | 1 | 1 |
| 3 | 4 | 0 | 7 | 3 | 2 | 3 | 5 | 4 | 4 | 7 | 9 | 7 |
| 9 | 6 | 1 | 8 | 2 | 4 | 6 | 9 | 1 | 4 | 1 | 3 | 5 |
| 4 | 0 | 4 | 2 | 7 | 1 | 0 | 7 | 1 | 7 | 1 | 0 | 7 |
| 1 | 8 | 7 | 8 | 5 | 8 | 0 | 2 | 3 | 1 | 8 | 9 | 8 |

(the numbers are taken from a random number table. their sequence is entirely arbitrary.) on a piece of white card, DIN A2, show the following in tempera colors (or gouache):
1 1 the gray scale with the numbers 0–9.
1 2 the colored tonal series with the numbers 0–9.
2 1 a diagram showing the numerical scheme.
2 2 a diagram showing the tonal values from 1 1.
2 3 a diagram showing the tonal values from 1 2.
the two tonal series and the two corresponding diagrams are to be arranged on the sheet in such a way as to facilitate comparison.

Anthony Fröshaug: course assignment for color theory, 1960

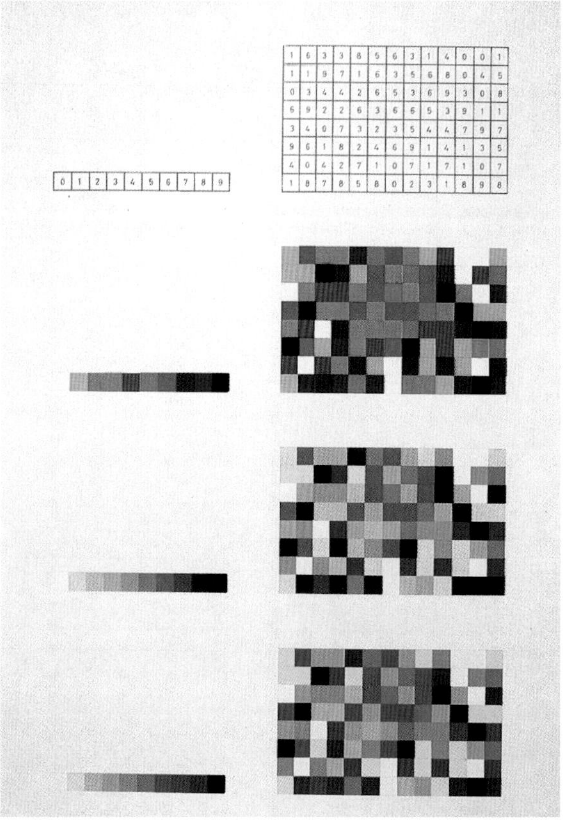

Random and purposive arrangement
Combination of colors using aleatory methods
Tom Dalley, 1960 – 61
Instructor: Anthony Fröshaug

for Hegel, as for the Romantics, "modern" was a word with well-established negative connotations; the Impressionists tied their own obsession with modernity to nature (which they thus helped to colonialize); Dada took arms against itelf; the Constructivists sought to justify their visual abstractions through many-layered religious or social motives; the champions of modern architecture offered gymnasia and sanatoria as model examples of their ideas; and the HfG took the case for modernity, quite justifiably, back to the last phase of the Bauhaus, and carried abstraction forward into systematization. The HfG set out to be on the side of the modern age and found itself, as a consequence of the very same impulse, subscribing to humanistic principles and so resisting the truth of its own modernity. Ulm encountered the situation that Novalis formulated in 1798: "Everywhere we seek the unconditional, and all we ever find is things."

On the other hand, "culture in modern times has brought forth structures of rationality from within itself"; and "without the guidance of the critique of pure and practical reason the sons and daughters of modernity learn how to pursue their cultural inheritance by dividing it, according to these same discrete aspects of reason, into truth issues and issues of justice and taste" (Jürgen Habermas). Alongside the "refractory nature of the aesthetic" (Peter Bürger), there exists that of the modern; but, as Tomàs Maldonado said in an interview in 1986: "Modernity as a project

*left*
William S. Huff teaching

*right*
Friedrich Vordemberge-Gildewart and
Horst Rittel

*left*
Screen pattern
Peter Fischer, 1964 – 65
Instructor: William S. Huff
Chosen for the permanent collection of the Museum of Modern Art, New York

*right*
Balancing contrasts ("Primadonna")
Achieving equilibrium through quantitative variation
Erich Rufer, 1960 – 61
Instructor: Friedrich Vordemberge-Gildewart

is not only a philosophical issue but also a highly practical one. . . . If we do not succeed in resolving this issue, this project, this great rendezvous with history, then there will probably be no such thing as a modernist project any more."

It is — to take an unreliable image — as if the modern age were a twin stream of continuity and change, running parallel, inextricably combined but never touching. Something is going on, and we have to intervene, although we know we are helpless. But how are those who seek to intervene — and how was the Hochschule für Gestaltung — to maintain clarity, uniformity, and unidimensionality under such circumstances? Which means, ultimately, that the final collapse of the HfG in Ulm merely marks its absorption into the ongoing process of the modern age and marks its place in history. "For design, as the self-definition of man on earth with the aid of a collective consciousness, is still with us as a positive legacy of the Enlightenment." (Kenneth Frampton.)

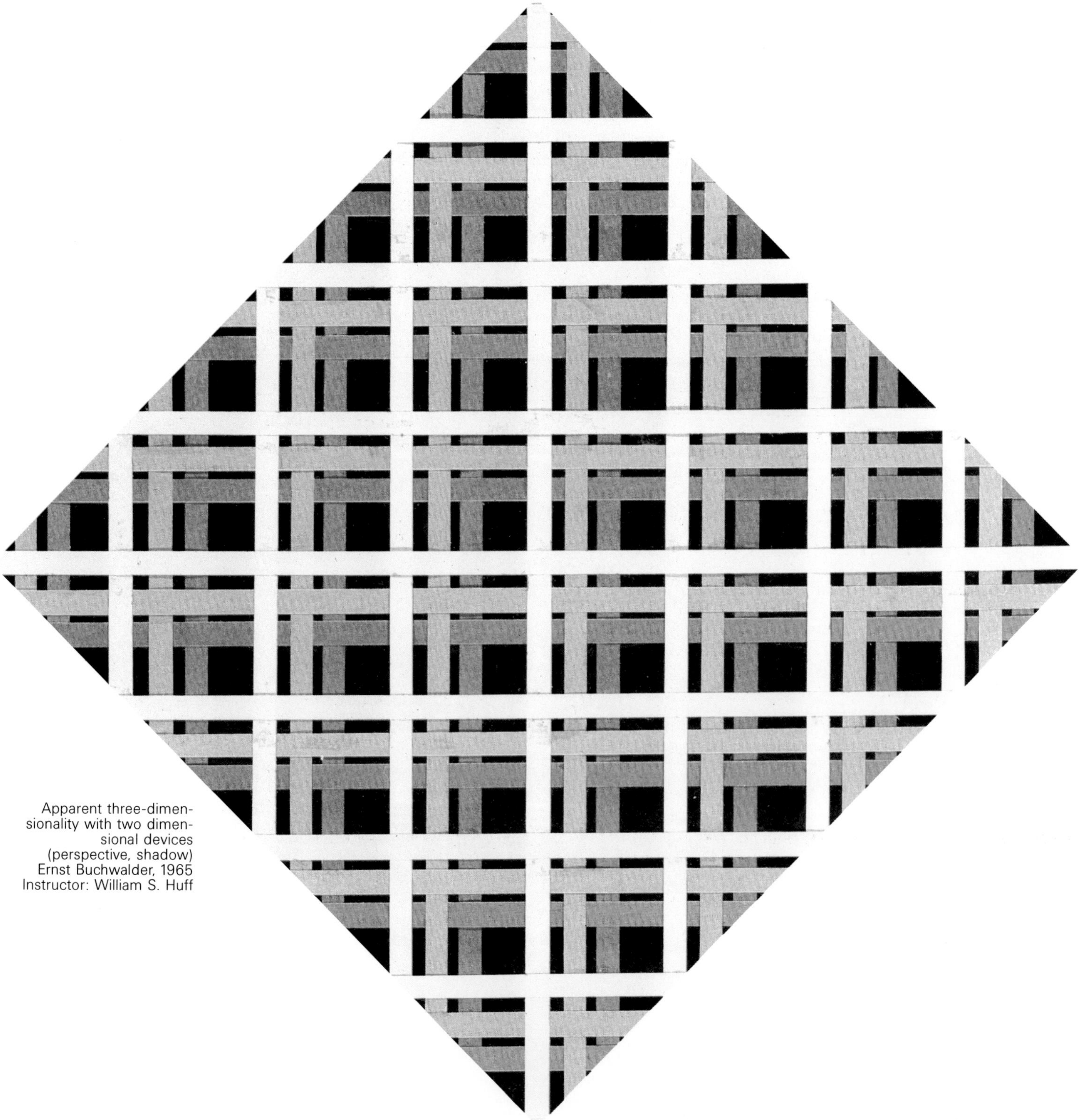

Apparent three-dimensionality with two dimensional devices (perspective, shadow)
Ernst Buchwalder, 1965
Instructor: William S. Huff

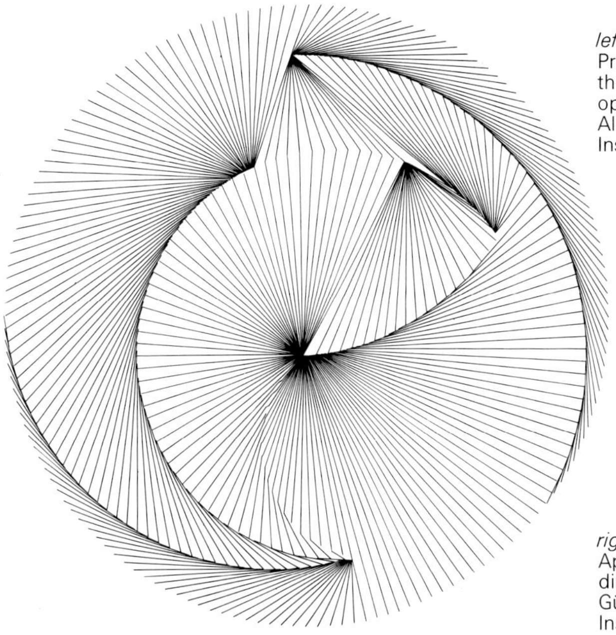

*left*
Programmed structures through symmetry operations
Albrecht Hufnagel, 1966 – 67
Instructor: William S. Huff

*right*
Apparent three-dimensionality
Günter Elsner, 1965 – 66
Instructor: William S. Huff

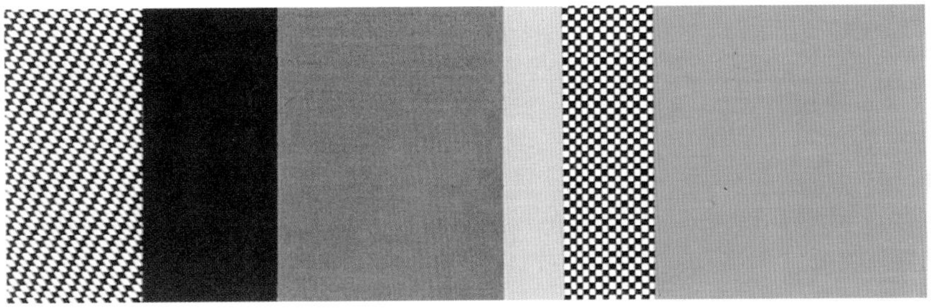

*above*
Positive-negative-ambiguously flat
Giovanni Anceschi, 1962 – 63
Instructor: Tomàs Maldonado

*left*
Interferences
Lothar Spree, 1962 – 63
Instructor: Otl Aicher

Reyner Banham

# HfG Ulm in Retrospect

What was it that took that devoted band of Britons to Ulm – those who stayed long, like Maurice Goldring and L. Bruce Archer, and those who made one or more shorter visits in the late 1950s and early 1960s, like myself or Richard Hamilton or Joseph Rykwert (to name only those that come immediately to mind)? In retrospect it seems to me that it was a peculiar combination of interests that were mostly transient on the British side, but more permanent at the Hochschule, more concerned with its basic orientations and its long-term patterns of intellectual growth. Both sides sought a way out of the local impasses into which design theory had fallen, but whereas the British impasse was one of almost total vacuity, the exhaustion of the "gentlemen's agreements" inherited from the 1930s and the lack of anything but the arts and crafts to fall back upon, the German situation seemed to us, as revealed by Tomàs Maldonado, to be one of excessive rigidity, a cast-iron system of "categorical imperatives."

But this was impressive to us visitors because of its relentless logic and intellectual clarity. I used to wonder why anybody bothered to speak to me at Ulm; I felt so stupid in my lack of dialectical method, and my head ached from having to find intellectual justifications for observations like "Yes, I would like another bread roll." Yet it was profoundly exciting to be in a milieu where issues about design could be discussed so intensively, especially if one

Closed form network of
catametric elements
Klaus Schmitt, 1961 – 62
Instructor: Tomàs Maldonado

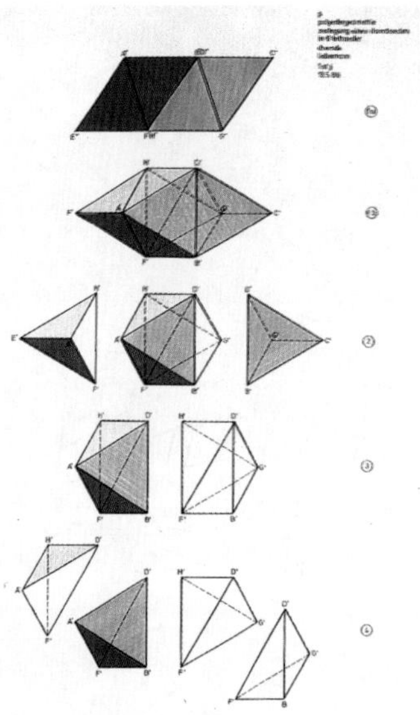

*right*
Sign transformation
Peter Polland, 1963 – 64
Instructor: Tomàs Gonda

*left*
Geometry of polyhedra
Erik Liebermann, 1965 – 66
Instructor: Horst Emde

had just come from London where there was at the time no intellectual discussion of design at all. And what we had to offer in return, in our unmethodical way, was, I think, a variety of pragmatic perspectives on design, based upon everything from industrial experience to Erwin Panofsky's celebrated essay on the movies.

These *ad hoc* methodologies never added up into a coherent system, but they were highly applicable in a piecemeal fashion — the Panofskian iconology that I had learned in art history and then applied to the study of pop art and product styling proved valuable and useful at the HfG: valuable as an escape from their rigid systems, useful as a tool for analyzing the visual rhetoric of advertising. Useful, but not immediately used — many entrenched ideological positions at the HfG were threatened by any proposal to give serious attention to pop art or advertising, and these disorderly Anglo-Saxon ideas were received with great caution. But in the end they were accepted — the *Systemleute* at Ulm were more intellectually flexible, more willing to take intellectual risks, than the supposedly eclectic or compromising leaders of Anglo-Saxon design, who could not abandon their ancient shibboleths like "Form follows function," or "A good design is forever."

Under the *Rektorat* of Maldonado, even the most outrageous ideas would be examined, skeptically but objectively, and tested against both real-world situations and strin-

Grid of catametric elements
Jan Thylén, 1961 – 62
Instructor: Tomàs Maldonado

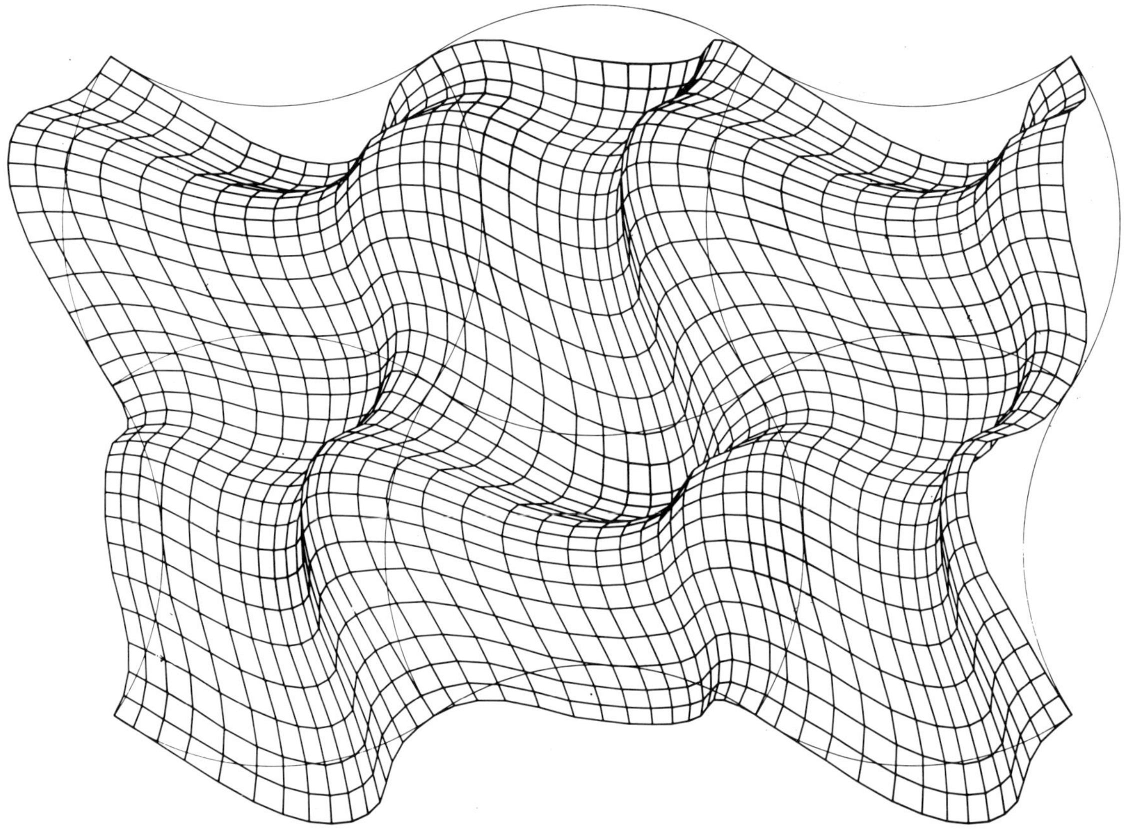

Network transformation
Kurt Christen, 1965 – 66
Assistant: Günter Schmitz

gent philosophical schemata, before being accepted or — more often — rejected. The skeptical rigor of HfG thinking — largely derived from the Frankfurt School, still almost unknown in England — was like a breath of painfully fresh air blowing down from the snowy Kuhberg.

Few of the designs actually produced by the students and faculty could match the quality of the thinking, alas; good as they might be, all were still too deeply involved in the stylized version of the Bauhaus inherited from Max Bill and every day reinforced by the example of teachers like Hans Gugelot. Deeply held cultural values were embodied in that visual tradition of *Die gute Form* in Germany, and it could not lightly be abandoned; but for Anglo-Saxons who agreed with Lawrence Alloway that "Good design is just another iconography," the hard-driving *Semantik und Semiotik* of Maldonado's "Information Seminar" was exactly the kind of intellectual reinforcement that we needed.

And just when the London design establishment had finally found what it believed to be the right cliché with which to dismiss the HfG as irrelevant — "all theory but no sense" — Maldonado cut the ground from under their feet with his timely and revisionist dictum of 1963: "For the design of simple objects like spoons and forks, good taste and common sense are sufficient."

Santa Monica, 1987

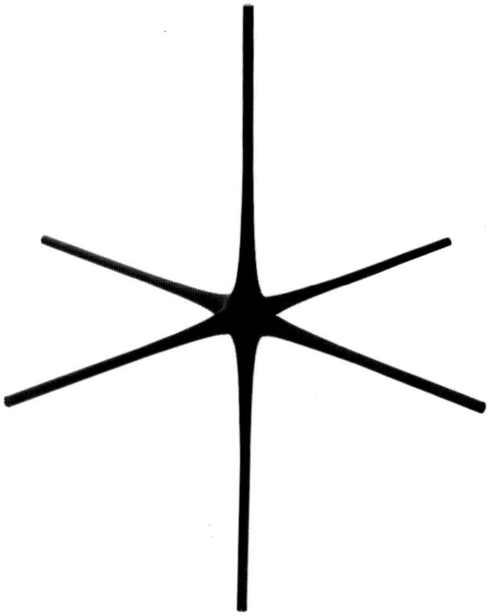

Nodal point
First Year, department of Product Design, 1962 – 63
Instructor: Walter Zeischegg

Branching element of two networks
Niklaus Hufenus, 1963 – 64
Assistant: Günter Schmitz

Since October 1962, the one-year introductory course has consisted of some twelve exercises. These are selected from a constantly growing repertoire of different problems. They are formulated in such a way as to give the students sufficient room to engage in formal creation within the assigned framework. A maximum spread is aimed at. The assignments sometimes sound relatively abstract, in order to stimulate the student's imagination; but they all refer indirectly to the problems of building. The complexity of the tasks increases, so that the student progressively learns to master the handling of different factors.

In setting the assignments, the following are the objectives pursued: to stimulate the use of imagination in designing; to convey the basics of designing prefabricated structures; rational decision making; the practice of a methodical mode of working, including purposive experimentation; and the practice of precise techniques of presentation.

All the exercises are based on the rational application of planar or three-dimensional grids: network and grid structures of various kinds are used as geometrical devices for creating order, through which new forms can be evolved. In this the emphasis is on the purposive arrangement, connection, and distribution of units. The parameters involved may comprise: correlations, forces, structures, directions, intensities, functions, extents, techniques.

The structuring of systems has long since become the pressing task of architectural design. Practical work with these basic exercises provides the student with an essential basis for the design problems of industrialized building.

Günter Schmitz, in *Ulm,* 19/20, 1967

## An Editorial Discussion

Through all the years of research and planning, the present Ulm project has been based on discussions: all those conversations and arguments from which much in this book is derived and that served to structure the book but that in themselves remained private. Private, that is, until we took the decision, when the work was almost complete, to take down one of our conversations and print it in the book. We were at the stage where we were having to consider what had been adequately covered in the book and what had perhaps not; so the conversation gave us the opportunity to fill a few gaps and also to give some pointers to a wider study of what went on at Ulm.

Like all the departments of the HfG, the department of Visual Communication has had a departmental first-year course of its own for four years now. The basic idea of a preliminary course has survived in several respects, but the assignments now concentrate on the work of the visual designer. In addition the original principle of a Basic Course has been abandoned in one respect: in the last trimester of the first year, practical design problems are tackled. This step has been taken to facilitate the transition to the applied departmental work of the second and third years.

The student begins with analytical studies of those phenomena that distinguish a sign or make it stand out from its background: figure/ground relationships in the dimensions of brightness, sharpness, quality, quantity. There follow studies of the relationships between visual elements: contrasts and minimal differentiations, similarity and affinity, geometrical and chromatic flickering, representation of processes, transformations, interference patterns. Then, in combination with a course on color, there are studies of ordering principles, i.e., symmetries. These exercises and studies are accompanied, in addition to the instruction in theory in general, by seminars on the theory of symmetry, structure, gestalt psychology, and color theory.

In the last trimester, tasks are undertaken that retain a general character but still carry over into the practical, applied departmental work: basic typographical principles, type areas, page layout, and photographic exercises.

Herbert Lindinger, *Ulm*, 17/18, 1966

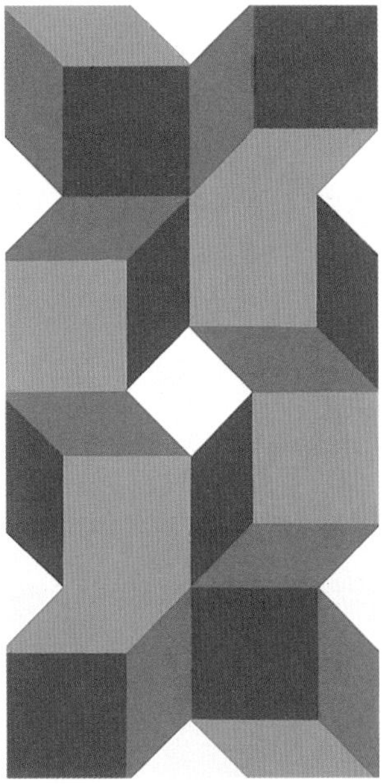

Ambivalence effect
Erhard Schreiber, 1965 – 66
Instructor: William S. Huff

Transparency of color and ordering principles
Lothar Spree, 1962 – 63
Instructor: Herbert Lindinger

## Ulm Curiosity

Editorial Discussion

Erlhoff:
I can't help getting the impression that the theoretical debates at Ulm were at least partly governed by chance factors. What was then propounded as rational often seems highly marginal now. One after another, with equal energy, the most various theories were professed, and then simply set aside.

Staubach:
That is what Maldonado describes: someone had heard something, more or less by chance, had taken it in, and straightway there was an attempt to put it into practice. So even in the theoretical disciplines there was no continuity: it all depended on who happened to be there at the time. The next instructor would come along and argue the exact opposite with equal vehemence. Ulm had no consistent theoretical foundation. And that was another of the ways in which it set out to be really avant-garde.

Lindinger:
It was a period of endless curiosity, and not only at Ulm. Maldonado in particular encouraged this: for him, intellectual

Avoidance of spatial illusion
Erhard Schreiber, 1965 – 66
Instructor: William S. Huff

unrest was the one outstanding characteristic of the College. New things were absorbed with enormous enthusiasm, and then often they couldn't be followed through, because the College lacked the equipment, the organization, the finance, and the personnel. It was never possible to do any long-term research in any field, whether ergonomics, perception theory, or semiotics. There was no personal continuity, either: the way the guest instructors outnumbered the permanent faculty led to a constant influx of new ideas, but for many students it also led to overloading and confusion. The volcanic Ulm atmosphere is inseparable from this mix of permanence and change. A lot of promising ideas came to grief that way.

Chemaitis:
All that endless curiosity, that tendency to pounce on anything new and play around with it till something else came along to grab your attention: there's something naive about all that.

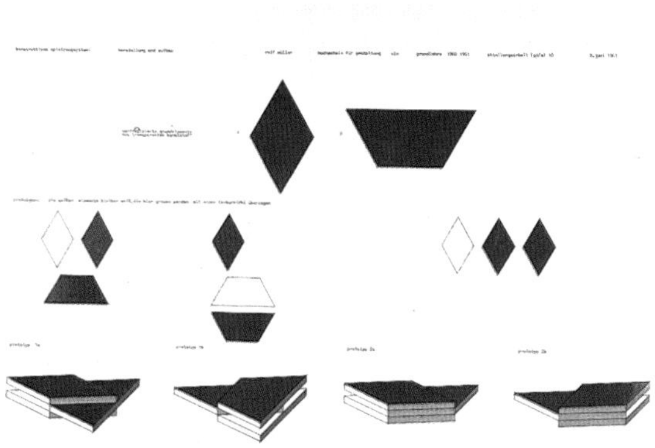

*above left*
Constructional toy
Rolf Müller, 1960 – 61
Instructor: Anthony Fröshaug

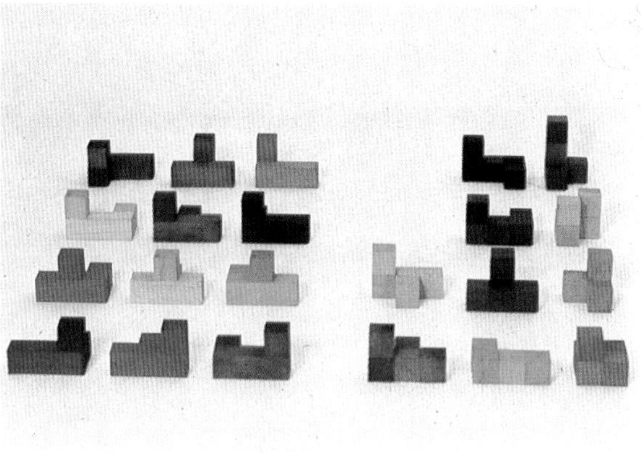

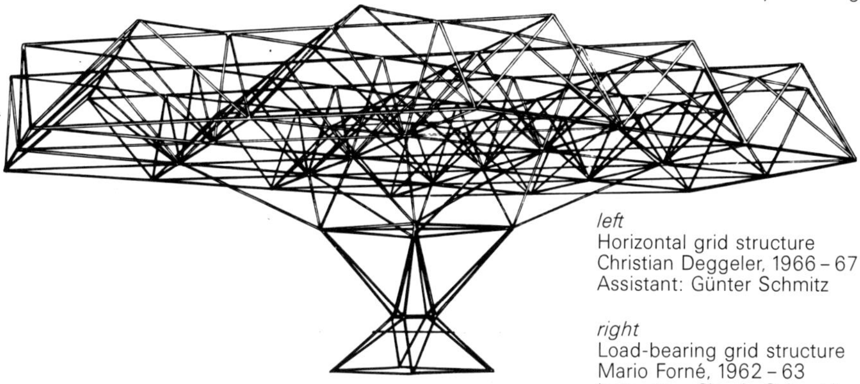

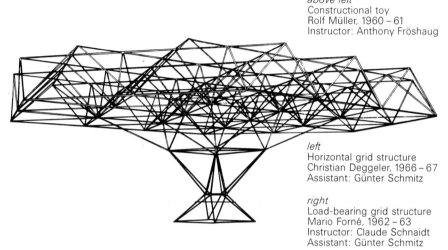

*left*
Horizontal grid structure
Christian Deggeler, 1966 – 67
Assistant: Günter Schmitz

*right*
Load-bearing grid structure
Mario Forné, 1962 – 63
Instructor: Claude Schnaidt
Assistant: Günter Schmitz

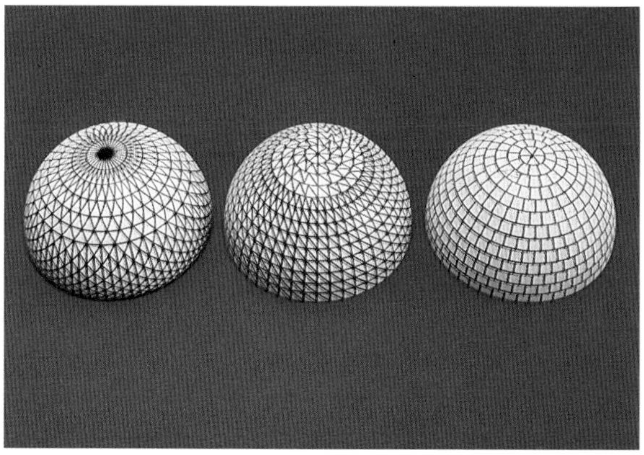

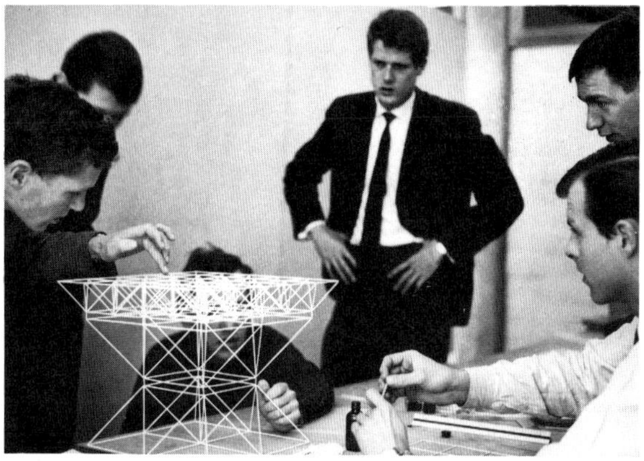

Shell structures
Eric Boss, 1965 – 66
Instructor: Herbert Ohl

Students of the department
of Building, 1962-63

U-L-M spells H.f.G.
A visit to Ulm means only one thing in our circle. Hochschule für Gestaltung doesn't roll off the tongue like Bauhaus – Ulm, however badly pronounced, is easier than college for the untranslatable . . .

Richard Hamilton, 1966,
in *Collected Words, 1953 – 1982*,
London, Thames and Hudson, 1982

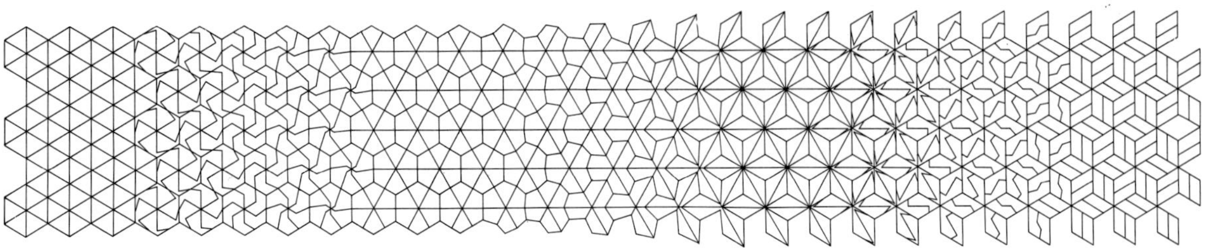

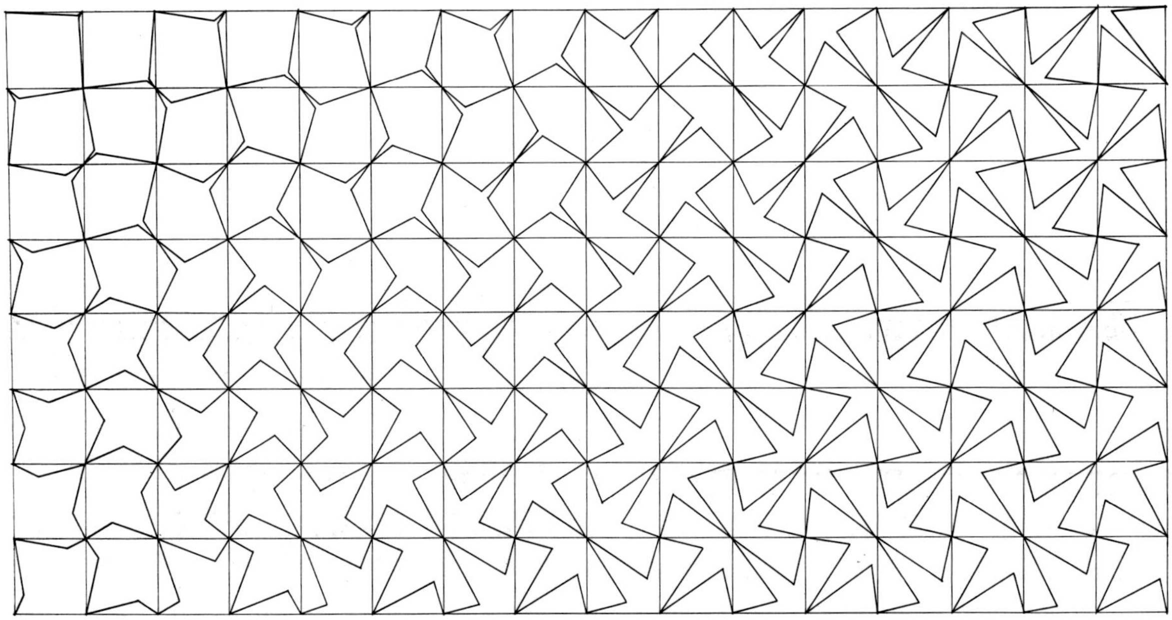

*this page*
Network transformations
Arno Caprez, 1965 – 66
Rolf Glasmeier, 1965 – 66
Pietro Rusconi, 1963 – 64
Instructor: William S. Huff

*facing page, above*
Nonorientable planes
Dirk Schmauser, 1965 – 66
Instructor: Gui Bonsiepe

*facing page, below*
Automobile warning lamps; in the background, typical 1963 model
Werner Zemp, Peter Westenfelder, Peter Hofmeister, Ciril Cesar, Uwe Rutenberg
1963 – 64
Instructor: Herbert Lindinger

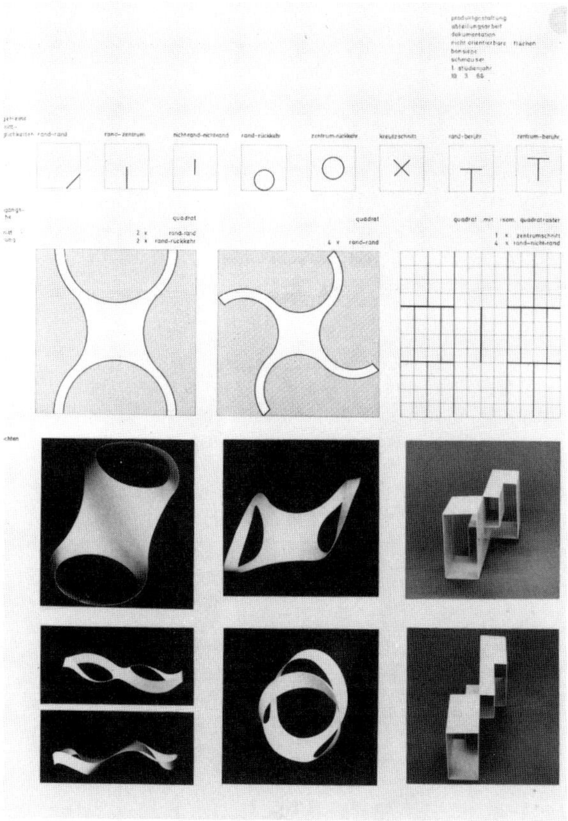

Michael Erlhoff:
There's that saying of yours, that often gets quoted — no doubt in very much abbreviated form — to the effect that design has to go "from spoon to city." So design is something that integrates, something that orders, the creation of order.

Max Bill:
That is a highly compressed presentation of the issues involved; of course, it is all about universal issues of the human environment. Objects, which are — as it were — added day by day to what is already there, have to be interpreted according to ecological and economical criteria, but also aesthetic ones. So that one doesn't simply produce just to stimulate consumption, or to activate social mechanisms, but to create things that stand in a certain harmonious relationship to the circumstances in which we live.

Michael Erlhoff:
But doesn't that mean, in your view, that a concept has to be presented in a wider context, not only as an isolated piece, but through a concern with every detail — as the building, the architecture, of Ulm already shows?

Max Bill:
Yes, you have to take on the responsibility for seeing that everything works, and that no detail gets overlooked. So that things, in a certain sense — economic as well as technical, aesthetic, or whatever — coordinate with each other. But it's not possible for this coordination, this agreement, to be unilateral. For instance, the Ulm buildings were there for a specific purpose, the coordination was automatically there, because it was one person who thought the whole thing through. I draw the conclusion from this that others ought to think in the same way, and then similar things would be created. Of course, you need an underlying ideology before you can even envisage such a thing. But without an ideology the result would be the chaos that we see in all the department stores today.

## Interview with Max Bill

Michael Erlhoff:
An idea that also involves a statement about the designer's social responsibility.

Max Bill:
Absolutely. That is so from the start. Without that it doesn't work. Otherwise design will never concern itself with ecology or economy or anything else. It has to be prepared to take responsibility for the whole thing.

Michael Erlhoff:
These theories, as you are now formulating them and as you have practiced them, have given rise to the objection that this is all an attempt to intervene in society from the top downward. It has been argued against Ulm, for instance, that it was no more than an attempt to use consumer goods to democratize a consumer society.

Max Bill:
There is absolutely no question of dictating from above. The idea that designers — it was a mistake to call it design — should take responsibility, does not mean rule by *Diktat* but moral responsibility. Just as doctors must take responsibility for keeping people healthy, not just patching them up when they get sick.

Michael Erlhoff:
There is also a criticism that Ulm produced uniformity and standardization, so that everything looked the same, and variety was lost.

Max Bill:
That is a mistake that comes from standardizing things instead of evolving them out of their own inner nature. And this so-called Ulm Style, with its tiny little labels, these scraps of paper: all that is absolutely wrong. It can't be done that way. No one can say I ever did any of that.

Michael Erlhoff:
But you lay stress on aesthetic categories, in design as well as in architecture.

Max Bill:
That's not what I start out with. What I have is an idea of how a thing has to work, then, when the thing is built, the necessary potentials have to be used. Let's take a window, for instance. A window is a necessity, and I need so much light in this particular room. Then I have to analyze the window: what part of it is to be opened, what has to stay closed, where can I admit light without the rain coming in, or whatever else there is. When I sort out all these things, together they ought to produce a window, and this window has a particular form of subdivision, and this subdivision depends on how much I use of each individual part. And finally I can still say: Whatever is subdivided in this window I can still regulate. The free options are very small. I have always maintained that the areas of freedom in architecture, if you take them seriously, are very small; but these areas of freedom are things that you have to master completely. You have to make something out of them, transcend all the compulsions, to produce a result that is as aesthetically good as it is fit for its purpose.

Michael Erlhoff:
So harmony between things — and that refers to disagreement and controversy, too — produces a beautiful appearance, aesthetics in fact, of its own accord. Would it be possible to coin the formula "Fitness for purpose is beauty"?

Max Bill:
It isn't, not in itself. It becomes beautiful when, in that free area that always exists — in every decision, whether it is a machine or a door or a window — aesthetic issues arise, and the final decision is about those. If those do not then conflict with fitness for purpose, then you can go ahead and make it that way.

Michael Erlhoff:
That would mean that a training in architecture and in design ought still to include a training in aesthetics?

Max Bill:
That's what I always hoped the Basic Course would do. We wanted the Basic Course to provide practice in aesthetics divorced from function, so that each individual, faced with purpose-free aesthetic problems, is enabled to take the decision whether to proceed in this way or in that way.

Once, at Ulm, I gave the same assignment to a group of people, then compared the results and started a debate as to which result was the best. Let's say: place four different-sized squares on a plane surface in such a way as to bring out the tension and the harmony between the four squares, or to create some additional expression. It became clear that these exercises lent themselves to general debate, and what emerged was a kind of aesthetic workout. Curiously enough, people never believe, to begin with, that there's going to be any difference. Then they sit down and have a try, and then there's a discussion. We had a class at Ulm — in the Basic Course — in which I carried this through: it was very striking that by the end everyone agreed on one piece of work that was simply the best solution to the problem that had been set. I think this proves that you can give aesthetic training. You can train for a quality that is basic to design work. Design must be carried out according to purely technical criteria, but areas of freedom can or must remain; and they can't just be disposed of by taking something out of the latest pattern-book. This area of freedom, which exists, must not simply follow the latest product from Braun, for instance, but must involve knowing what is inside the product.

Michael Erlhoff:
And when Ulm began there was some discussion of changing the form of production and exchange.

Max Bill:
You can't change the form of production, you can only change economic forms. By the time Ulm was actually open, there was overproduction. The country had survived the war, and there was a great need to catch up. In many cases it was imperative to make something new; and objects had to be produced that were good, or at any rate adequate, and usable. That was the situation we had started out with; but while we were still in the preliminary stages the situation changed somewhat — so much, in fact, that for a time there were doubts as to whether we should start up the College at all, because there was just too much available. There were already rumblings of the economic miracle, Germany was remilitarized, and we faced such problems that on one occasion I nearly said: "We won't do it." Although I was convinced that it was necessary.

It then became apparent, as the economic miracle got under way, that the growth of productive capacity brought with it a mounting pressure to produce and sell more and more. This was happening even then, and it became more and more evident as time went on. But at the outset there was a quite different situation, and at that phase of the reconstruction of Germany there were so many young people who wanted to do something; we wanted to give them the opportunity to play their part in the reconstruction. And I think it was this that gave the College its necessity and its justification; whether it would be justified now is quite another question. I would say that what is needed now is some kind of economic police force.

Michael Erlhoff:
That was a time when a college of design, or the designers and the architects, could still express a moral sense of responsibility through a product.

Max Bill:

All these things have to be looked at within an economic and political context, otherwise there is confusion. I mean, it was an unhappy situation at that time. A bit earlier, I made a speech, just at the time when we were having discussions in Ulm — I was discussing things with Ulm from 1946 or 1947 onward and at that time there was just nothing anywhere, there were no buildings standing. There were no ordinary consumer goods to buy; everything was hard to get. It was a war situation, which brought with it a dearth of material objects, but also made it possible to take the risk of starting up something like this. There was nothing to suggest that the economic situation would change so rapidly.

When we founded the Hochschule für Gestaltung in Ulm we were confronted by a scene of utter devastation, and there were a lot of problems to be faced. For example: since then it has become apparent that everyday utilitarian objects represent a large part of our culture, and that we now look on ordinary objects, used in the past, as cultural relics — even the very simplest things, such as simple tools — we regard them as the wonderful achievements of a long-past culture. And so there was a question, when the HfG was founded, whether anything found today was in any sense worthy of a place in a museum. This idea, which was present in a sense at the very inception of the College, meant of course that we intended to found a culture of the industrial age and create the component parts for it.

Berlin 1986

# Product Design

The students in this department were trained to design objects for everyday use in the home, in offices, in production, and in science.

On one hand there was a need to integrate all the factors involved in the design of a product: functional, cultural, technological, and economic. On the other hand there was a need to invent and determine new and rational patterns of use. Design assignments were keyed not only to individual, self-contained products but also to groups or systems of products.

Instructors: Max Bill, Gui Bonsiepe, Hans Gugelot, Georg Leowald, Herbert Lindinger, Tomàs Maldonado, Walter Zeischegg

Something for the students to sit on. That was the program that produced the stool. We made it ourselves in the workshop. The idea was that we had a normal height for sitting at a desk, a more comfortable height for sitting round for a discussion and so on, and we could take it into the canteen. So people wandered around with these stools. And when you had your things with you, books or implements or whatever, you put those into it to carry them from one room to another.

Max Bill, in a West German Radio broadcast, 1981

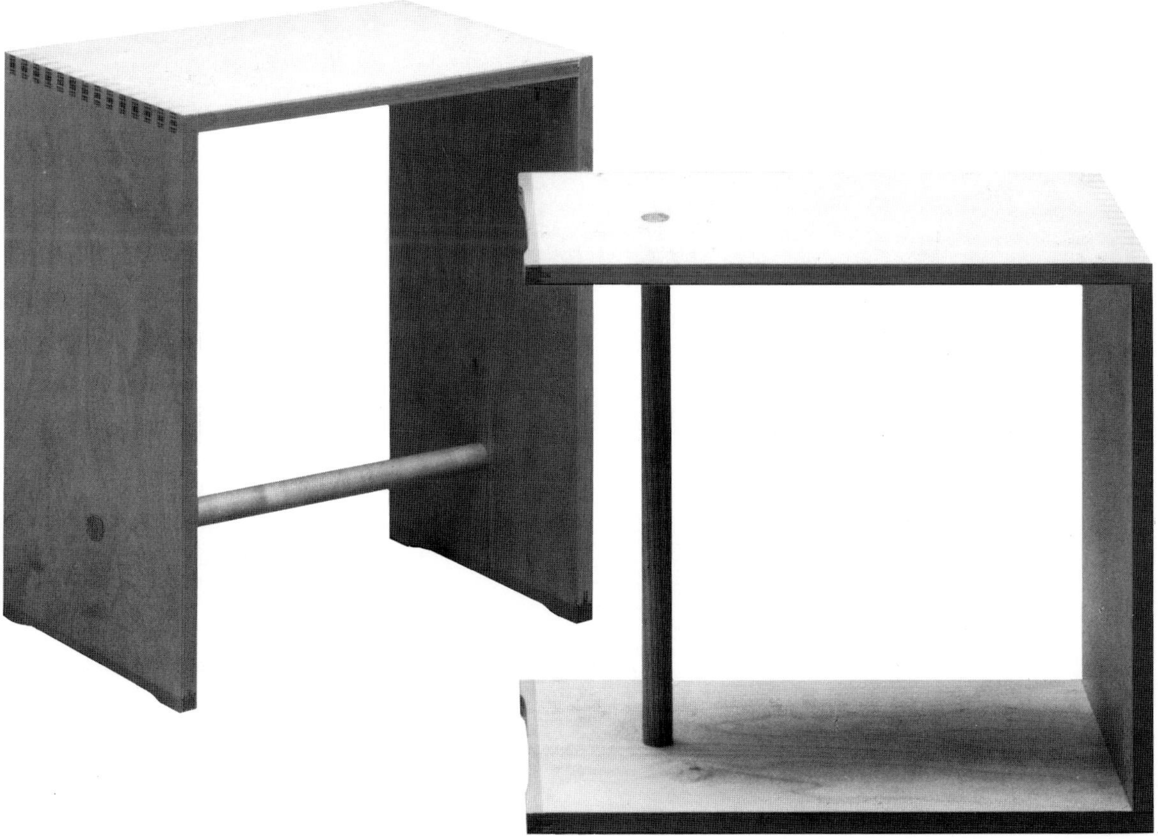

The HfG stool
Max Bill, Hans Gugelot
1954

# Alone in Midstream

Michael Erlhoff

## The Hochschule für Gestaltung and 1950s and 1960s Culture

The HfG is one of those things that are easier to define as a clear-cut negative than as a proliferating positive. It was the College's enemies who gave it a common cause; and as the open confrontation receded, or — which is really the same thing — as the HfG became integrated into normal life and acquired a position in society, the common cause tended to get lost. (The weakness of the resistance to the shutdown in 1968 was one of the consequences of this process of assimilation.) Ulm was a blast against the revival of German coziness; against a new conservatism that wanted "no experiments"; against stupidity and sentimentality; against blank looks and letting sleeping dogs lie. It was also against kidney-shaped tables masquerading as organic form; against showy automobile design, fins and all; against all the "styling" that was nothing but packaging, putting a glittering facade on contents that were outmoded, sloppy, and useless. It was against K 2 r, Odol, Brisk, Rexona Soap, and the Colgate Smile; against cliquishness and reaction; against those who banned Bertolt Brecht, born in Augsburg, from the German stage, and reserved their praise for Kotzebue; against snug, smug hometown movies; against the knitting patterns that passed for policies in conservative circles; against the Borgward Isabella and the Volkswagen Beetle; against record-players in walnut chests; against arts

A student's room in the dormitory tower

HfG door handle
Max Bill and student
Ernst Moeckl, 1955

Max Bill teaching

Door furniture: handle and escutcheon. The hand movements on the door handle when it is used, such as reaching out, touching, sliding, pressing, pulling, or pushing and sliding off, all movements that run into each other without a break, create the arched basic shape of the handle and a cross section without sharp angles, evolving into an ellipse. The free end of the handle curves slightly inward toward the door, so that the hand has a good grip even at the end of the handle and does not slide off too soon. The door can also be opened and closed with the arms (when both hands are full), without their slipping off. The incurved end also helps to prevent garments and other objects from getting caught.
Ernst Moeckl, 1955

and crafts; against "of course, in an ideal world . . . "; against the alleged death of geometry and the anarchy of urban reconstruction; against nylon shirts, bicyclists, garden gnomes, and lotharios; against four-leafed clover: against, in fact, the official, quantifiable majority of the population of the new Federal Republic.

The HfG was so avant-garde that it made a double leap into the future: it was criticizing the critics. The people who gave us kidney-shaped tables, Odol ads, and "styling" in general, regarded *themselves* as the avant-garde, as heralds of the new age and critics of the old. And, in a poll held in 1955, only 8 percent of the West German population were in favor of the kidney-shaped table, while 60 percent voted for German Renaissance, and 20 percent preferred German Rustic. However hard this majority may have pressed their noses to the windows of the new car showrooms and furniture stores (in 1952 the American firm of Knoll International opened its first German store), what they bought was the stolid petty-bourgeois comfort of the reconstruction years: chests of drawers, the "Victor" three-piece suite, long johns, rubber plants, and "tropical fruits." The majority voted Christian Democrat, lent an ear to the Friedland "Liberty Bell" and the clang of the rifle butts of the new German army, shouted "No Partition!" and supported "Operation Clean Sheet."

In the mid-1950s, it was only those who thought themselves modern — and who could

The designer of consumer products now has a greatly increased responsibility to the public at large, especially since mass production has more and more displaced the individual handcrafted article. The design of good forms, that is to say technically correct, functional in use, and aesthetically irreproachable products, is a cultural and an economic necessity.

The development of a product for mass production presents demands quite different from those involved in the design of a craft object. The designer must deal with the technical parameters, learn all about the production process, and be in constant contact with technicians, business people, and users.

From *HfG-Info,* 1955 – 56

Building Interior
by Max Bill

Lighting
by Walter Zeischegg

afford it — who spoke out in favor of the things that by 1960 had become universal: streamlining, canned food, plastic, behavior manuals, the rules of a new society, modernity in general, towns laid out for the automobile with gentle curves; things that were later followed by functional furniture, sofa beds, practicality, and — above all — middle-class do-it-yourselfism.

In fact, it was not until the late 1950s, or even the early 1960s, that the doctrines of Ulm ceased to be ahead of their time and came into direct collision with a prevalent taste. Instead of being two leaps ahead, Ulm was now only one leap ahead. Now the Ulm morphology came out of the shadows, and it became possible for the Braun company to bring out the radios that created a sensation and a scandal at the 1955 Berlin Radio Show. This was the very company that as recently as 1954 had sent to one German design group, the Kronberger Werkstatt für Gestaltung (W. Schwagenscheidt, T. Sittmann, and H. Dormauf), the following horrified letter of rejection: " . . . and would venture to say that in our view the German public is not prepared to accept modern cabinet designs for radio apparatus. Germans are too attached to tradition, and no radio manufacturer can afford to put a major range of unconventional cabinet designs on the market. We would never take such a risk, especially as our sets are manufactured only in long production runs. We regret, therefore, that we can make no use of the designs you offer."

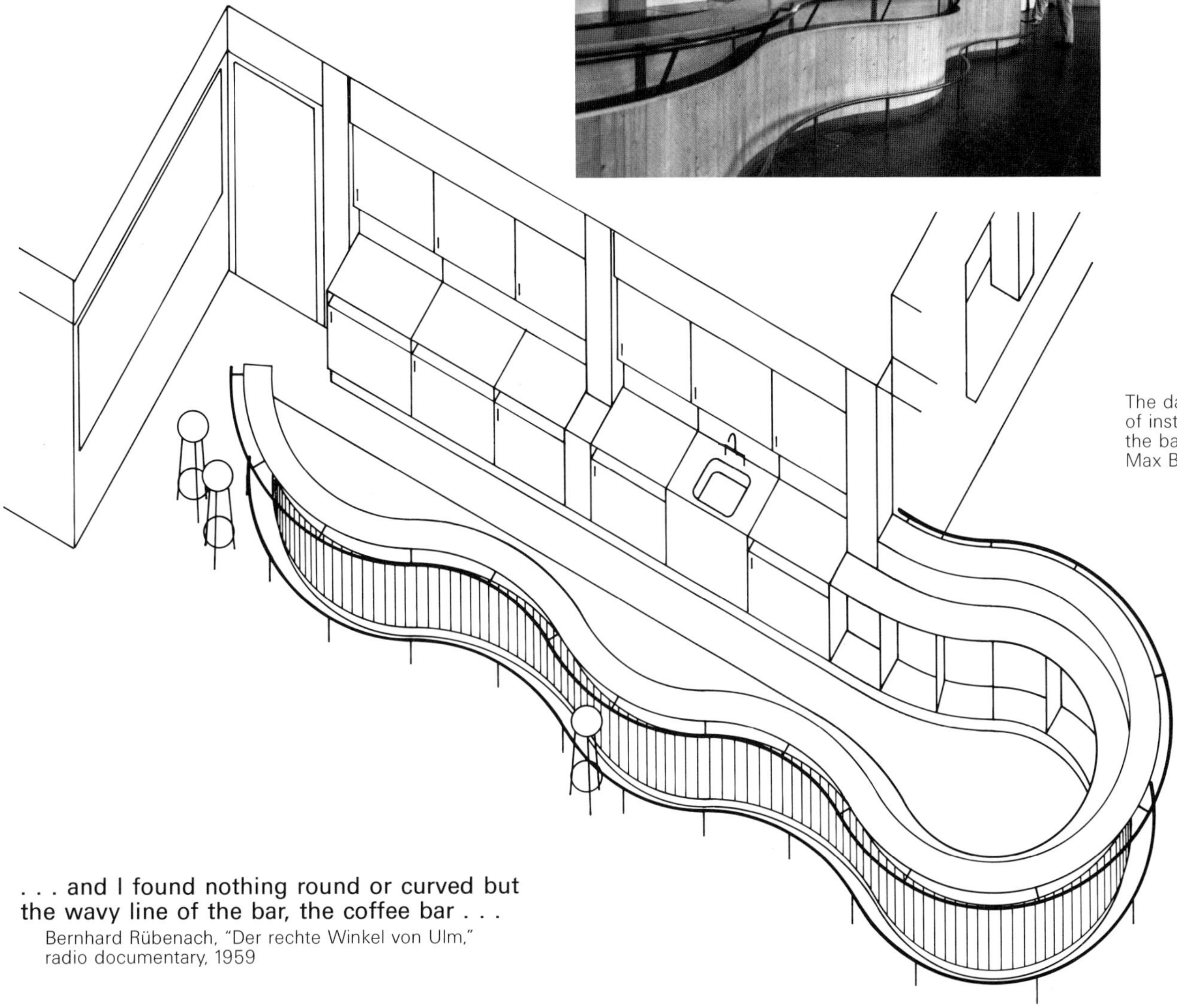

The daily meeting-place of instructors and students: the bar
Max Bill, 1955

**. . . and I found nothing round or curved but the wavy line of the bar, the coffee bar . . .**
Bernhard Rübenach, "Der rechte Winkel von Ulm," radio documentary, 1959

The HfG as a political entity was there as an alternative model of higher education, with alternative learning models and explicit political concepts. It was not, however, quite so isolated within West Germany as one might think today. The Easter Marches (in which some members of the HfG joined) were just beginning; there was a militant element in the trade union movement; within the Social Democratic Party there was a Trotskyist wing which produced the magazine *Sozialistische Politik;* and there were the Socialist students of the SDS (Sozialistischer Deutscher Studentenbund). Then there were one or two theoreticians – Ernst Bloch, Hans Mayer – who were publishing in East Germany. Theodor W. Adorno and Max Horkheimer, among others, were working in Frankfurt, as was Alexander Mitscherlich; and there were the beginnings of a press that was at least liberal.

Nor was the HfG alone within the context of aesthetic debate and practice. Yet it took surprisingly little overt notice of what was going on around it: of the radio rebels Ernst Schnabel and Alfred Andersch, for example, or of Hübner, the director of the Ulm theater. Since 1954, too, there had been a Rat für Formgebung (Design Council), conceived in highly idealistic terms as the custodian and patron of good German design: *die gute Form.* There was the Brussels Expo in 1958, with the pavilion by Le Corbusier that received great publicity even in Germany; there was a mass of architectural and planning work that

HfG bed
Hans Gugelot, for Dunlop
1954
Bill, Aicher,
Administrator Schlensag,
Hildinger, Gugelot

Kitchen clock
Max Bill, Ernst Moeckl,
for Junghans
1956

Electric plug
Max Bill
1955

Washbowl for dormitory
and studio use
Max Bill, Walter Zeischegg,
Otto Schild
1955

Telephone switchboard
Max Bill, Karl-Heinz Bergmiller
1956

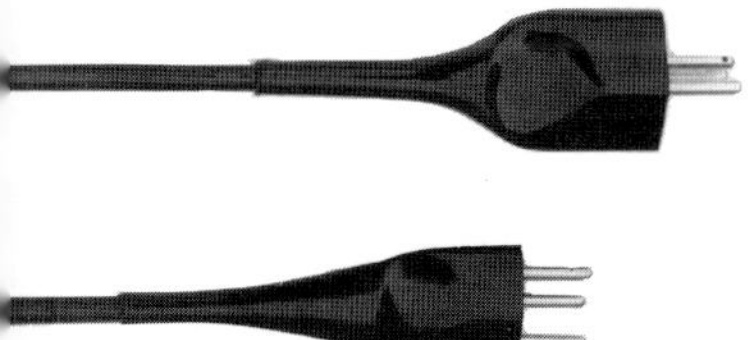

had gone on in Germany since 1945, and all the activities of the German Werkbund; there were the idols of a rebellious younger generation, rock 'n' roll, James Dean, and Marlon Brando; in art, there was the impact of Allen Kaprow's first happenings, the founding of the Fluxus movement in Germany in 1962, the actions of the Living Theater, the texts of "concrete poetry"; and later the advent of the Beatles. Very different from all this, and yet historically inseparable, there was the coming of "direct action," the "Viva Maria Commune," and the politics of student socialism and the protest movement. Those 1950s and early 1960s were also the years of the controversies surrounding Sartre and existentialism, and of French films that brought with them new behavior patterns and new kinds of dreams. And in the mid-1960s: "Marx and Coca-Cola."

And then, in 1955, there was the first of the *documenta* series of art exhibitions in Kassel — founded and initiated by one man: a Kassel academy professor, painter, and exhibition designer, Arnold Bode.

"As has often been pointed out, *documenta 1* was a reaction to a vacuum. After twelve years of Nazi rule and ten years of reconstruction, it enabled people to catch up with both information and ideas about twentieth-century art. After a barbarous interregnum, it set out to mark the return of German art into the continuum of modern European culture.

The HfG was swept along, both mover and moved, within

On opening day, Gropius and Bill shake hands. And there they are, on opening day, standing in front of one of the white Braun radios. For the handshake they keep smiling: the Bauhaus and the HfG shake hands. In front of the radio, Bill stoops, gesturing animatedly. Gropius stands upright, one hand in his pocket, critical and reserved . . .

Bernhard Rübenach, "Der rechte Winkel von Ulm," radio documentary, 1959

G 11 Super radio and record player
Hans Gugelot,
for Max Braun AG,
Frankfurt, 1955

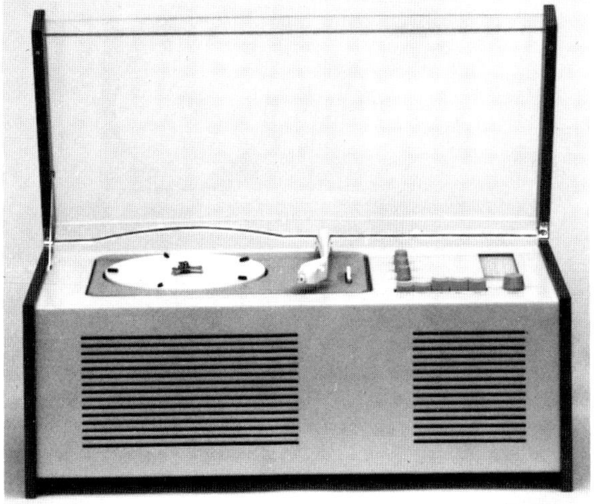

Walter Gropius and Max Bill
1955

SK 4 radio and record player
Hans Gugelot,
for Braun, 1956
*(also overleaf)*

the contradictory and paradoxical stream of German history after 1945, in which — as in the past — self-styled majorities made policies and formed taste; and in which, at the same time, there were rival attempts, as varied as they were contradictory, to redesign society and its forms of self-expression. This was not, by any means, "peaceful competition": it was the anarchy of the market structure of economic liberalism (on which the "social market economy" superimposed itself as an ideology).

An unholy mess, a complex process, in which design — *Gestaltung* — was an urgent necessity.

# The Pursuit of Reasons and Systems

Editorial Discussion

Erlhoff:
Ritual and system: both are elitist, both create only a group form of democracy, both offer security and means of identification. By virtue of its inner structure, the HfG actually stood for ritualization, and in its theoretical statements it stood for systematization: for verifiable criteria of objectivity and order. To find something that can be justified and is real: this combines Goethe's hope of finding an *Urgestein* or primordial rock form, the anarchists' hope of finding a primordial form of society, and the positivists' hope of finding a primordial explanation. Was the Ulm desire for a system just the desire to have something to take home in black and white?

Lindinger:
There was a whole series of movements, such as De Stijl, that pursued the ultimate ideal of finding colors and forms that would be indivisible and irreducible. The early Bauhaus was affected by this too. This idea of operating, as it were, with the primordial elements of design did not exist at the HfG. What did exist was the belief that chaos could best be countered by rationality, and that the mounting flood of new objects could be contained by systematizing them in modular building-block units. And as for methods, work at the HfG concentrated exclusively on their instrumental character.

Erlhoff:
There's one implication that may have been unthinkable then, but seems very bothersome now. It's this: all that faith in rationalization, objectivity, and operational control

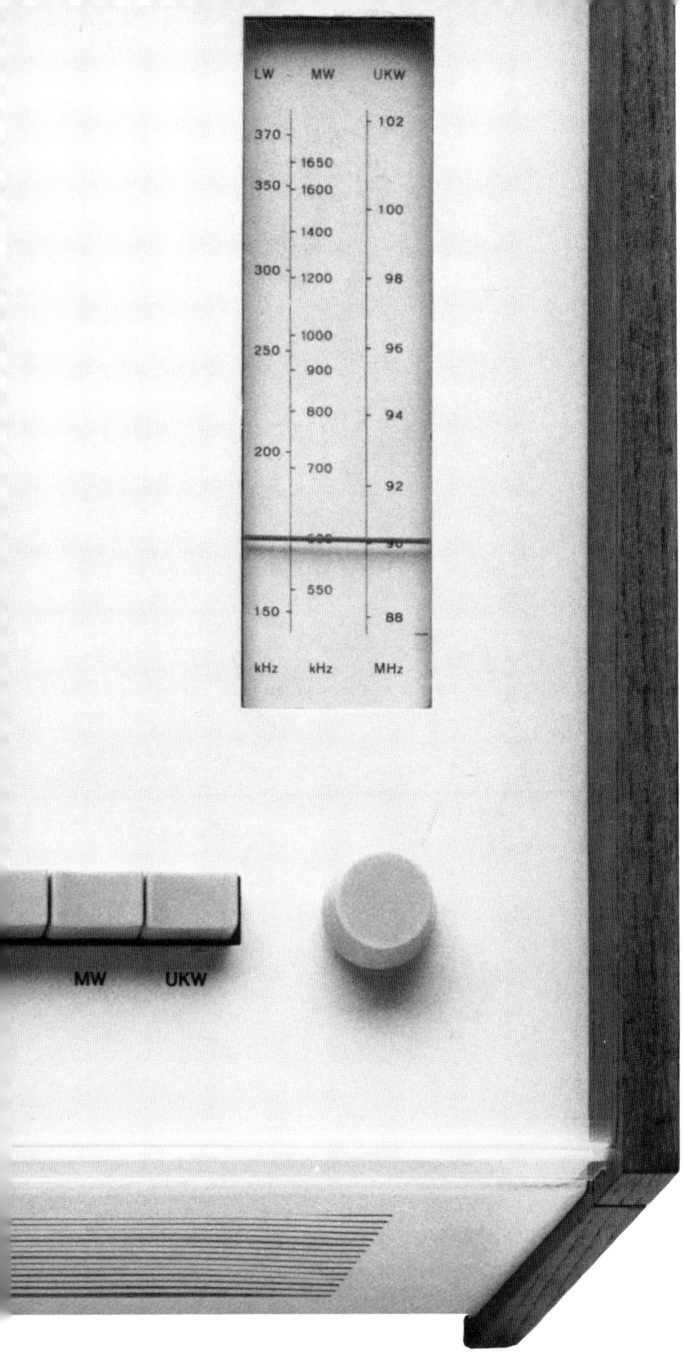


completely undermined the dialectic of the Enlightenment. And once Ulm ceased to subscribe to that dialectic, it was identifying itself with the common 1950s aspiration to be "value-free," and with what later came to be called "helpless antifascism": the belief that there are objects that lie outside politics.

Lindinger:
But that wasn't typical of the HfG at all.

Erlhoff:
That's just it: the HfG lived in a context of undeclared contradictions.

Lindinger:
The dominance of rationality at Ulm has a number of sources: firstly, we could all remember Fascism, with its attempt to rob human beings of their reason, to make deliberate use of symbols and unreason to enslave them.

In total contrast to this, we believed that this world could be made a better place; we believed in reason, and we believed that there was a place for us within the Enlightenment tradition. Bense and Bill, in particular, reinforced this attitude. Bense brought Hegel, Kant, and Descartes into almost every one of his classes. Bill, who as a person was by no means an extreme rationalist, stood for a theory in which argumentation, the ability to justify your own work, was strongly emphasized and indeed glorified. The whole Ulm approach is basically an Enlightenment one: it is an endeavor to find an organic

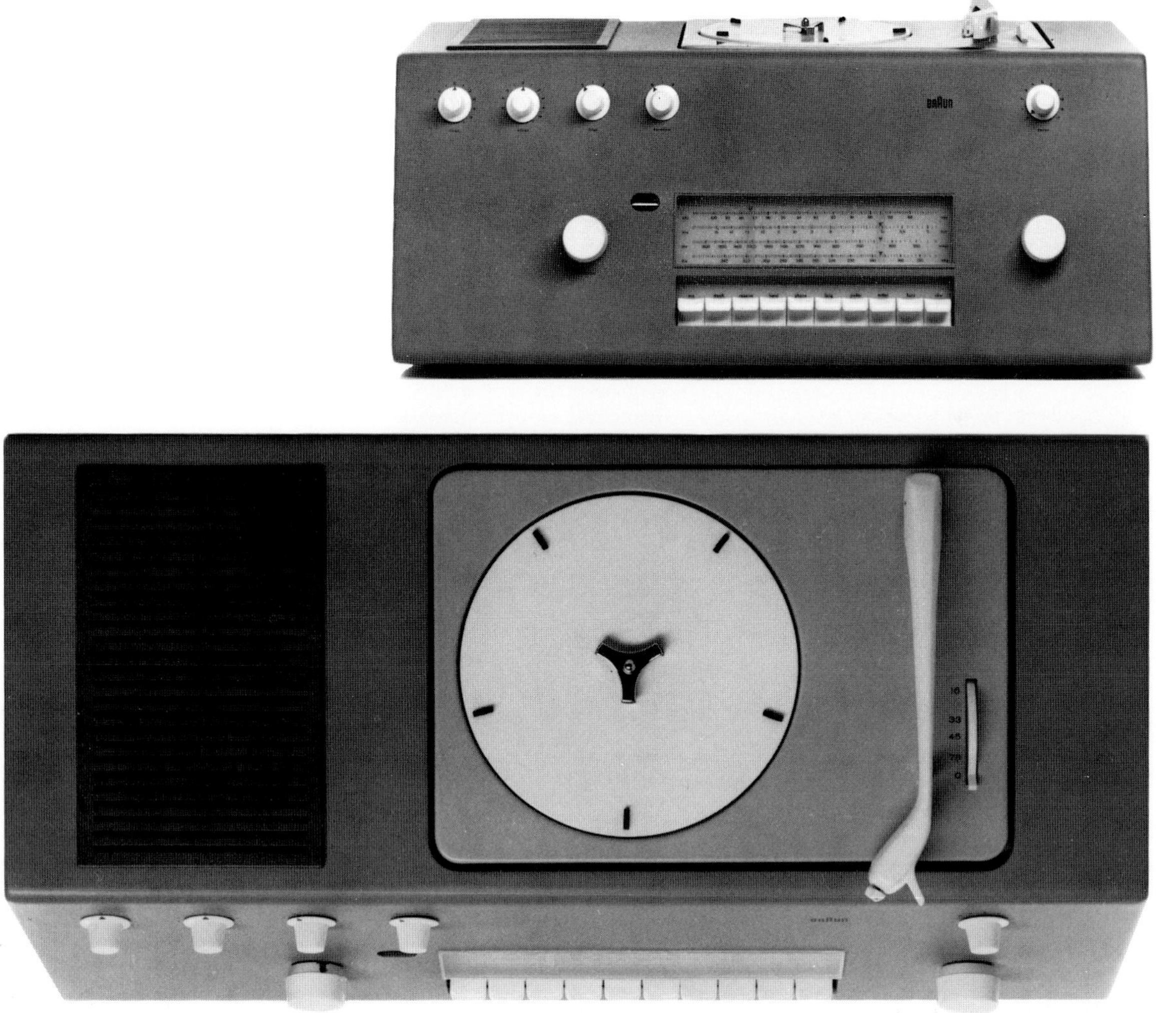

Studio 1 radio and record player
Hans Gugelot, Herbert Lindinger, for Braun, 1956
Casing in synthetic resin reinforced with glass fiber

unity between society and culture, on the one hand, and science and technology, on the other. There seemed to be no prospect of doing this if the founding premise was an emotional one. Inevitably, we were virtually forced to go to extremes, right down to the abandonment of value judgments; and there later had to be a course correction. It should not be forgotten, either, that our present-day critical approach to progress and technology, after Chernobyl and the whole ecological catastrophe, was not on the cards then.

It is certainly true that at Ulm there was a fixation on geometry as a visual language. The emphasis on rationality inevitably favored mathematical thinking in design. All this was very largely set off by Max Bill in his essay on "The Mathematical Thought Processes of Art in Our Century."

Erlhoff:
It probably wasn't yet possible to realize that Fascism largely pursued a rational course, and that — however horrible it may be to contemplate — the murder of Jews, Gypsies, and political opponents partly stemmed from plain economical facts.

History just does not consist of a dualistic relationship between Reason here and Unreason or Emotion there. Between these two poles there exist a variety of contradictory relationships. Those were the relationships that Ulm denied, as can be seen in its inflexible pursuit of systematization. I know that all those ideas about the systematization of building might have gained legitimacy

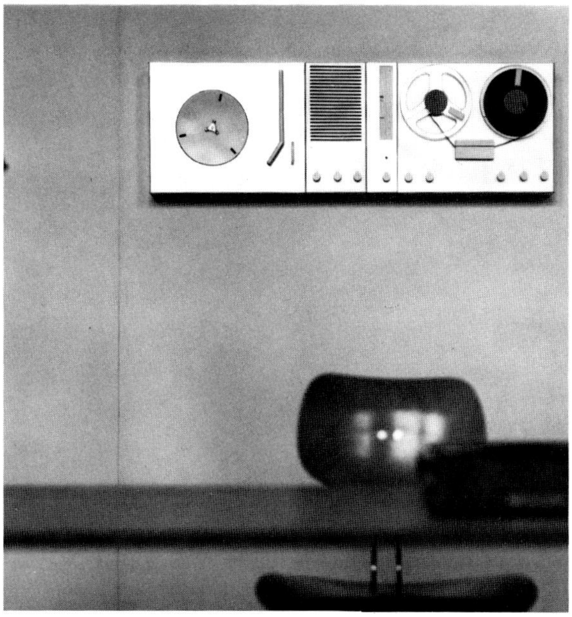

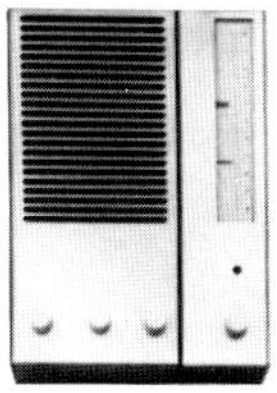

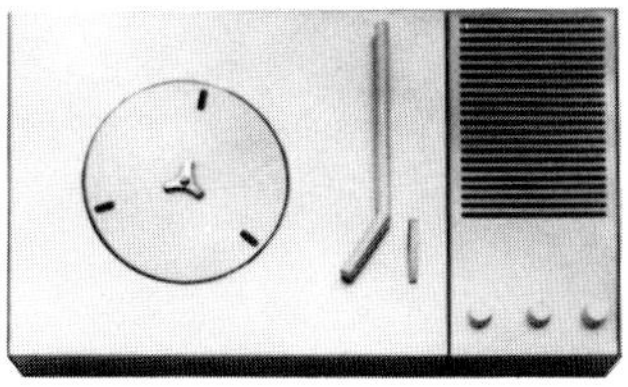

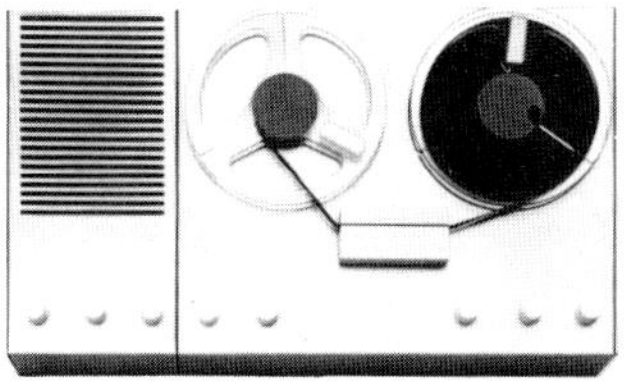

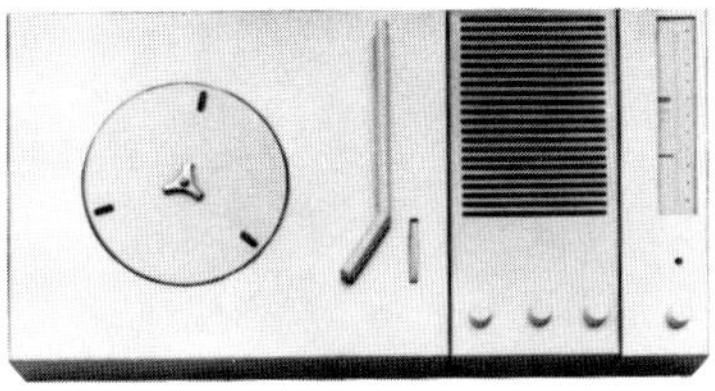

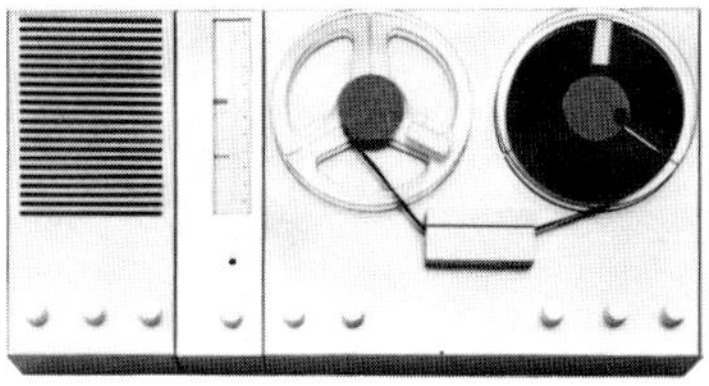

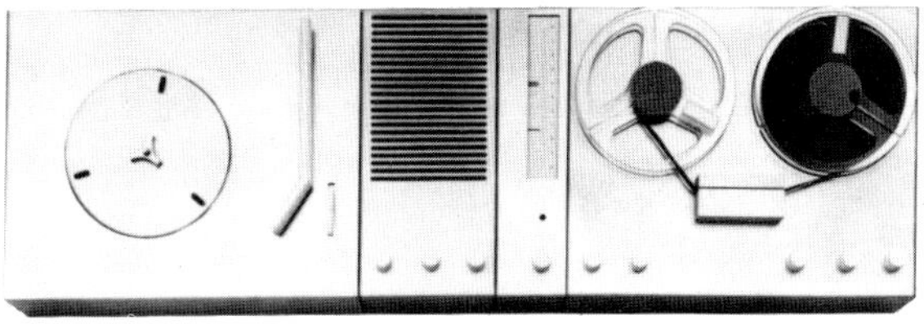

Hi-fi component system
Herbert Lindinger,
for Braun AG, 1958
Diploma work within the Gugelot product development group:
The first comprehensive Proposal for a Hifi-System.
A forerunner of the hi-fi systems of the 1960s that went into production in 1962 as the Braun audi 1 and audi 2

through the utopia of a democratic construction industry; and systematization did offer to make everything cheaper and more accessible to people. But even this democratic ideal is subject to a dialectic process.

Lindinger:
By excluding from our teaching, from the very beginning, not only art but taste and fashion, we freed ourselves to some extent from the emotive and irrational characteristics of these fields of activity. We set out to work in areas where we could indulge our craving for rationality.

That is one reason why Ulm people never went into the design of decorative items or interiors, or not with any great success, anyway (this is still true now, with a very few exceptions), and why Ulm teaching and Ulm theory have stayed alive and been practiced only in those fields in which the technological, economic, and latterly also ecological interests of society have been dominant. Not in the middle-class living room, not in fashion, but in objects of everyday use, in the office, in instrumentation, in machines, in hospitals, in the nursery, the bathroom, the kitchen, and the street.

Chemaitis:
But in people's minds Ulm does still play a stylistic role, even today, curiously enough. If you ask people today what they think is good, or what good design is, then very many Ulm objects tend to get mentioned, directly or indirectly. Even if these same people have all sorts of convoluted, mind-boggling objects in their own homes.

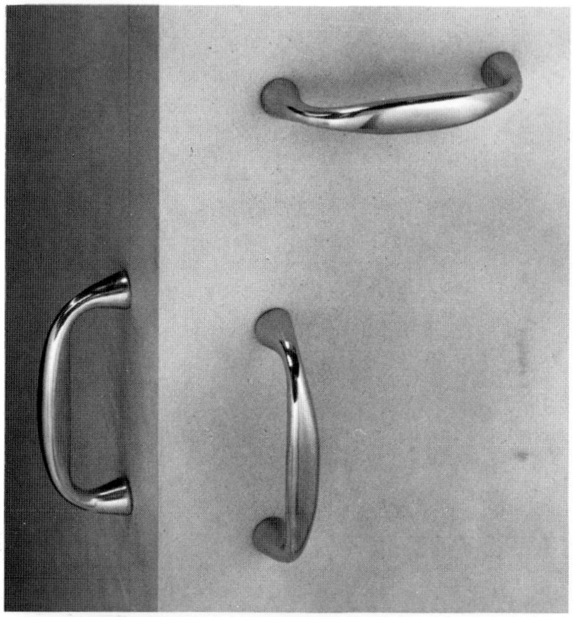

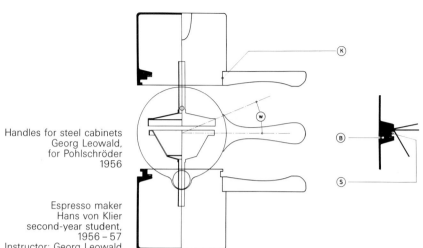

Handles for steel cabinets
Georg Leowald,
for Pohlschröder
1956

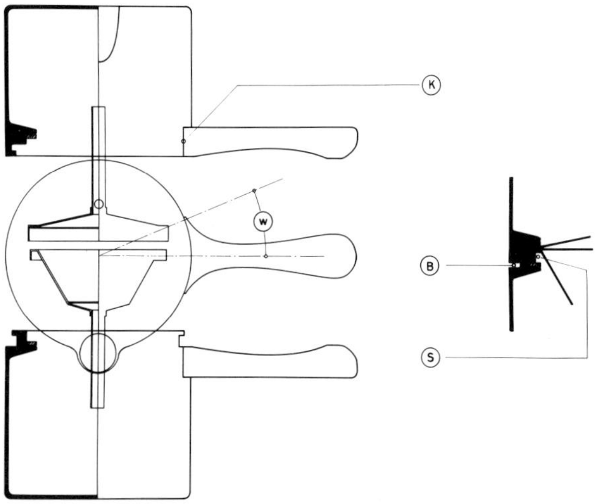

Espresso maker
Hans von Klier
second-year student,
1956 – 57
Instructor: Georg Leowald

Georg Leowald with
students Andries van Onck
and Hans von Klier

Writing implements
Hans Roericht
Third-year student,
1957 – 58
Instructor: Hans Gugelot

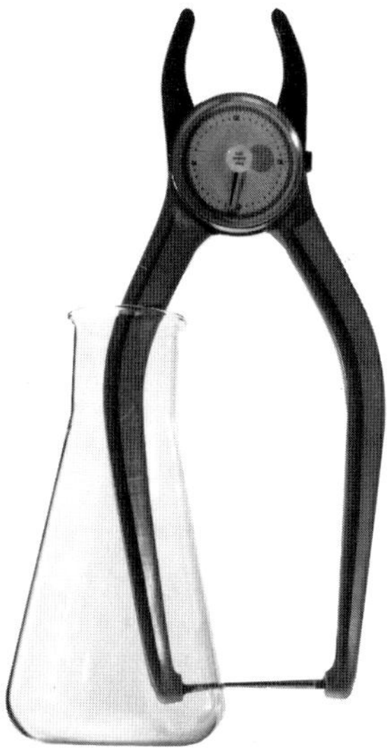

Double-ended calipers
Klaus Krippendorf
Second-year student,
1957 – 58
Instructor: Dieter Oestreich

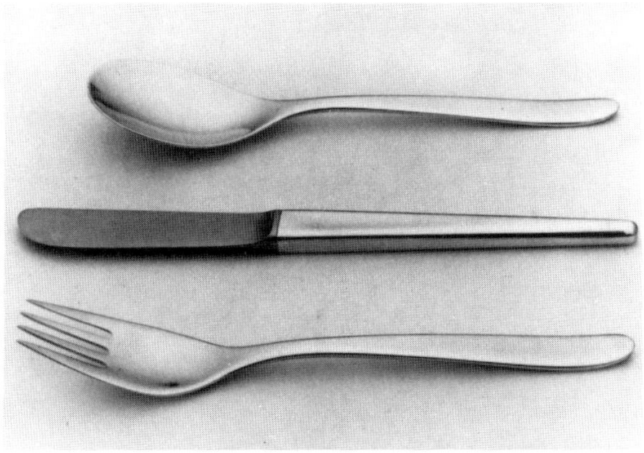

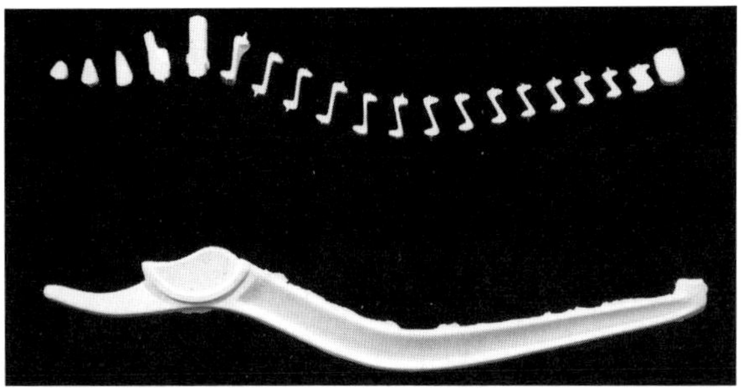

Flatware
Ernst Moeckl
Diploma work, 1957 – 58

Development of a pair
of spectacles
Herbert Lindinger
Second-year student,
1956 – 57
Instructor: Hans Gugelot,
for Angerer, Linz

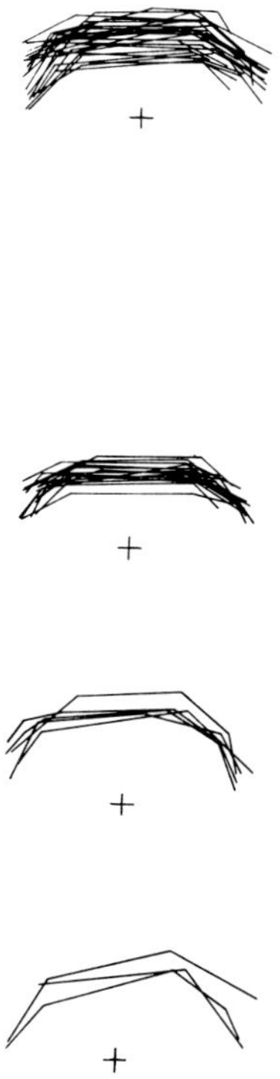

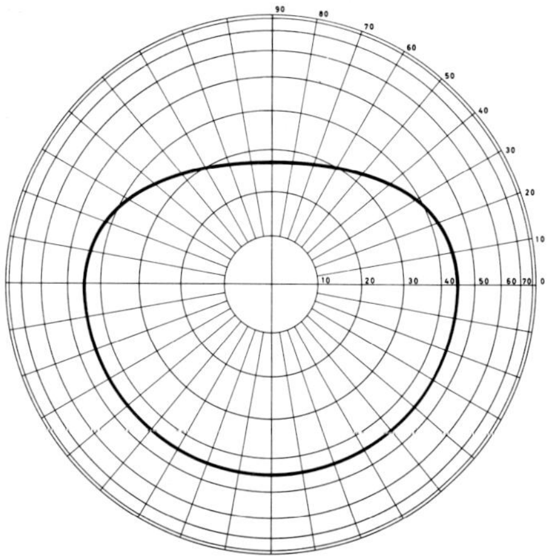

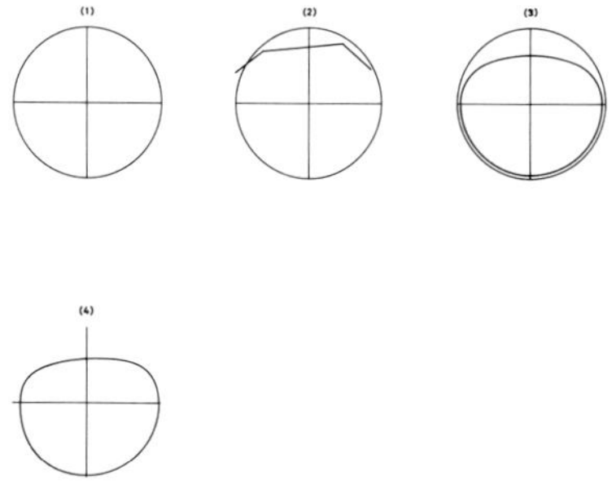

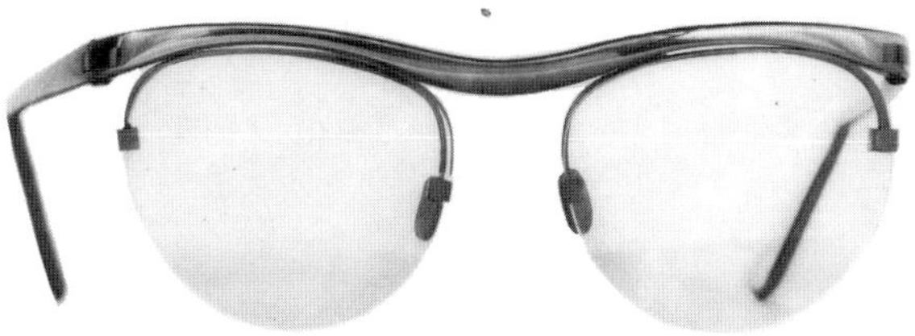

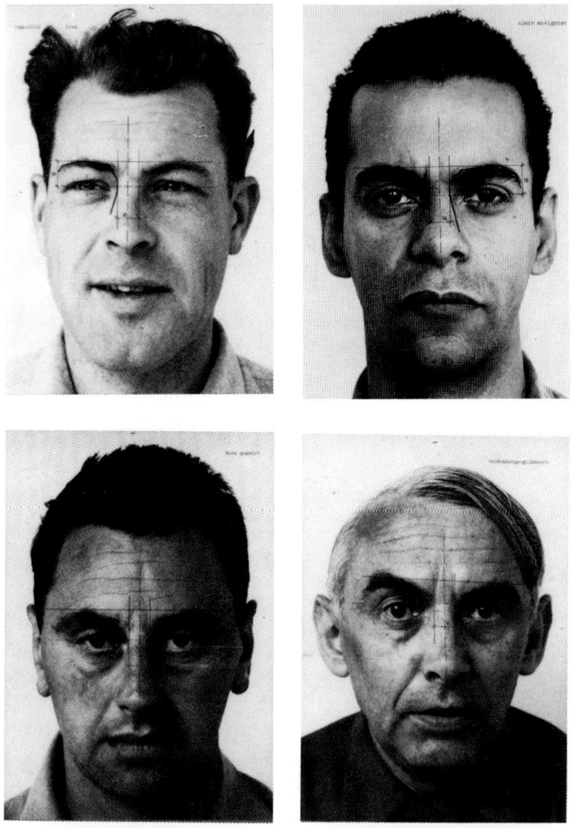

Eyebrow studies of
Paul Hildinger, Almir
Mavignier, Hans Gugelot,
Friedrich Vordemberge-
Gildewart, Hans Conrad,
Deborah Sussmann

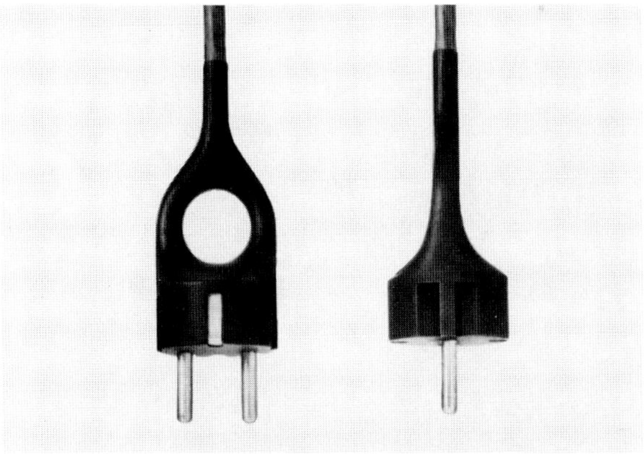

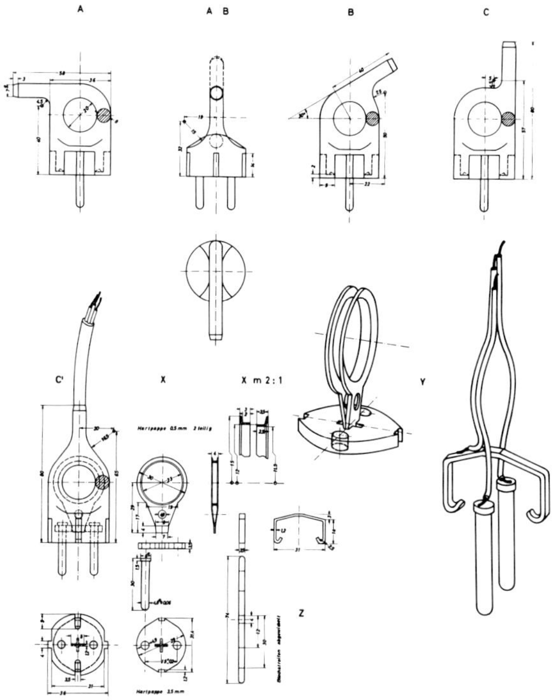


Electric plug in soft PVC
Aribert Vahlenbreder
Second-year student,
1958 – 59
Instructor: Walter Zeischegg

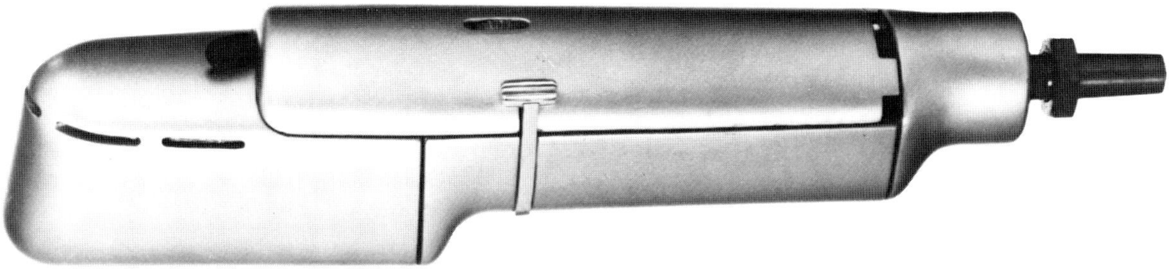

Multipurpose drill
Richard Fischer
Diploma work, 1958 – 59

M 125 furnishing system
Hans Gugelot,
for Bofinger, Stuttgart, 1957
The first comprehensive
flat-pack laminate furnishing
system

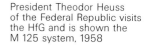

President Theodor Heuss of the Federal Republic visits the HfG and is shown the M 125 system, 1958

The idea of thinking in systems, and breaking down large complexes into units that can be produced using simple techniques, has found numerous adherents in the field of furniture manufacture, and the large number of current offerings in this sector only goes to prove the basic correctness of Hans Gugelot's conception. His main aim was always to develop connections, because any system presupposes a breaking-down into elements that have somehow to be put together again. It was this kind of analytical thinking, too, that led him to the 12.5 cm (4.9 inch) module, one-eighth of a meter, which the architect knows how to handle, and the multiples of which cover almost all the objects that people have to find room for in a cabinet or closet. . . .

Gugelot declared that as a matter of principle he would thenceforth develop only furnishing systems, not pieces of furniture.

Rudolf Baresel-Bofinger, in *Hans Gugelot – System Design,* exhibition catalogue, Munich, 1984

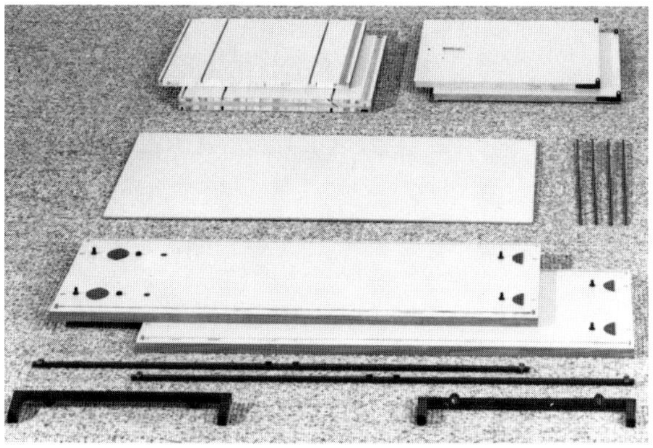

M 125 components

**The contribution of the industrial designer will be to enhance the use value of an article. Through his coordinating activity, his constructional ability, and his specialized concern with the relationship between human beings and the equipment they use, he, alone among the members of the team, determines the final structure of the product.**

Hans Gugelot, lecture in Tokyo, 1960

Animal construction toy
Hans von Klier
Third-year student, 1957–58

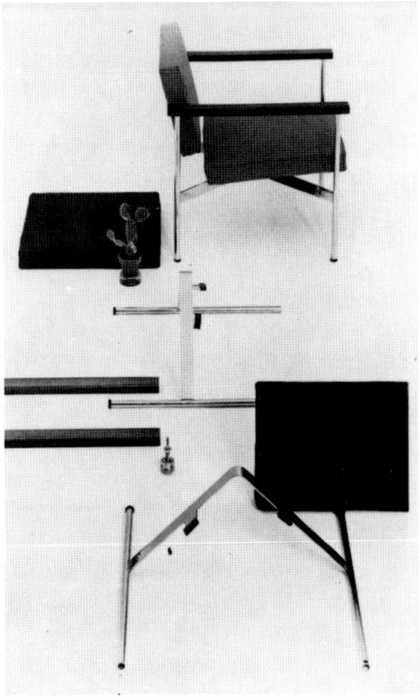

*right*
Gugelot development group:
Helmut Müller-Kühn, Hans
Gugelot, Herbert Lindinger
1959

*left*
Knockdown chair
Hans Gugelot
1959

Knockdown chair on the
*Kolonialstuhl* principle
Karl-Heinz Bergmiller,
Ernst Moeckl,
for Wilde & Spieth, Esslingen
1958

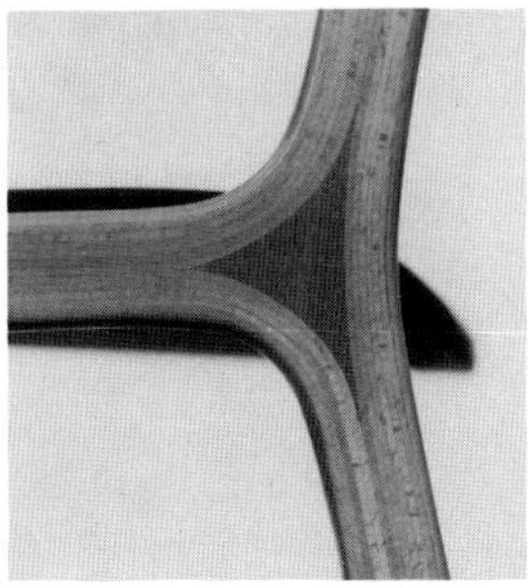 

Laminated wood chair
Wilhelm Ritz,
for Wilkhahn, Bad Münder
Diploma work, 1958 – 59
A chair using an innovative patented wood lamination technique; still on the market after 30 years. Selected for the permanent collection of the Museum of Modern Art, New York

Systems and modular component systems
Sets of compasses, unit furniture, kitchen machines with one power unit and a series of attachments, are common examples of modular component systems. Like systems in general, of which they form a subclass, they consist of elements. These elements must relate to each other, whether in their dimensional, qualitative, formal, or other properties. A system comes into being only when its elements are coordinated. Office papers, for example, belong to a system only if their dimensions form part of a series, and if typographical constants are established.

Gui Bonsiepe, in *Ulm*, 6, 1962

TC 100 stacking tableware
for hotel use
Hans Roericht,
for Thomas/Rosenthal AG
Diploma work, 1958 – 59
Selected for the permanent
collection of the Museum of
Modern Art, New York

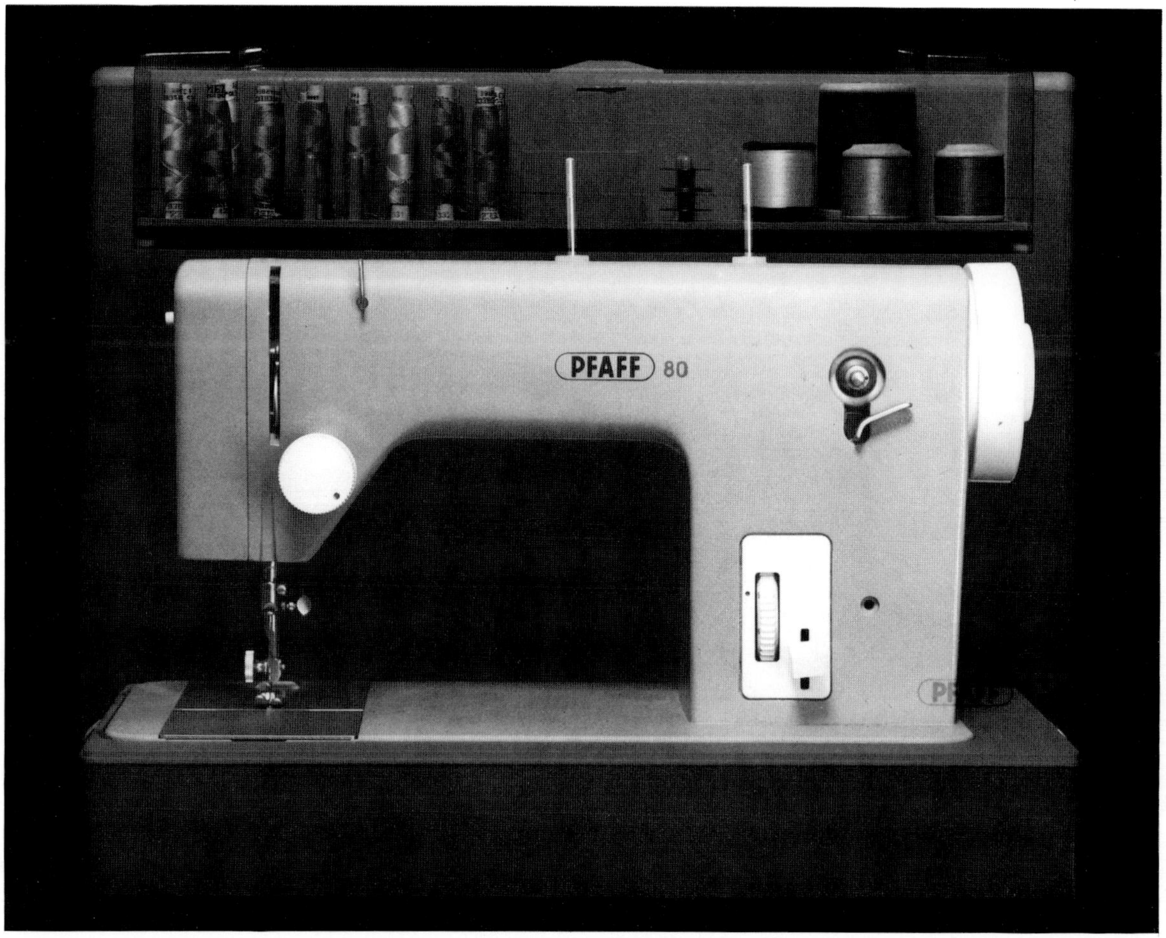


Sewing machine with case
Double exposure
Hans Gugelot, Herbert Lindinger, Helmut Müller-Kühn, for Pfaff AG, Kaiserslautern
1959 – 60

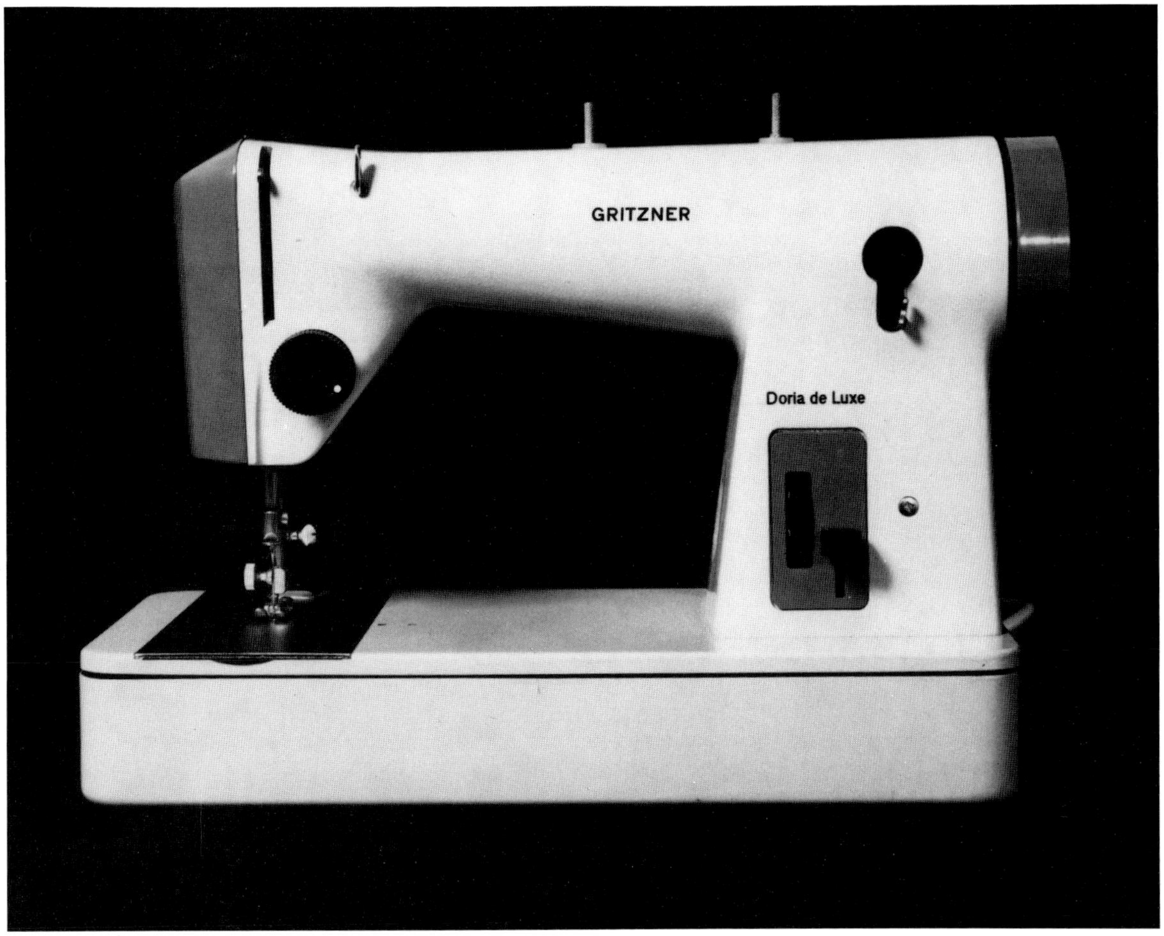


Sewing machine
Hans Gugelot,
Herbert Lindinger,
Helmut Müller-Kühn,
for Gritzner, Karlsruhe, 1959

U-Bahn subway system, Hamburg
Hans Gugelot, Herbert Lindinger, Helmut Müller-Kühn
Graphics: Otl Aicher, Peter Croy, for Hamburger Hochbahn AG; manufacturers, Linke-Hoffmann-Busch, Salzgitter, 1960 – 62

*above right*
Herbert Lindinger and Helmut Müller-Kühn at work on the development of train seats

*above left*
Interior of a car, U-Bahn, Hamburg

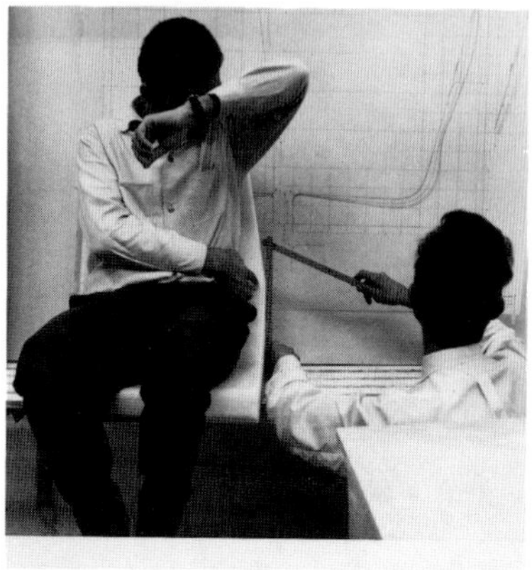

# The Ulm Lifestyle

Editorial Discussion

Erlhoff:
Life at the HfG, for students as well as instructors, was marked by a set of specific attitudes and gestures that became famous as the Ulm mystique: you weren't allowed to ride a bicycle, everyone wore black, you had to throw your bobby sox away and cut each other's hair.

Lindinger:
That's true in principle; anyone who did not abandon every vestige of bourgeois costume within the first two weeks was a marked person. The hairstyle changed first, then the shirts and the suit materials. It's not true about black, though: the great idea then was to get as far away as possible from bourgeois materials. The only cloth you could get in those days was a kind of drill called *Schwesterndrillich* — nuns' veiling — and otherwise only certain coarse materials that workers wore. We put our own clothes together out of these materials. In those days there were still plenty of people who did tailoring work cheap. Old suits were dyed, as well — one of the busiest people in Ulm was the dyer. The instructors and assistants mostly wore smart grey suits, and the students sometimes wore sweaters. There was a certain leveling effect at the HfG, which wasn't made explicit anywhere. There were unwritten rules that were simply in the air. You caught the smell of them, and anyone who failed to detect them soon became an outsider.

Chemaitis:
And the story goes that there were parties every weekend, and smaller ones every night.

*left*
Front view of train

*right*
Side view of train

The designers evolved a series of ideas that set new European standards in rail vehicle engineering: the consistent use of the new material of fiberglass, the one-piece floor-to-ceiling fiberglass side walls, the floor curving up to meet the side wall, and the uninterrupted strip lighting.

Erlhoff:
All this was colored by the close relationship that existed between students and instructors. It's hard to imagine nowadays how close the relationship was then.

Riemann:
It's interesting that people who were in different years used the formal form of address, *Sie,* to each other, and people didn't normally use the informal *du* inside the College.

Erlhoff:
In the photographs, life at Ulm tends to look a bit like the French films of the period: haggard-looking types, in rather loose, baggy suits that look as if they ought to be neat and tidy but never are.

Lindinger:
Shoulder pads were absolutely out.

Chemaitis:
There was a glimmer of existentialism showing through.

Lindinger:
A lot of things came together: the campus idea, indissolubly linked with the isolation of life up there on the hilltop, out of town; the revolutionary attitudes to life and design, which did not fit in with the small-town ideas of Ulm at all; the alien terminology; the wild parties. It all hung together.

Erlhoff:
But this drove the faculty and students of the HfG — like other intellectuals of the period — into a curious paradox: they had to come across as an elite, although what they really wanted was to hit the world with a grand democratic gesture. They were forced into an anomalous position. Up there on the Kuhberg, with all the

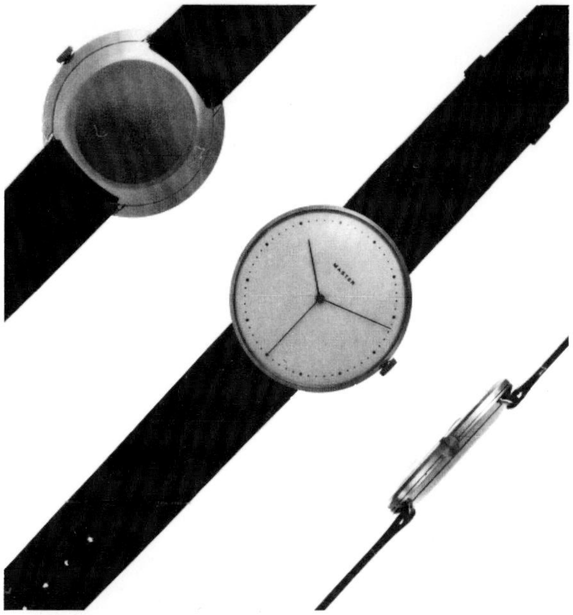

Wristwatch
Reinhart Butter
Third-year student, 1960 – 61
Instructor: L. Bruce Archer

Of course, all these things are also tasks for the engineer, or the architect, or the scientist. But the designer is not competing with all these specialists; on the contrary, he complements their work, and also controls an important area of no-man's-land between all these specialties. He takes as his starting point the purpose and function of the thing to be designed; his design has to be optimized from many viewpoints at once; his point of departure is the planning of the whole creation, in its complex interaction with its environment. For example, he looks into the suitability of the design for its eventual user or recipient, into its ease of production, its characteristics in use, and so on. Whereas the engineer sets out to achieve technological fitness for function, and the artist, perhaps, to achieve aesthetic expression, the designer has to go further and consider other factors: those that belong to no-man's-land. A car that technically works well need not look good at all; and a good-looking one can be bad. So the designer's work does not by any means end on the outer surface of things. He is not a cosmetician of external form, or a casing specialist. An object can be brought into line with the complex demands that are made of it only if what is under the surface is thoroughly understood, and in most cases this can happen only in collaboration with the appropriate specialists. The designer is not an engineer, and he is not an artist: his professional qualifications are of a new and unique kind. . . .

The contemporary urgency of the problem is obvious. We live in highly technological environments that consist almost entirely of industrially produced components; we are plugged into complex communications networks such as radio, television, press, and transportation; at the same time, we are actors in the various scenes of our complicated social structure: as consumers, as taxpayers, as purchasers, as voters, etc. With the growth of immense systems of production and communication, the design of a chair becomes a difficult problem. Even 150 years ago this was a matter between the craftsman and the client, who was also the user. Now a product run of 20,000 chairs has to be planned for an anonymous mass of users who defy description except perhaps in statistical terms; the making is done in a largely mechanized production facility; and the distribution process passes through a number of intermediate stages. And then there is a market with wide ramifications and a powerful dynamic of taste. Faulty planning would have far-reaching consequences.

Horst W. J. Rittel, in *Werk*, 8, 1961

rituals that went on, and with the Basic Course as a fully fledged initiation rite, it's hard to keep away from the image of a monastery.

Lindinger:
Yes, but all this changed a great deal over the fifteen years of the College's existence. The students' whole aspect and demeanor underwent a transformation. We mustn't forget that in the initial phase many of the students were still feeling the effects of the pre-1945 situation in Germany, and many would have been studying much earlier if they had not been prevented by the war. So they came to Ulm as adults, many of them with bitter experiences behind them. These were people who had all the enthusiasm to try something completely new, and they didn't much care where they went. For instance: in the design field, in the mid and late 1950s, there were no employment prospects whatever in Germany. The occupation of a designer just did not exist. And yet young people came and studied it — in the fanciful expectation that they could make the world move their way.

Riemann:
But all that changed around 1957.

Lindinger:
Yes, as West Germany became increasingly prosperous, a generation of students arrived who cared about diplomas; the first students had not given a thought to such things. Suddenly, the diploma became an issue, and so did the formalization of the curriculum, which arose from student pressure.

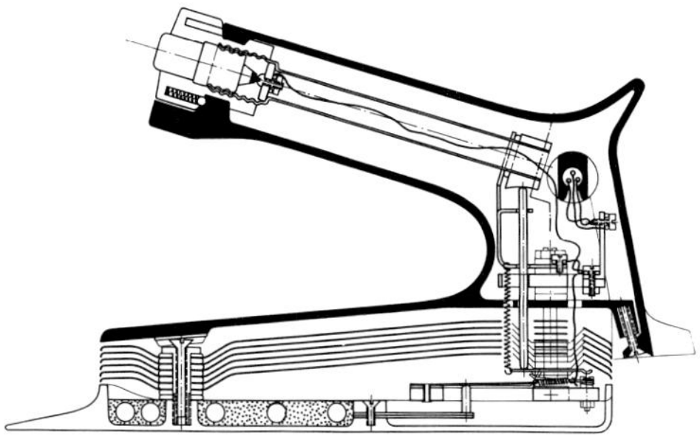

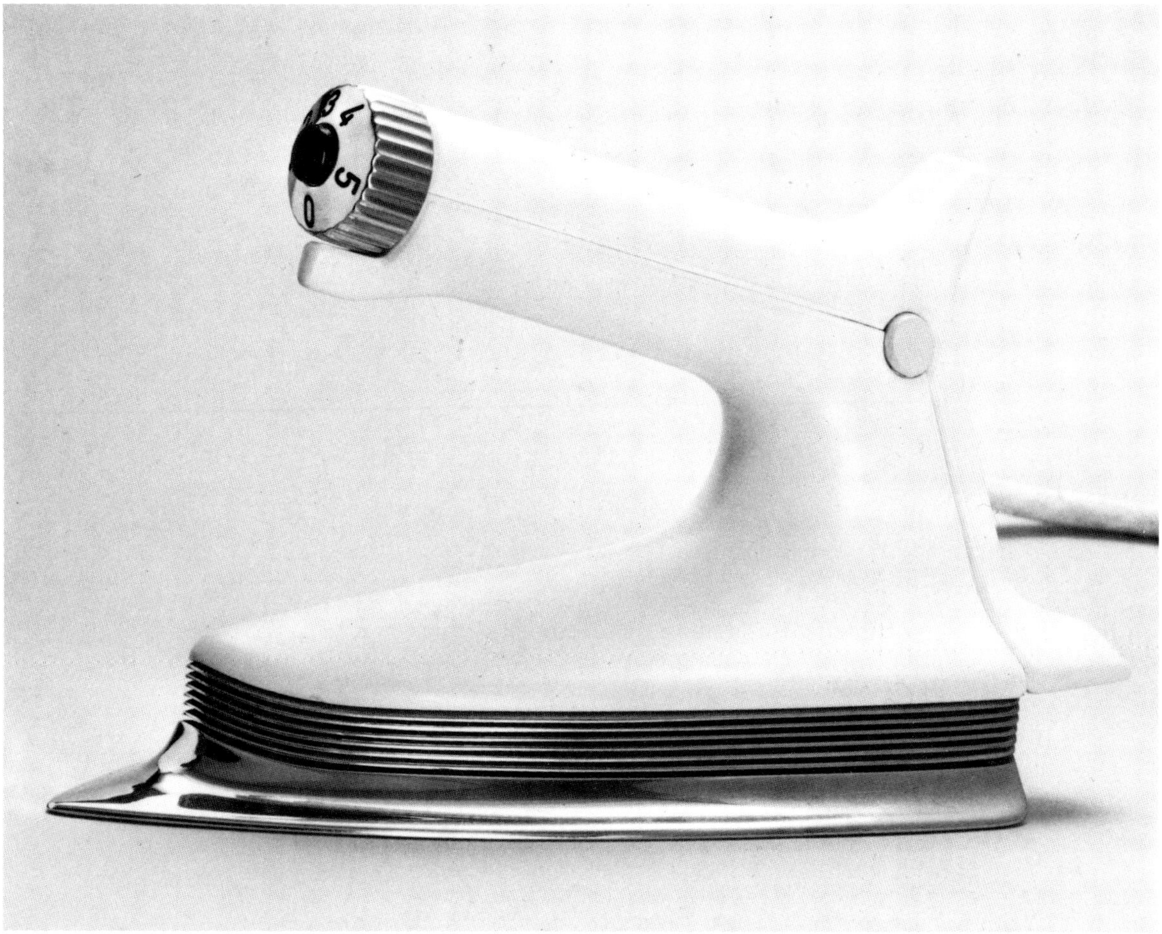

Iron
Reinhold Weiss
Diploma work, 1958 – 59
Open handle in contrast to the then customary closed handle, to improve convenience in use.

Naturally it all started in the Building department, because it's hard to get to build anything without a recognized diploma. But the other departments came under increasing pressure as well, and the College tended to become more academic, and the freedom of choice in teaching methods was reduced. The HfG had made promises, and it now had to keep them. The transition from the planning stage, the utopia, to the real world leaves marks behind it.

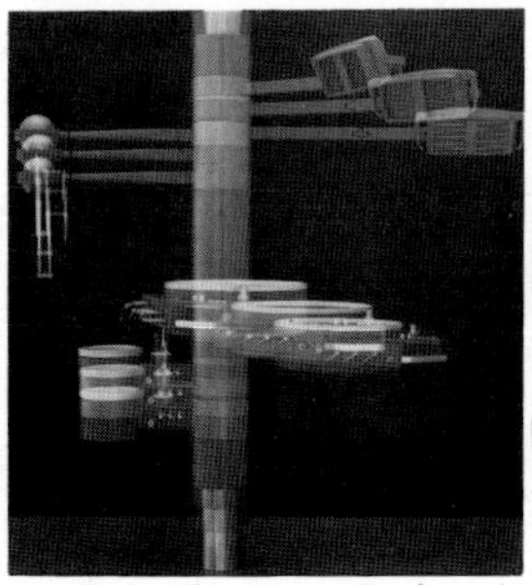

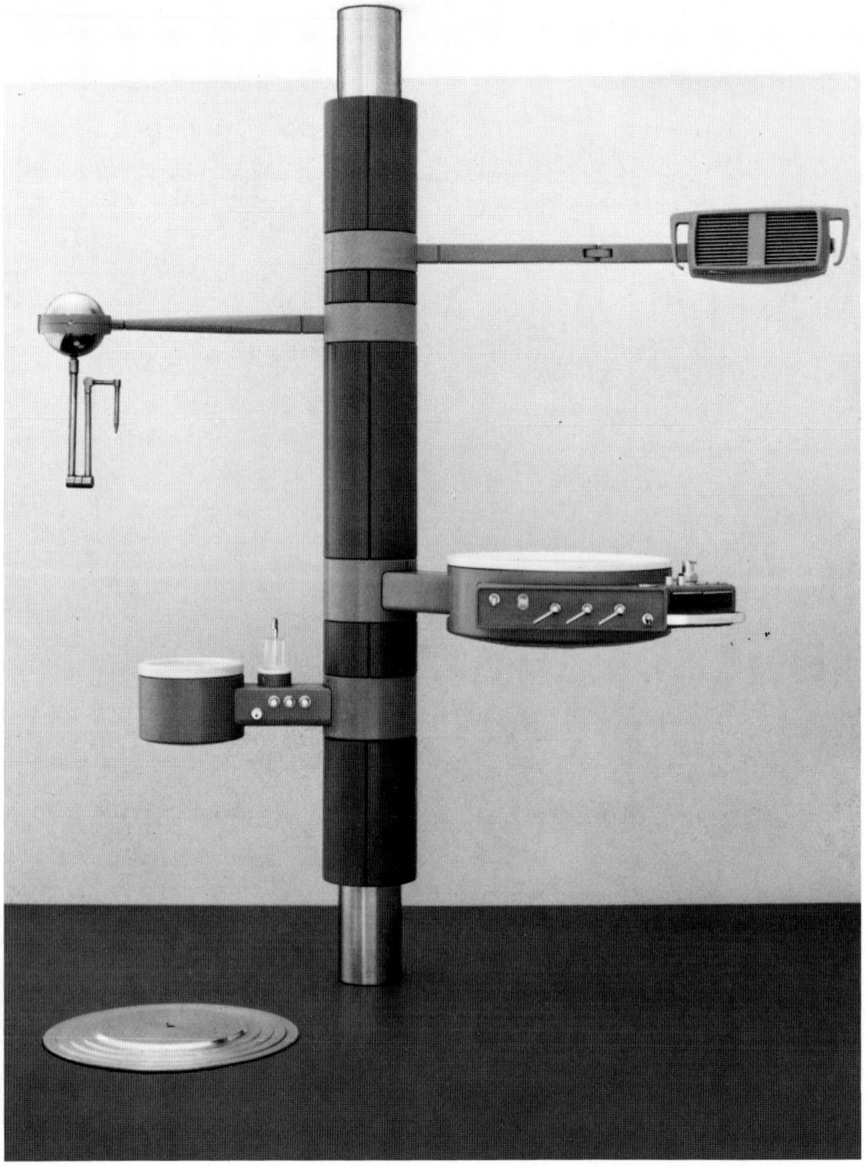

Dental unit
Peter Emmer, Peter Beck, Reinhold Deckelmann, Dieter Reich
Third-year students, 1961–62
Instructor: Walter Zeischegg

# Ulm Students

Editorial Discussion

Erlhoff:
How did you get to be a student at Ulm? Did you send in an application? Was there a selection process?

Lindinger:
There were those Spartan-looking Ulm prospectuses, and those were very unusual for those days, both in their graphic design and in their rhetoric. It was all printed in lowercase, too. Everything about it was a signal — a signal, in my own case, to drop everything and rush to get there. At the College you naturally had to show that you had some aptitude. Religious and political views were disregarded on principle, as were high school diplomas and all other evidence of graduation. This was another way in which the College set out to break with the past, both educationally and politically. There were, however, questionnaires on culture and society that had to be filled out.

Chemaitis:
What they were looking for was clearly a certain general education in the sense of "cultural integration," and some involvement in political, social, and cultural life.

Lindinger:
What was called for was not deep or thorough knowledge but an unconventional attitude to the questions that were asked.

Erlhoff:
What were the chances of admission?

Lindinger:
As far as I can recall, about a third of the applicants were accepted. But a good third of those usually had to leave after

Course assignment: sanitary installation for residential use
Instructor: Walter Zeischegg
Students: Robert Graeff, Walter Kiehlneker, Heinz Wäger

Detailed brief

Appearance
Sanitary equipment and fittings must not have unduly hard edges; upper surfaces smooth; colors light and inviting.

Arrangement
The hygiene system must be applicable to the greatest possible variety of ground plans; toilet separate from bath area as far as possible; controls within the user's reach; accessories built in; adequate provision for storage (toilet tissue, soap, etc.).

Cleaning
Hard-to-clean angles must be avoided; rounded transitions between surfaces, with adequately large radii; floor must be cleanable with a long-handled implement, with no need to bend down.

Construction
Plumbing to be built into the system itself; only the elements necessary for the user to be visible; system to be freestanding (not supported by walls); noise generation and noise transmission to be avoided.

Manufacture
The system is to be mass-produced. Use of new and appropriate materials; quick and simple assembly.

Climate control
Draft-free ventilation; wall and ceiling to absorb steam.

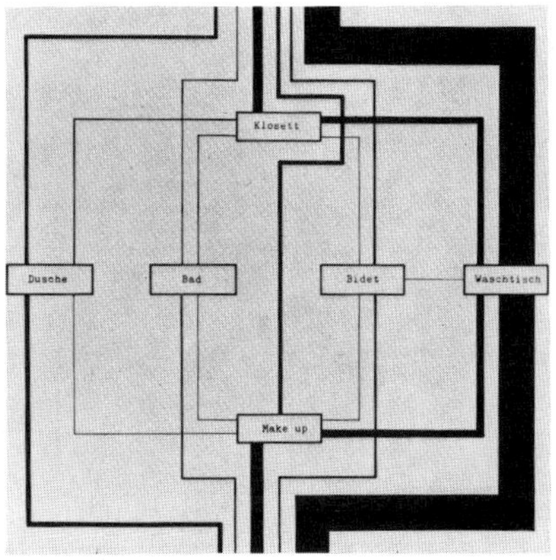


Diagram showing use patterns for hygiene area

For two trimesters in the academic year 1961–62, third-year students of the Product Design department under Walter Zeischegg made a series of designs for so-called *Sanitärzellen*. In contrast to current practice at the time, the individual components were to be looked at and developed not as separate pieces of equipment but as parts of a system. The brief covered the design of the individual items, both form and structure, and their assembly. The extensive functional analyses were carried out by the students as a team. Differences in the interpretation of the resulting information lead each student to put forward his own set of solutions.
From *Ulm*, 8/9, 1963

Project conference: Walter Zeischegg and students

the first year. There was another weeding-out in the second year; and so there was quite a radical selection process – in the course of which, naturally, some glaring errors were committed.

Ulm is a beautiful and memorable city in my youthful days and is also my "Heimat". . .
Sennosuke Shimizu, Tokyo, 1987

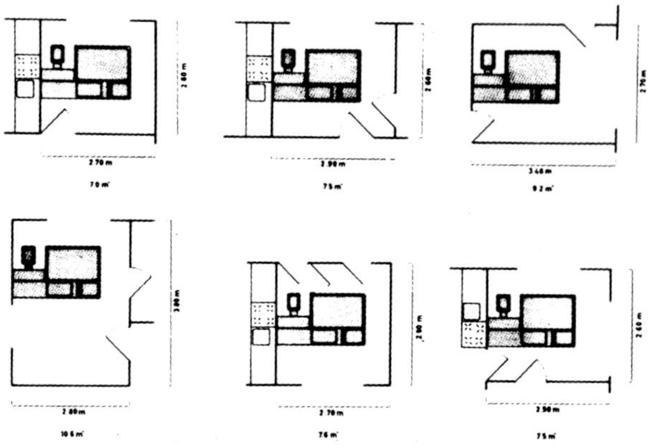

*above right*
Examples of room plans incorporating the sanitary block

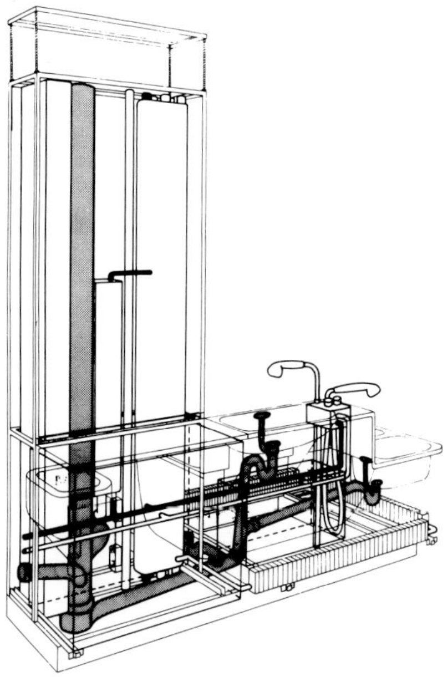

*left*
Diagrammatic view of plumbing, showing room heaters, toilet cistern, and space for washing machine and water heating

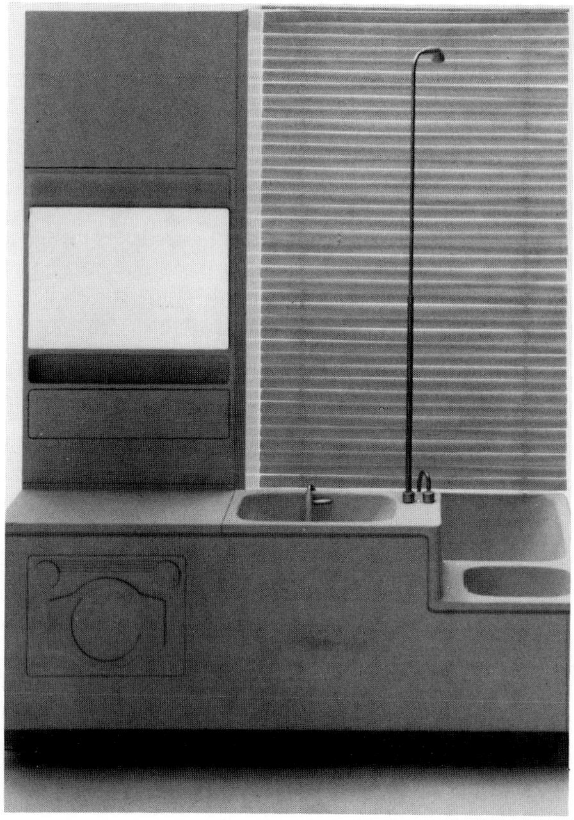

*below right*
Sanitary block
Walter Kiehlneker
Third-year student, 1961–62
Instructor: Walter Zeischegg

All of the plumbing is combined in a single block, which stands free of the walls of the hygiene room. This design was the result of the following considerations: because the system is free of the walls, the windows and doors can be sited on any wall of the room; because the unit is relatively independent of the shape of the room, it is suited for mass production; the internal spaces created by the individual items fit together, and better use can thus be made of them; the connecting pipework is short and can be installed in the factory; it is easily accessible for repairs; at the structural stage of building work, all that needs to be done is to leave a hole in the floor for the plumbing. The washing machine, which forms part of the "wet area" of any dwelling, is incorporated in the hygiene unit.

From *Ulm*, 8/9, 1963

*left*
L. Bruce Archer in a seminar

*right*
Charles Eames visits
the HfG, 1955

Grader
Klaus Krippendorf
Diploma work, 1960

Braun Sixtant shaver
Hans Gugelot,
for Braun AG, 1961
First use of matt black

## The Kuhberg as a Male Domain

Editorial Discussion

Erlhoff:
It's very noticeable that there weren't many women around at the HfG.

Lindinger:
Yes, that was one disastrous thing about Ulm. When people say that it was like a monastery, that's not only because of its isolated position but also because there was just not enough of a female element.

Erlhoff:
So Ulm became a male society.

Staubach:
The women at the HfG: were they treated as equals? Or wasn't it the case that they needed a whole lot more dynamism than the men, even to avoid being relegated to so-called feminine roles and duties? Was the image of women at the HfG a traditional one, and did the women conform to the traditional patterns?

Lindinger:
No, it was rather the other way around. If I look at it from the viewpoint of today's liberation movement and present-day ideas about the role of women, it becomes clear that there were many women there who had fought for their new roles and who had got there by sheer professional competence; and that was something that was not customary at all. So there were the first stirrings of the new consciousness.

Chemaitis:
Ulm people were anything but representative. An avant-garde can't be representative.

Erlhoff:
But how many of the women became full professors, for instance, compared with the

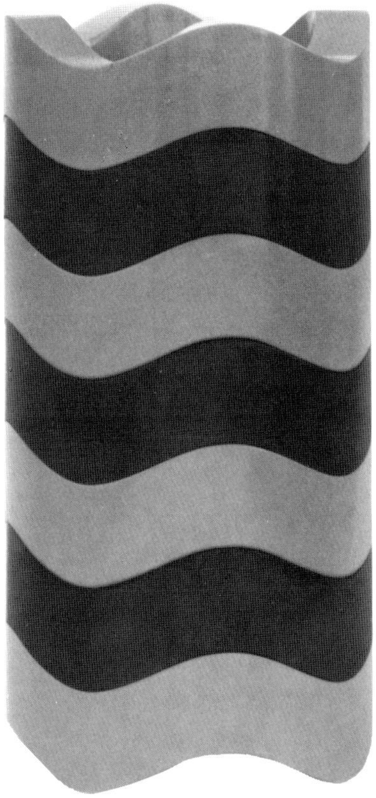

The department lays particular emphasis on the design of apparatus, machines, instruments – products largely untouched by the craft tradition. The design of ornamental or luxury goods is not within the brief of the department of Product Design.
*Ulm,* touring exhibition catalogue, 1963

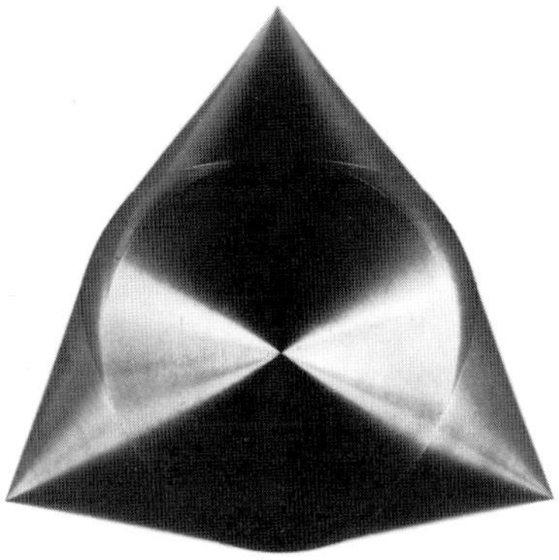

Stacking ashtray
Walter Zeischegg,
for Helit
1966 – 67

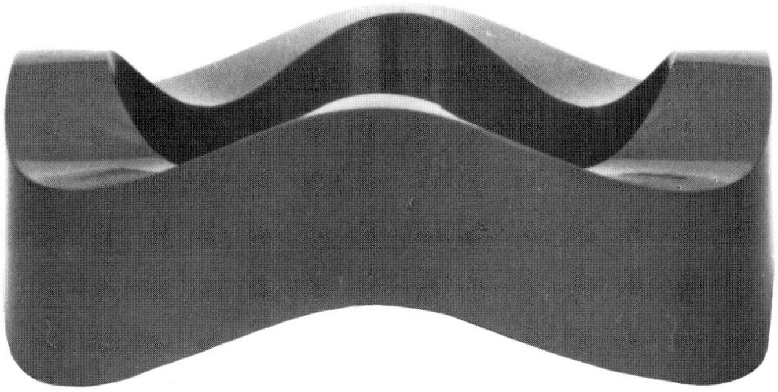

Tetraconosphere
Walter Zeischegg
1966

10 percent of Ulm graduates who became full professors somewhere or other?

Riemann:
Not many. Cornelia Vargas taught in Chile, and Eva Pfeil at Auburn University.

Lindinger:
Even so, quite a few had successful careers: Ilse Grubrich was Editor-in-Chief at S. Fischer Verlag for a long time and edited the complete edition of Freud; Margit Staber was Director of the Kunstgewerbemuseum in Zürich; Deborah Susman did the color design concept for the Los Angeles Olympics; Ulla Stöckl and Janine Meerapfel made some interesting films and won major prizes.

Chemaitis:
In the College itself there were never many women instructors. Helene Nonné-Schmidt . . .

Riemann:
. . . and Käthe Hamburger, Helge Pross, and Elisabeth Walther.

One sometimes comes across the idea of design as a magic formula that will miraculously solve all the problems of our environment: design as a philosophy of life. But any attempt to see the whole of human experience in terms of design, as a theory of existence, is doomed to failure. From the coffee spoon to the city: yes, but not if it means falling prey to a uniform, processed image of reality. There is no such thing as a perfect, definitive coffee spoon – any more than there is a perfect, definitive city. What can exist is a series of approximations to the best conceivable coffee spoon, and approximations to the best conceivable city, within a given culture and a specific social order. But where design takes on the function of a restrictive ideology, the designer takes on the guise of a Grand Inquisitor, handing down justice and mercy. This attitude is exemplified in the relationship between designers and industry. It turns the designer into someone who dictates the form of the products from outside, in accordance with the motto: "The designer gives the orders, the engineer obeys."

Tomàs Maldonado, from a speech at the HfG, 1957

UniData system
Tomàs Maldonado,
Gui Bonsiepe,
Rudolf Scharfenberg,
for Alex Linder GmbH
1963 – 64

Tekne 3 electric typewriter
Ettore Sottsass in association with Tomàs Maldonado,
for Olivetti, Milan
1960

Tomàs Maldonado and
Ettore Sottsass

Typewriter keys
Preliminary studies
for Tekne 3
Tomàs Maldonado,
for Olivetti, Milan, 1960

## Inge Aicher

Editorial Discussion

Erlhoff:
Let's talk about Inge Aicher: her part in setting up the HfG, her total commitment to the creation of such an institution, and the intentions on her part that released such a volume of energy.

Lindinger:
I don't think we can do justice in this book to her importance for the HfG as an institution. And if anyone now asks where all that strength came from, I think that Inge Aicher – or Inge Scholl as she still was then – derived from her experience under the Nazis, and from the execution of her brother and sister, the unstoppable will to find something new that she could set against all that.

However, this alone does not explain her part in the development of the HfG. The fact is that without her the College would never have existed. To get together a million marks, at that time, when everything was in ruins: that was an unbelievable achievement. No intellectual, given the time and the circumstances, would ever have taken it on. I think it's all in Inge Aicher's character. There's something unshakable there, and at the same time relaxed: she has the faith that moves mountains.

Chemaitis:
That relaxed quality was no disadvantage; in fact, without it the College would never have been set up. Given a moderately clear notion of what was in store, no one would ever have got started.

Lindinger:
Inge Aicher's persistence is also shown in her relationship with the city of Ulm: there

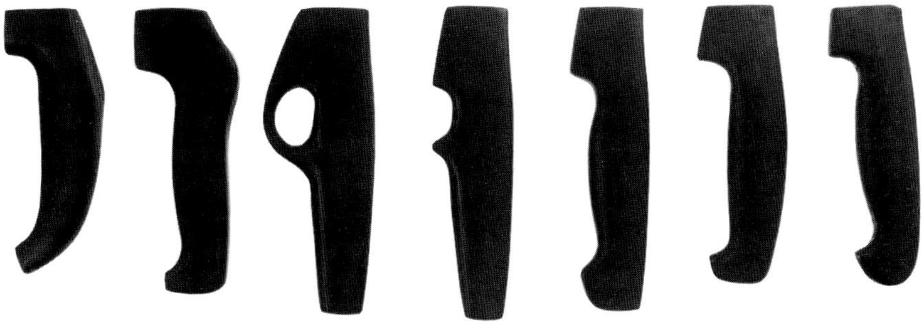

Handles for butchers' knives
Peter Staudacher
Third-year student, 1964 – 65
Instructor: Peter Raacke

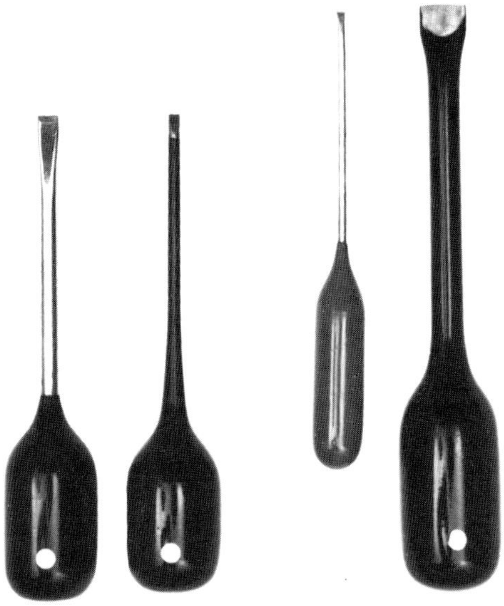

Screwdriver
Frank Hess
Third-year student, 1963 – 64
Instructor: Peter Raacke

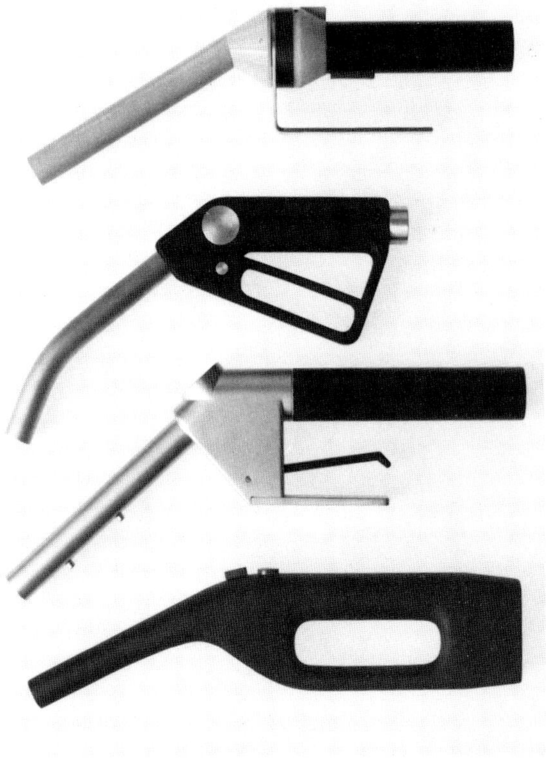

Filling station nozzles
Peter Hofmeister, Werner Zemp, Verena Loibl, Edith Ross
Second-year students, 1964 – 65
Instructors: Gui Bonsiepe, Peter Raacke

have been only two occasions in the history of Ulm since 1945 when the citizens have organized a petition for a referendum under the city constitution: one is happening now, over the Richard Meier building planned for a site right in front of the cathedral, and the other was in 1952, when the old guard of Ulm inhabitants petitioned against the founding of the HfG.

Erlhoff:
It may be important, nevertheless, to point out here that in 1945 Inge Aicher was not alone in trying to foster democracy in Germany. It's well known how many magazines were started in the first few years after 1945, with titles that voice aspirations toward a democratic utopia; and how many organizational forms for a democratic Germany were tried out. This was the context in which Inge Aicher worked — with her unshakable determination to make democracy a reality, and her faith in the feasible.

Lindinger:
That faith in the feasible, the faith that things could be changed, was everywhere at that time, and it was linked with a limitless belief in progress. This was reinforced by Inge Aicher's infectious sense of conviction. For years on end, she succeeded in maintaining the finances needed to keep the HfG running, and she got money for the institution from all sides, inspired by the conviction that this idea, and this college, could not and must not be under state control. After the experience of the Third Reich, Inge Aicher could guess how soon denazification

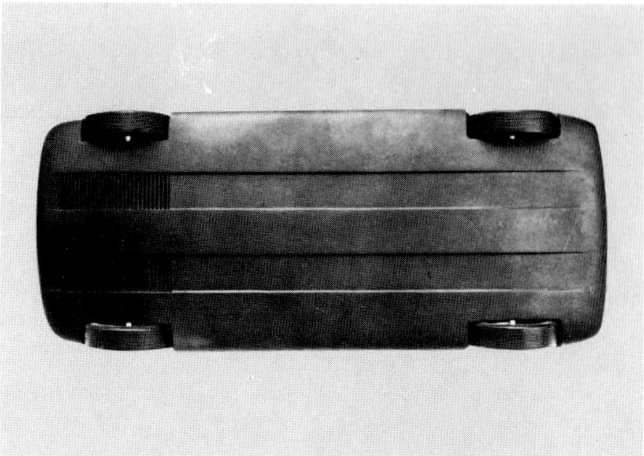

Bodywork for a private car
Pio Manzú
Third-year student, 1962 – 63
Instructor: Rodolfo Bonetto

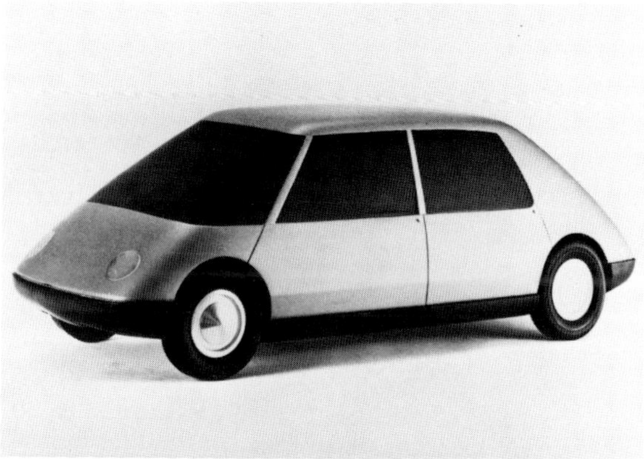

Small delivery truck
Kerstin Bartlmae,
Peter Kövari, Michael Penck
Second-year students,
1963 – 64
Instructor: Rodolfo Bonetto

Rodolfo Bonetto and
second-year students at
a project conference

would be quietly shelved, and how many people who had been in responsible positions under the former regime would hold them again, in education and culture as elsewhere. The new freedom must be made possible through an institution that was independent of the state. This was and remained one of the great, even the tragic, utopian ideals of the HfG.

Chemaitis:
That was an attitude that certainly helped her in dealing with the Americans, because they were used to private initiatives and, especially after Fascism, they had more faith in a private initiative than in any state-run body.

Staubach:
In this connection it ought to be pointed out that Inge Aicher was involved first of all with setting up the Volkshochschule in Ulm, the adult education organization that became one of the sources of the HfG. This particular Volkshochschule was notable for its political activities, fostered by returned former émigrés and democratically minded intellectuals. Clearly, the initial idea for the Ulm college was not at all that it would be a new Bauhaus, but that it would be a college of politics, a *Hochschule für Politik*. There was a sudden change of course, and the HfG was conceived. It's not clear how the change came about.

Lindinger:
It mustn't be forgotten that a large number of the instructors whom Inge Aicher recruited for the Volkshochschule were also present or future guest instructors of the HfG. For instance, in

Steamroller
Frank Hess
Fourth-year student,
1964 – 65
Instructor: Walter Zeischegg

Tractor
Pio Manzú
Diploma work, 1963 – 64

the lectures of cultural integration, as it was called: the Wednesday seminars. There was a certain coordination between the two schools, which can be explained through Inge Aicher's position as Director both of the Volkshochschule and of the Scholl Foundation.

It also needs to be said that Inge Aicher, throughout the whole lifetime of the HfG and since, has never lost sight of her original role: witness her effective campaign after the Chernobyl catastrophe, or her commitment to the peace movement.

Erlhoff:
I don't think the HfG summed up all her activity, by any means; she went right on insisting that political education was vital. She represents the element of continuity in the whole story.

Staubach:
Frau Aicher represented the link between the self-governing institution of the HfG and the Foundation. The personal link between her and Otl Aicher meant that much of what Otl Aicher stood for inside the HfG was taken up by the Foundation.

Lindinger:
That is rather a sore point, of course. This personal link – as you rightly say – was a potential source of conflict from the very start; and the more strongly the HfG structured itself as an institution, the more critical this factor became. In other words, the more Otl Aicher became involved in the running of the HfG, the more Inge Aicher's position at the head of the Foundation came

Autonova fam
Fritz B. Busch,
Michael Conrad, Pio Manzú
1965

When this prototype was shown at the International Automobile Exhibition in Frankfurt in 1965, it was nearly twenty years ahead of its time in its logical, integrated overall design, in its structural conception, and in a whole series of constructional ideas.

In contrast to the automobile design of the period, which was modeled largely on American styling and oriented mainly toward prestige requirements, the designers here felt free to concentrate on enhancing the car's practical usefulness in the context of a new conception of the car.

The compactness; the maximum use made of the interior space; the increased internal headroom with a consequent greater ease of entry and exit, better visibility, and greater seating comfort; and the impact voids integrated into the bodywork in place of the then customary fenders and tail fins: these ideas slowly became standard features, but not until the 1980s.

Herbert Lindinger, 1987

to appear invidious. Regardless of the expectation, which everyone went on taking for granted, that she would take care of raising the necessary funds. It's obvious that this caused her a great deal of unhappiness, and that is why she later resigned from running the Foundation.

The Autonova fam prototype
in the 1965 street context

Carousel S slide projector
Hans Gugelot,
for Kodak, 1963

## HfG and Industry

Editorial Discussion

Lindinger:
Where did Ulm really find its way into people's homes, in between four walls, into living rooms? Through Gugelot, a little, but basically only among intellectuals — even "Snow White's Coffin" never really became a mass item.

Staubach:
But in offices, in administration, in organization.

Erlhoff:
And even there only later. I don't think Ulm ever became general currency in its own time. German industry didn't even really participate.

Lindinger:
On the contrary. Take the U-Bahn in Hamburg, which was used by millions of people every day from the day it opened; or Gugelot's slide carousel.

Erlhoff:
OK. We have the Hamburg U-Bahn, we have Braun objects, we have Pfaff, Kodak, and Lufthansa. That's not a lot. The great push for Ulm, the event that made Ulm into a school for everyday living, came later. I find it infuriating — in face of the relationship between American designers and American industry, or between Italian designers and Italian industry — that in the Federal Republic of Germany, from the very beginning and almost up to the present day, there has been practically no recognition of good design. Ulm — in the full sense, in which it can now be seen and in which it seems to dominate the shapes of the things around us at this time —

A Catechism for Design Engineers
The nature of the design engineer's work long remained clear, to his and others' satisfaction. The nature and manner of his contribution to the form and structure of products remained unexplored, so long as "industrial design" had not established itself. But, over the past four decades, the technological product – or, if you like, the general attitude to that product – has undergone a qualitative change, in such a way as to make it clear that the work of the design engineer does not cover every aspect of the product. The impulse to cultivate civilized values was, if not foreign to his nature, at last not a pressing one to him. Things were designed as they had been designed before, because that was how they had been designed before. The rightness of this remained unquestioned right up to the moment when people started to question the quality of the results. Design engineering in this sense involved no wide-ranging program. Industrial design, on the other hand, made great claims for itself. One of them was the demand that the technological environment must be humanized. Another was the claim that technology was in itself a legitimate form of cultural expression.

Within industry, industrial design was regarded with suspicion in some quarters, in others welcomed with high hopes, and in others simply ignored. Middlemen were quick to appear on the scene (although probably only in Germany), with a mission to educate and convert industry; they included unblushing apostles of a specific morphology. The result was that large sections of German management were left with totally erroneous ideas of the aims and methods of the industrial designer. The design engineers were no less uncertain about their new partner. Was he an irresponsible amateur? An unwanted rival? A commercialized artist? A cosmetician?

Gui Bonsiepe, in *Ulm*, 7, 1963

Ulm never was really absorbed by German industry, with a very few exceptions. I get the impression that in Germany an unshakable belief in brand identity and the "Made in Germany" label, and all the enthusiasm of the reconstruction period, led people to forget about the whole idea of design. And this is equally true of the German public. A college that had been working with that kind of commitment and that kind of quality for fifteen years ought to have found public acceptance and social integration.

Lindinger:
I don't agree. First you have to remember that Ulm was not alone. The ideas that Ulm stood for were taken up in other schools in Germany, England, Japan, and South America, admittedly with somewhat varied results. The multiplication effect on this level should not be underestimated. That means that the concept spread, changed, and maybe improved.

Secondly, I would say that the Ulm approach has established itself very widely, extending as far as everyday home life. Very certainly in the field of consumer and capital goods.

Staubach:
But in what form? Bowdlerized by industry and for industry.

Lindinger:
I'm talking about Ulm attitudes and not about a specific morphology. I mean the way people think, the way they plan and shape an object, how they

Street lighting
Peter Hofmeister, Thomas Mentzel, Werner Zemp
Third-year students, 1965–66
Instructor: Walter Zeischegg

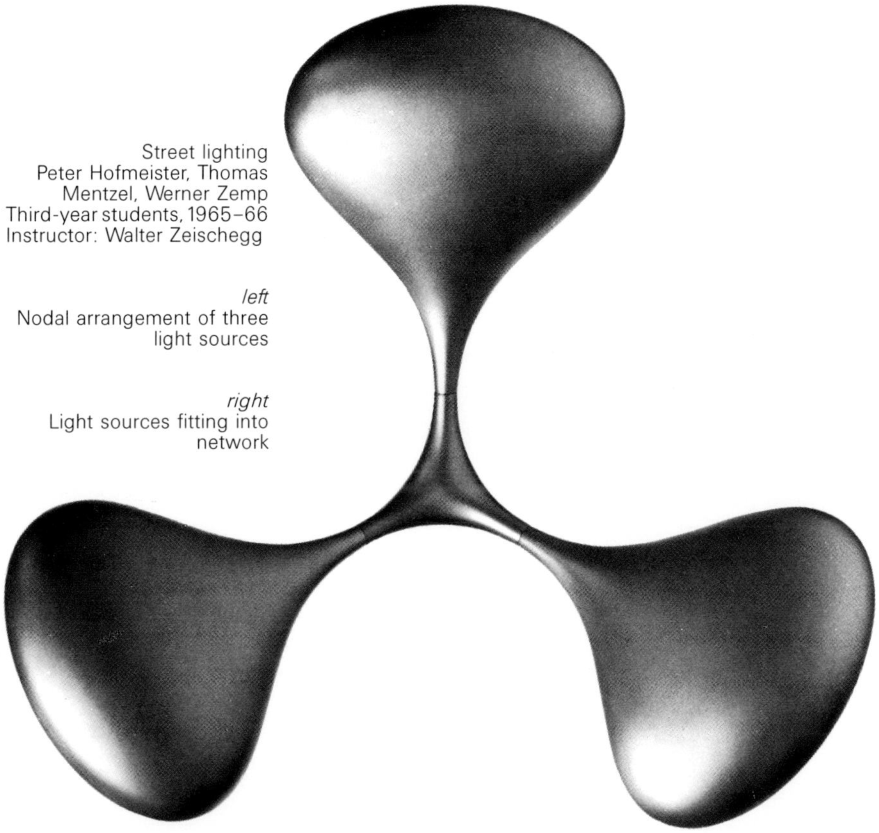

*left*
Nodal arrangement of three light sources

*right*
Light sources fitting into network

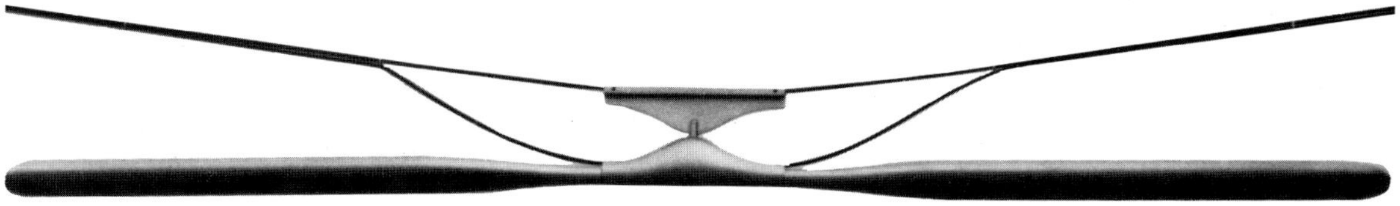

Hanging light

approach it, through analysis, evaluation, synthesis. This set of principles has established itself. Morphological concepts have changed; and it has to be stressed that Ulm always rejected the idea of a single definitive morphology. Rounded forms were used where they were needed for ergonomic reasons, and rectilinear ones where the priority was compactness.

Staubach:
By bowdlerized I mean purged of the element that's in our title: purged of morality.

Lindinger:
This is one of the contradictions. That morality remained in the realm of utopia, because, to the extent that industry took up the Ulm concept, the moral superstructure was tossed overboard. Design was reduced to the level of a marketing tool.

Erlhoff:
So the question is this: did Ulm succeed on a fundamental level, or is it all window dressing?

Staubach:
No, not on a fundamental level, certainly. But in some specific areas of life — production, domesticity, administration — we've already talked about that.

A second great sphere of influence is the whole business of design training. The curriculum structure is still basically set by Ulm, and by the people from Ulm. I do think, however, that Ulm ideas have quite often frozen into doctrines. Society has changed since

The incipient process of rationalization – something that designers cannot avoid, unless they are prepared to risk being marginalized in the future – has not had uniformly encouraging results to show thus far. Often it yields more deformation than formation, more distortion than creation. There is a sarcastic saying in American schools of architecture that anyone who has no talent for architecture can go into planning. (It may not be very many years before this has to be reversed.) It is said, and not wholly out of spite, that those who are most captivated by the glitter of design methodology are those who lack what is called the ability for *Gestaltung:* formal creation. They thus use design systems not as a practical tool but to mask the deficiencies of their own design ideas.

The adoption of rational methods, the incorporation of scientific techniques and knowledge into the design process, can be the result of a variety of often contradictory motives. On the one hand, there was and still is the desire to make scientific achievements practically useful for the humanization of the environment – a task that has been criminally neglected thus far. On the other hand, the word "science" has performed – and still performs – an essentially suasive function in the process of consolidating the social status of

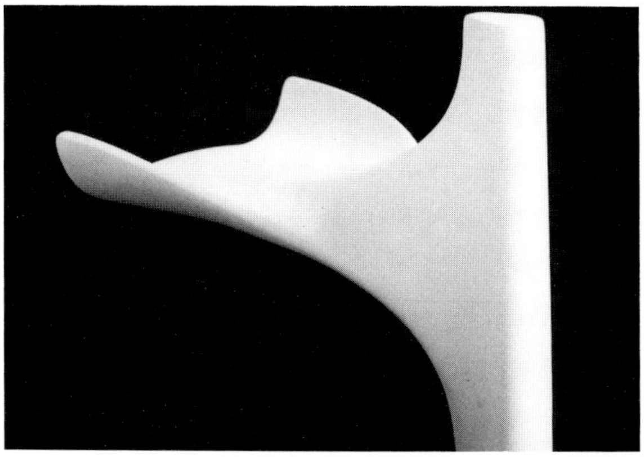

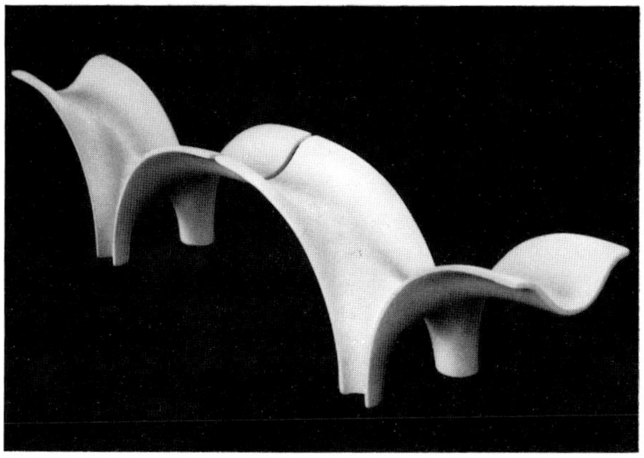

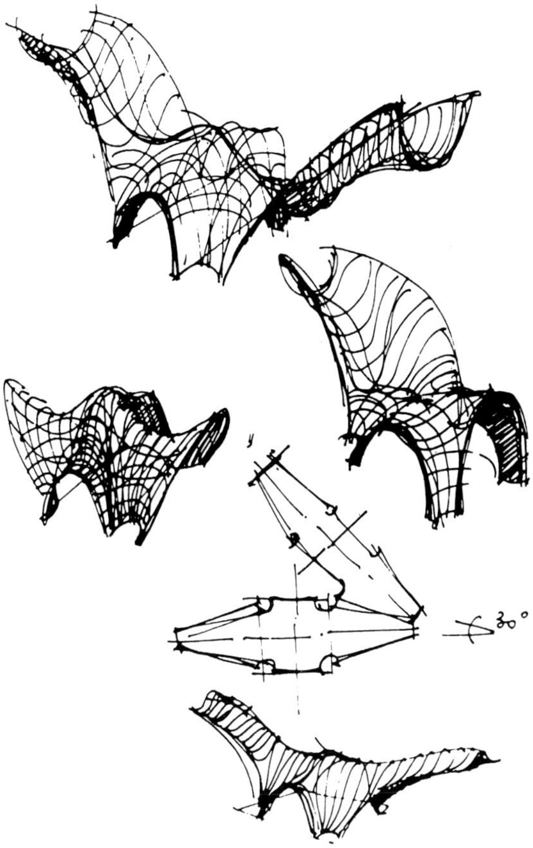

Details from the diploma work "Nature Studies and Abstraction"
Werner Zemp, 1967

then, politically, socially, culturally. Thinking of the Ulm attitudes to consumption and technology, for instance, one might say that society has evolved away from Ulm.

Lindinger:
To some extent, yes. But that is because the image of a future humanity, which is something that Ulm had, never became anything more than a utopian vision. It was a dream – even though there were individuals, groups, subclasses in society who matched up to that dream, and who still do so. The immense thrust to differentiate, the pluralistic society of the 1970s and after, was something no one had any idea about in the 1950s.

Erlhoff:
So we're talking about the Enlightenment dream, and the modernist dream – the dream that keeps this bourgeois society going, although society never allows it to come true.

Lindinger:
That's it, exactly. It's just that it's all too easy to blame all of this on industry. Ultimately, industry just produces what there's a demand for. The average Italian family has no more designs by Bellini in its living room than the German family has by Gugelot or Rams.

Erlhoff:
So maybe we could agree that Ulm is still there as an irritant, a thorn in reality's flesh.

Lindinger:
Yes, and maybe that also explains why people still tend to find Ulm impressive.

the designer. Adaptation to prevailing values is not always an unmixed virtue, even when these are the values of "science," whose conservative uses all too easily displace their original critical impulse. Anyone who works out rational criteria of decision making, and who prides himself on optimizing design solutions, is by that very fact presenting himself as a solid, serviceable sort of person, just what an industrial system requires. In design work, a scientific emphasis may mean one of two things: on one hand a practical concern, and on the other a craven capitulation to science – or to what designers themselves fondly imagine to be science.

Gui Bonsiepe, in *Ulm,* 19/20, 1967

Development of all-terrain vehicle on a modular construction system
Alexander Neumeister
Diploma work, 1966 – 67

Proposal for an unsupported plastic car body
Hans Werner
Diploma work, 1963 – 64

Traffic sign system
Richard Schaerer
Second-year student, 1965 – 66
Instructor: Herbert Lindinger

The idea of education has begun to fall apart; and so has the idea of taste. Because taste went along with education – even if it was more often the other way round. German bourgeois culture, in particular, had no time for aesthetics in working hours, and relegated culture to the position of something exceptional, reserved for the evenings. Taste has ever since been a somewhat ambivalent thing. Originally, it simply meant the general capacity to make aesthetic distinctions. Now it has become less ambitious: it mostly manifests itself in "interior" and "personal" design, the presentation of one's rooms and of oneself, which is essentially a matter of arrangement. Dexterous copywriters are at pains to extract and exploit the last vestiges of its once-aristocratic aura, and to promote that aura where taste may be a little lacking. But the target of the advertisement is always the one who gets taken in. He is always just too late. Taste, when taken over by industry, loses the very spontaneity without which it can never manifest itself. Taste exists at the point where the artificial transcends its own artificiality and suddenly becomes natural. Taste is something that you only have if you can turn your back on it. People who proudly proclaim their own lack of taste are failing to appreciate this fact; but then so are those who try to make it the sine qua non of design. Whenever an industry tries to promote one form of taste as a norm, this shows that its eager propagandists have shut their eyes to the realization that taste, although always dependent on a group sanction, signifies conformity with a consensus concerning finer distinctions. Give taste a twist, and it will pass through the eye of a needle: it has no oscillations.

Gui Bonsiepe, in *Ulm,* 10/11, 1964

Town bus
Herbert Lindinger,
Michael Conrad, Pio Manzú
1966
First prize in international competition

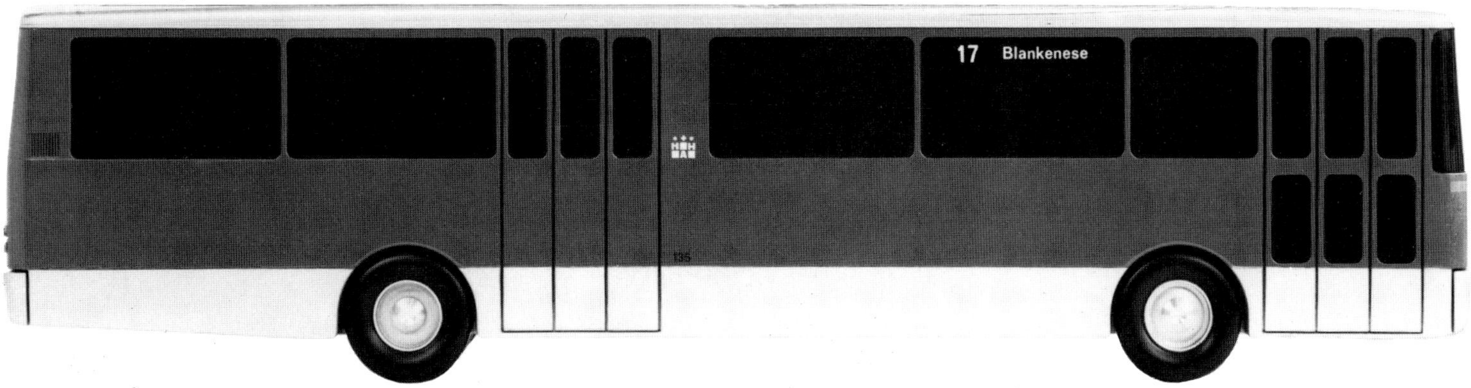

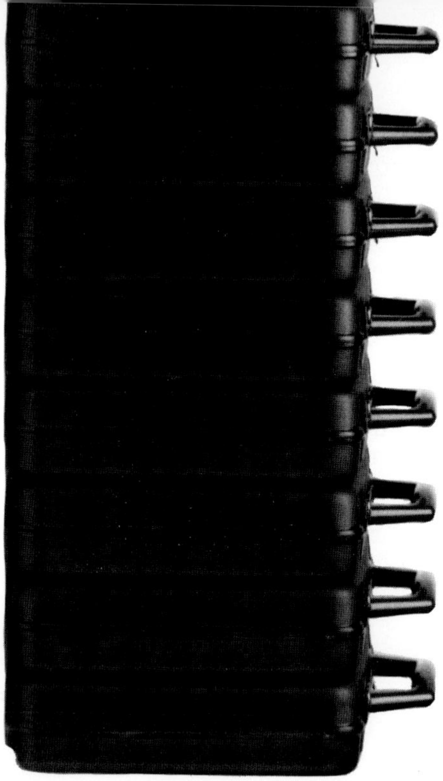

Plastic shell suitcase
Peter Raacke, Dieter Raffler,
for Hanning, Paderborn
1965 – 66

*right*
Maldonado development group:
Gui Bonsiepe,
Rudolf Scharfenberg,
Tomàs Maldonado

*left*
Peter Raacke, guest instructor

Herbert Lindinger and
Claude Schnaidt

BO 105 helicopter
Herbert Lindinger, for MBB,
Ottobrunn
1968 – 70

Bus stop
Karl Gröbli, Jean-Claude
Ludi, Rudolf Schaerer,
Michael Weiss
Fourth-year students,
1967 – 68
Instructors: Herbert Lindinger
and Claude Schnaidt
An interdisciplinary project
linking the departments of
Building, Product Design,
and Visual Communication.
First prize in a German
competition. Principal West
German exhibit at the Milan
Triennale, 1968

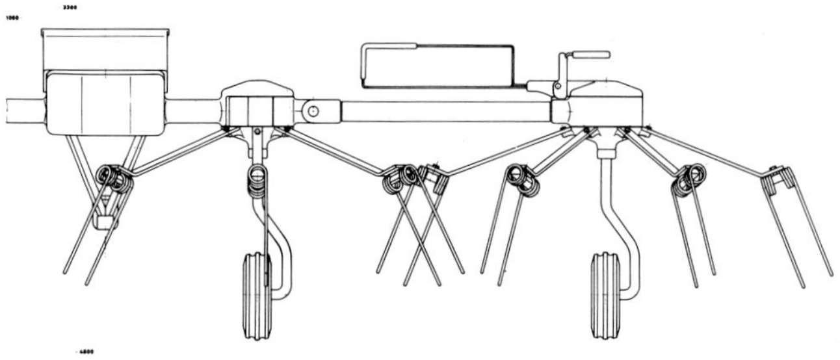

Tedder Eberhard Wahl
Diploma work, 1965 – 66

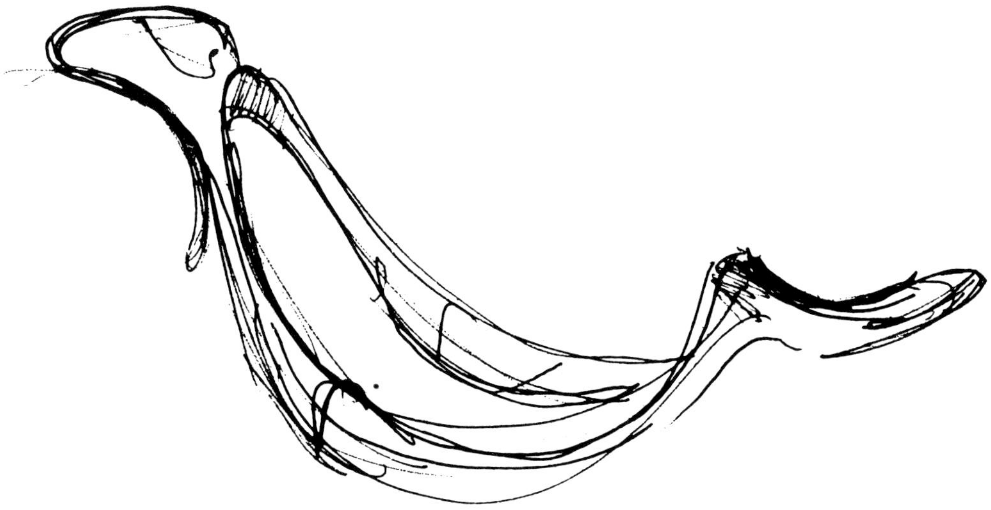

Children's seesaw in fiber-glass-reinforced plastic
Peter Hofmeister
Diploma work, 1967

Horst W. J. Rittel

# The HfG Legacy?

In those far-off 1950s, in the rosy haze of the Adenauer years, in the heyday of Gelsenkirchen Baroque and the kidney-shaped coffee table, the inhabitants of Ulm considered the products of what they called the "hochschule für ungestaltung" to be a standing provocation — or at best a farrago of incomprehensible aberrations. To the good people of Swabia, the "Ulm Right-Angle," the studied economy in gray and white, the lowercase mania, the abstruse language, the exotic behavior, and the constant rows were an abomination.

At the same time, a numerically small but cosmopolitan elite was already admiring, debating, analyzing the "Ulm Miracle" as the vanguard of a whole new culture, as the leaven of a more enlightened future world.

At home and abroad, the Ulm experiment enjoyed some success, for a time. Stackability, ergonomic efficiency, sign systems with "semiotic" appeal became highly marketable, especially for elite consumption: after all, "today's luxury is tomorrow's mass demand." And so the Ulm style eventually found its way into department stores and home kitchens, often as a watered-down and misconceived copy. People became "design-conscious," in that they learned to see travel alarm clocks and washing machines as aesthetic objects and began to look for good and functional design.

A Pyrrhic victory? Times have changed, and people and ideas have changed with them. "Design consciousness" is more widespread than it has ever been, but design as understood at the HfG has become Design, by which is understood the imposition of a modish, symbolic added value onto some otherwise entirely ordinary product (like pants or coffee machines). Design has turned into styling, in other words product cosmetics. It delivers status; it signals membership in a subculture. What you buy is the image: the object's mundane functions are a secondary matter. In extreme cases, what you pay for is the right to display the designer's name stitched onto your shirt. The orthodox HfG graduate is horrified; Thorstein Veblen would be pleased to see the vindication of his concept of "conspicuous consumption"; and Raymond Loewy was right after all. The Ulm Look has become a straightforward episode in the history of fashion and style: a bit old hat, useful for specialized markets at best.

As a cultural leaven, the HfG had a lasting influence. Its teachers and graduates went out into the world and proclaimed their own versions of the right way to shape the environment. Countless schools assimilated Ulm ideas in their own various ways. And long after its demise the HfG was a favorite theme for the cultural critics of the feature pages.

In retrospect it becomes apparent that the HfG's most durable legacy was the endemic internal strife that kept the institution going. It is widely believed that the HfG was destroyed by the "policy conflict" between "designers" and "theorists." The truth is the exact opposite. The HfG stayed alive just so long as it remained a hotbed of discord.

The rows began far back in the legendary prehistory of the College, starting with those between the "founders" and Father Bill. Things first came to a head when the assistants rose up against their master, and he left. There followed a long "theory buildup" in which a heterogeneous assortment of guest lecturers in new subjects were entrusted with the task of providing theoretical justifications for the new approach to design. This soon led to a chronic crisis, in which the "real designers" resisted the rise of what they saw as subversion and negative thinking. And then, when the "founders" and the "real designers" had things to themselves once again, they in turn fell out spectacularly, and dispersed in all directions to sulk. The ensuing comic, anarchic finale led to governmental intervention and final demise; the Baden-Württemberg government had never liked the HfG from the very start.

A sectarian farce, a storm in a provincial teacup? Partly. An undigested Bauhaus residue, an echo of the sandaled progressivism of long ago, and a number of pseudoscientific doctrines of salvation were all part of the mix. And yet: this conflict was a lesson in the crucial issue of environmental design as a purposive effort to improve the human condition. It may be described as a confrontation between two attitudes; let us call them X and Y.

Type X is sensible of the moral duty of quickly and thoroughly making the world a better place. He has a clear vision of what it ought to be like. His world is clear, orderly, and comprehensively designed; it is peopled by positively minded people whose lifestyle and attitudes enable them to live in joyful harmony with each other and with their environment. Their utopia is maintained by rationality, common sense, and a sense of mission. Type X has no doubts as to what an "ideal world" might be; he knows what is best for all of us, and how to achieve it.

Type Y people are those who have not been blessed with this enviable self-confidence, this ultimate certainty. They too consider the world to be imperfect. They would like to find out how the world ought to be. The more they think about it, the harder the problem gets. The more skeptical they become about claims to have found permanently valid solutions and eternal truths, the stronger is their inclination toward pluralism and tolerance.

The remarkable thing about the HfG was that both types (and, in spite of all the purges, Type Y outnumbered Type X) so long coexisted under one roof in a state of blatant mutual incompatibility. What an almighty scene took place whenever Type Y could find no scientific evidence for the assertion that Akzidenz-Grotesk was objectively the most legible and thus the best of all typefaces; or denied that there is any such thing as a universally comprehensible "language" of matchstick-men signs! Or said that a "universal integrated industrially prefabricated constructional element" can in some circumstances be inferior to the brick! Or that status use (conspicuous consumption) does constitute a form of use, important for some people in some circumstances and completely unimportant for others! The Type Y people failed, in fact, to comply with the role that was expected of them, which was that of supplying the Platonic utopias of Type X with the quality label "scientifically tested."

The most influential thing about the whole controversy was the residue it left behind. This consisted of a definition of *Gestaltung* – or, in less mystical terms, of design or planning – on which there was actually (not without some gritting of teeth) a measure of agreement.

Design is activity that involves planning; it is concerned to control its own consequences. It is hard intellectual work and demands meticulously informed decisions. It is not always primarily concerned with appearances, but with every aspect of its consequences: production; handling qualities; perceptual qualities; also economic, social, and cultural effects. Objects to be designed must not be seen in isolation but in conjunction with the contexts in which they are to be placed. Above all, the designer should always step back and take a critical look at the thing he is working on: how you design has a decisive impact on the product. A critical "process consciousness" is recommended.

The HfG's "founders" never accepted that "their" school had taken on a life of its own as a forum for the discussion of themes like these. Day-to-day practice in the College was repressive. In spite – or because – of this, not many students left with their initial naiveté intact. It was the X/Y conflict that prevented the HfG from degenerating into a mere *Gesamtkunstwerk*, a "total work of art."

On balance? The avalanche of environmental disasters, the bankruptcy of large-scale planning, the arrogance of "postmodernist" architecture, all confirm that there is no lack of gurus even today. There is work enough for many HfGs, blessed with good, meaty X/Y controversies.

Berkeley, 1987

Guest Instructor
Mauricio Kagel, Argentine
Lucius Burkhardt, Suisse
Henry B. Bahrick, USA

# Visual Communication

The students in this department were trained to deal with the design tasks inherent in the visual field of mass communications. The department was divided into two sectors: Typography and Filmmaking/Television.

Subjects covered by the Typography sector were graphic design, photography, typography, exhibitions, and packaging.

An important part of the work was concerned with the problems of advertising. To this was later added the area of so-called technological communication, including information displays for machines and tools, technical sign systems, and representations of scientific data. The visual designer was expected to be in a position both to solve specialized problems himself and to exercise coordinating functions.

Instructors: Otl Aicher, Anthony Fröshaug, Herbert Kapitzki, Friedrich Vordemberge-Gildewart

The Visual Design department is based on the preceding Basic Course. It works closely with the Information department.

The department is structured like a wide-ranging graphic studio. In the course of time the students are thoroughly familiarized with the various techniques that are of use in the overall field of visual design.

Specialized knowledge of typography, photography, etc., is taught in self-contained time sectors. Theoretical specialties complement the studio training. . . .

The aim of the training in the department of Visual Design is the education of specialist professionals in every field of advertising design, including book design, exhibition design, etc.: specialists who are not merely trained in one limited field, such as typography, graphics, or photography, but have mastered all these fields and can combine them.

From the 1950 Curriculum of the HfG

Billboard pillar of the Volkshochschule Ulm, 1953

Otl Aicher in his studio 1954

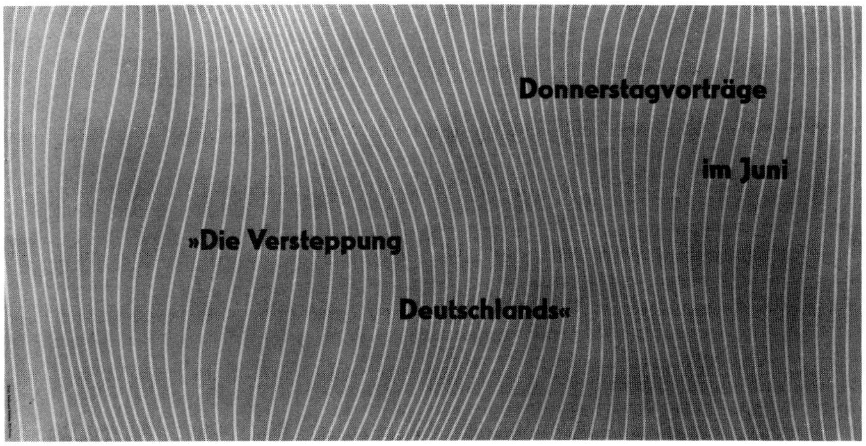

Posters
Otl Aicher, for the Volkshochschule Ulm, 1954 – 55
Aicher's Volkshochschule posters were awarded the Grand Prix at the Milan Triennale in 1954

Otl Aicher

# Bauhaus and Ulm

when walter gropius came to ulm and offered us the name of "ulm bauhaus" for our college of design, we turned it down.

looking back from today's vantage point, even i am almost surprised. in our present-day civilization, in which the packaging of a thing often means more than the contents, far more importance than ever before has come to be attached to cross-references, names, and connections. but anyone who had just come home from the war and was helping to set up a new college was in a different situation altogether. so little of substance had survived, whether material, political, or cultural, that the idea of relationships, affiliations, borrowings, or reciprocal influences was as good as meaningless.

of course we were quite aware of the prestige, in terms of cultural politics, that would adhere to a school that called itself the "ulm bauhaus." but the whole idea of "status" had rather negative connotations, as far as we were concerned. we wanted to do what was right in practice, without speculating on public response and recognition.

nor had we the slightest intention of setting up a second bauhaus, a repeat of the first. deliberately, we set out to distance ourselves from it.

when i say "we," this needs to be qualified; the basic assumptions were not by any means unanimous, especially as far as the relationship between art and design was concerned: max bill thought differently from walter zeischegg, tomàs maldonado, hans gugelot, or myself. max bill had no more desire to go over old ground than we had, but what he did want resembled a new bauhaus in several ways. we all agreed that in the ulm curriculum, unlike that of the bauhaus, there should be no place for painting and sculpture; but bill's plans for the new building nevertheless included artists' studios and a goldsmiths' and silversmiths' workshop.

walter zeischegg and i had both initially set out to work as artists, but we had soon left our academies, he in vienna and i in munich. we had made the break as a matter of principle. we had come home from the war and gone to the academies, where we were expected to practice aesthetics for its own sake. this would no longer do. no one with ears to hear and eyes to see could fail to realize that art represented an evasion of the many tasks that presented themselves, in the field of culture as elsewhere, now that nazi rule lay in ruins.

so we had to ask ourselves whether a culture and an art that ignored the true human problems of a postwar period were not simply asking for trouble. was art not basically an alibi, an excuse to abandon reality to those who were in command of it? was art not a bourgeois disguise, donned on sunday in order to keep practical matters all the more firmly under control for the rest of the week? had not those who had done the most for art been those who were interested in dominance?

i had no conclusive answers at that time; but it was clear that our interests had entirely changed direction. we were interested in the shaping, the gestaltung, of everyday life and the human environment; we were interested in the products of industry and the attitudes of society. we refused to accept any longer that creativity should be classified according to its object. should the highest human creativity continue to be that of purpose-free pure aesthetics, while practical life and the objects of daily use remained secondary? was this on the principle that the soul was superior to the body? this kind of dualism is outworn now, as far as psychology, medicine, or philosophy is concerned. but a poet's word is still worth more to us than a journalist's, and an object in a museum has a greater aesthetic value than one on the street. we still make the distinction between mind and matter, and we still need art as evidence.

what was needed, in our view, was not to enrich art by a few more works, but to show that culture today must have the whole of life as its theme. we went so far as to believe that the traditional culture business was a trick to distract attention from everyday objects, which could then be left free for commerce to exploit. art is written with a capital a by people who make a living out of trash. eternal values are proclaimed by those who would rather not be caught out in their own dirty dealings. we wanted no part of this platonism. culture must face reality.

we discovered the bauhaus, constructivism, and de stijl, and found what we were looking for in the writings of malevich, tatlin, moholy-nagy.

gestaltung, the shaping of everyday things and of the really real world — design, in fact — had become the locus of all human creativity. and if anyone was working with squares, triangles, and circles, with colors and lines, then this was a process of meaningful aesthetic experiment: that was all. it had to justify itself by mastering reality, however broken-down, dirty, and desolate that reality might be.

max bill was a survivor of the bauhaus, and had been able to preserve, through the schweizer werkbund, a little of what had been banned and destroyed in germany and austria. for us he was the authentic bauhaus, which we otherwise knew only from books. but bill had another experiential world behind him, as well. design had to draw its conclusions from the object: we all agreed on that. but for bill painting and sculpture retained a transcendental status, whereas we wanted to prevent design from slipping back into the applied-art mold and taking its cue once more from art. charles eames's chairs were already famous at that time. they were convincing models of the unity of technology, function, and aesthetics. this was design that sprang from the design brief, design that borrowed no forms from art. conversely, rietveld's constructivist chairs turned out to be no more than mondrians for sitting on, unusable art objects that claimed to be useful.

for plato, even for aristotle, matter was the veil that concealed mind. the world would be ideal if there were no matter. mind would be free if there were no body; love would be great if there were no sex. this view lay at the root of all bourgeois cultural activity; it was so universal that hardly anyone even thought about it. the dada movement made a start: the kitchen stool, the toilet bowl, the bicycle wheel, and the broomstick found their way into the museum as provocations.

a number of paradoxical propositions were now valid: he who has nothing to say looks for a style; he who lives by materialism worships the mind; he who does business patronizes culture.

hugo ball was the initiator of the dada movement in zürich and gave lectures on kandinsky. he was also the first to quit dadaism. for him it was not enough to reject bourgeois culture.

he drew the logical conclusion and condemned what was bourgeois in dada and the tendency to escape into a world of the mind. in 1919 he was elaborating an antithetical "philosophy of productive life": "respect and recognition of one's neighbor, love of one's neighbor, can give rise to an ordering of things in which the great concern of productivity constitutes the foundation of morality." in opposition to the art of pure mind, he propounded a humanistic philosophy of the shaping of concrete objects: a kind of philosophy of design. he was beginning to have his doubts about kandinsky's decorative curves.

for adolf loos, similarly, architecture no longer consisted of style and construction, divided in the same way as body and soul; and karl kraus no longer divided language into content and form. form was a form of statement.

in ulm we had to go back to the facts, to the things, to the products, to the street, to everyday life, to people. we had to turn right around. this was not to be an extension of art into everyday life, into practical application. it was to be a counter-art, a work of civilization, a civilizing culture.

we found architecture primarily in the building of factories, form in the construction of machines, design in the way tools were put together.

i first met walter zeischegg when he visited an institute of handle research near ulm in search of material for an exhibition called "hand und griff," hand and grip, in Vienna. i was doing posters. my principles were confirmed for me when soon after i left the academy one of my posters hung in the new york museum of modern art next to a painting by paul klee.

i created for the street as others did for the museum. because i did not sign any of my works, in order to avoid falling into the habits of the art business, the poster was labeled in new york "artist unknown." that suited me too. just as others sought to make their mark on the visual marketplace, i enjoyed the anonymity. craftsmen, draftsmen, and engineers do not sign their names.

the bauhaus had undergone a number of inner mutations, inner revolts, like the shift from craft design to industrial design, from the painting of an adolf hoelzel or a johannes itten to that of a theo van doesburg, from a werkbund ideology to a stijl ideology. but it never managed the final leap, away from art. on the contrary: its true princes were the painter princes, kandinsky, klee, lyonel feininger, and oskar schlemmer. and their concern, as always, was with the mind, the spirit. "the spiritual in art" is the title of the book in which kandinsky sets out his theories.

kandinsky and mondrian were adherents of theosophy, a doctrine of pure spirituality, which seeks to transcend materialism through union with the absolute spirit, with god. for both of them, painting represented access to pure spirit, and the transition to a world without objects was a renunciation of the hard, material world.

malevich in russia sought for pure form, pure plane, pure color, with a fervor that people otherwise reserved for icons. his goal was a rarefied aesthetic of pure form, of squares, triangles, and circles, of lines and color. klee spoke of the cosmos, of primordial ages, and of primordial forms of movement. in kandinsky, objects became nexuses of energy and complexes of lines. he sought in his painting for purely abstract beings to be equal citizens of an abstract realm. the quest was on for spirituality, the transcendental, the supraindividual.

but surely the world, as it is, consists only of the individual, the specific? is not the spiritual, the universal, simply a part of a human conceptual world, created in order to deal with the world in linguistic terms? william of ockham, an early forerunner of today's analytical philosophy, would have confirmed that.

but all this spirituality on the part of the bauhaus painters is not the whole picture. from the outset there was also a tendency toward the practical. the first bauhaus program demanded a return to hand workmanship, a new guild of craftsmen, work in the workshop spirit. it demanded the unity of the arts in building and declared art to be a higher form of craftsmanship. the arts and crafts idiom of this whole program is exemplified in the last sentence of the first manifesto of 1919: "let us therefore create . . . the new building of the future . . . which will one day rise toward the heavens from the hands of a million workers as the crystalline symbol of a new and coming faith."

there is something more to be inferred from this statement: that the age of craftsmanship had a lofty ethos of work. the thing was done for its own sake. manufacturing and service sectors fixated on profit are based on the delusion that the presentation takes precedence over the thing itself.

the architect gropius kept the bauhaus interested in worldly goods, buildings, tables, chairs, and furniture. but this was not for their own sake, only as elements in the painters' newly discovered faith in elementary geometry, in the square, in the triangle, in the circle, and in the primary colors, red, yellow, blue, black, and white.

and so the battle lines were drawn. is design an applied art, in which case it is to be found in the elements of the square, the triangle, and the circle; or is it a discipline that draws its criteria from the tasks it has to perform, from use, from making, and from technology? is the world the particular and the concrete, or is it the universal and the abstract? the bauhaus never resolved this conflict, nor could it, so long as the word art had not been rid of its sacred aura, so long as people remained wedded to an uncritical platonist faith in pure forms as cosmic principles.

of course, there were some dissenting voices. younger people, above all, like josef albers, mart stam, hannes meyer, and marcel breuer, questioned this submission to an ideal aesthetic. they regarded the results of their work as products of their working methods, of the properties of the materials, of technology, and of organization. as empiricists they stood in opposition to the idealists of pure form. hannes meyer had to leave the bauhaus. he said straight out that art was composition and therefore incompatible with function. life, he said, was inartistic; the aesthetic was something that emerged from economics, function, technique, and social organization.

the bauhaus was always dominated by a geometric style derived from art. this had more influence on art deco than on modern industrial production. the bauhaus had more impact in museums than in modern technology and economic life.

geometrical principles of design found some application in furniture or in typography, but even in chairs such formal preconceptions had to be approached with caution; how much more so in motor cars, machines, or tools. industrial production went its own way, and it was only with designers like charles eames that it became possible to see what it really means to develop products on the basis of purpose, material, means of production, and use.

we all had our reasons to have reservations about the bauhaus.

walter zeischegg and i were neither of us economists, the sort who might have seen aesthetics as a mere by product of a purely technical manufacturing process. it seemed right to us to apply aesthetic categories such as proportion, volume, series, interpenetration, and contrast, and to work with them experimentally: not as an end in themselves, and not as a transcendental, all-encompassing, spiritual discipline, but as a kind of grammar, a syntax of design. the result of a design had to correspond to the brief, and the relevant criteria were use and making. aesthetic experimentation was important to us; the conceptual control of aesthetic processes was as exciting as it was essential. but we did not consider newtonian physics to be more important than nature itself.

in hans gugelot the development group acquired an inventive technological brain; in maldonado another theoretician and designer who had originally been a painter. gugelot brought his technical ingenuity into the training in product design, while maldonado organized the scientific structure of the curriculum.

bill seemed to go along with all this for a time, as long as it was a matter of integrating art and design. the critical issue was whether he could follow us in our belief that painting or sculpture was an experimental discipline concerned with the definition of colors and volumes, and not in any way a superior one.

for gugelot the question came down to the hierarchic relationship between engineer and product designer. is the designer superior to the technician?

gugelot had never had anything to do with art and could make an impartial judgment.

for bill the designer remained superior to the engineer. for gugelot this was a nonissue: both, the designer and the engineer, came to a problem from different points of view, one concerned with technical efficiency, the other with use and appearance. gugelot took the engineer so seriously that he could not conceive of him as subordinate in any way, just as he expected the technician to take the designer seriously and not regard him as subordinate either. he did not see the world in terms of higher and lower but as a union and a network of differing activities, all equal in status. in addition, technology was the source of too many aesthetic qualities for him ever to regard himself as superior to it. so he himself became a technician, in order to explore to the full the aesthetic potential of technology.

the same went for zeischegg, who now read more books about mechanics and kinetics than about art, and gratified his intellectual curiosity more at engineering trade fairs than at art exhibitions. at the same time he delved into solid and positional mathematics, in order to master the laws of volumetrics and topology. he could never impose a hierarchy on something that he could see to be the result of position and viewpoint.

maldonado and i worked on mathematical logic, only to find in the end that the answers we received to the world's problems depended on the method we used to formulate the questions. here too a vertical world-order collapsed. mind was a method, not a substance. we experience the orderings of the world as orderings of thought, as information.

one of the first books i ordered for the library of the HfG was charles morris's theory of signs. the classification of information into semantics, syntax, and pragmatics gave us a theoretical foundation for defining criteria of design and interpreting art as a syntactical activity. we found this as crucial as others found sigmund freud's declaration that the psyche was the form by which the physical was organized.

i learned all over again how dangerous a purely syntactical art of squares, circles, and triangles could be if it lacked the awareness that it was divorced from the semantic basis of information. my posters had found their way into the formal area of so-called "concrete art," and i had to ask myself whether they were still primarily in the service of communication. a photographer, christian staub, who was in charge of teaching photography, pointed out to me, apropos of my own photographs, the danger that they might become formal "artistic" statements in their own right; i must not mistake syntactical exercises for information. where was the communication?

four years after the opening of the college, max bill left it. without him there would never have been a hochschule für gestaltung in ulm. we were eager to have the benefit of his bauhaus experience. his views on design showed us the way. but on a level of principle, as we saw it, he was still tied to the bauhaus. he remained very much the artist, and art for him retained a special status.

i myself had not found much in the bauhaus that was useful as far as typography and graphic design were concerned. on the contrary, in typography the exclusive reliance on the basic geometric figures, circle, triangle, and square, as for example in the design and the assessment of typefaces, was a disaster. a clear, legible typeface cannot have a circular O or an A based on an equilateral triangle. geometric letter design represents a lapse into aesthetic formalism. a legible and thus functional typeface sets out to do justice to the writing and reading habits of human beings.

the same goes for photography. the bauhaus did pioneer work, if you think of the syntactical aspect of photography. but photography as communication did not rank particularly high. what was wanted was perspectives, light and shade, contrasts, structures, viewpoints. photography as a means of communication was evolved by others, the photographers who did reportage for the illustrated magazines: people like felix h. man, stephan laurant, erich salomon, eugene smith, robert capa, or henri cartier-bresson. the photography of a man ray, or a moholy-nagy, was primarily formal aestheticism, a self-sufficient aesthetic form, or at best a syntactical experiment. reality was rendered as a signal, which did at least enhance the value of this kind of photography for advertising and graphic design. this interpretation is supported by the fact that this kind of photography is nowadays bought and sold as art. its formal ambitions were in inverse ratio to its function as communication.

finally, today's postmodern design can also point to bauhaus precedents. furniture is disintegrating, just as it did in rietveld's time, into cubes, cones, and cylinders; and the colors are the primary ones of elementary formalist design. the spherical teapots and cylindrical vases of the bauhaus will remain immortal just so long as elementary geometry remains marketable as art; witness aldo rossi.

it has to be said that the true commercial potential of kandinsky is now being exploited. his lines, dashes, waves, circles, dots, segments of circles, half-moons, and triangles are being exploited at the moment as the very last word in fashion, now that mondrian has been dismissed as too cold and klee as too poetic. visual fashion today relies on massive transplants and transfusions from kandinsky. even architectural plans plunder his syntactical repertoire. the age is once more in thrall to a higher, universal reality. art is back in favor: not responses to a situation, or results of a case study, or solutions to a problem, but art. art offers eternity.

The Bauhaus has made its own way in many parts of the world, but I have only to walk into this building to realize that here is the place where it has most truly taken root.

Walter Gropius, in a speech at the opening in 1955

we are back to the controversy that split the dadaists. they fell into two camps: the aesthetes and the moralists. hugo ball, as a moralist, withdrew and left the aesthetes to their own devices. even marcel duchamp, who was getting over his expressionistic painting, stopped making provocative objects to upset the bourgeoisie.

the moralists refused to give up their indictment of the world as trash, lies, and deceit. the modern market principle was based on profit, and neither factory-made goods, nor chemical techniques, nor the products of the food industry sprang from a sense of responsibility to the product and the matter in hand. the moralists had to abandon the aesthetes.

this situation has not changed much to this day. the world is not a very different place. most designers have gone over to the camp of the stylists, the aesthetes, in order to present products in accordance with aesthetic sales criteria. presentation is all. it's a shame there's no ulm any more.

Rotis, 1987

The first prospectus of the Scholl foundation (Geschwister-Scholl-Stiftung, Ulm, Bahnhofstrasse 1) for the Hochschule für Gestaltung, 1953

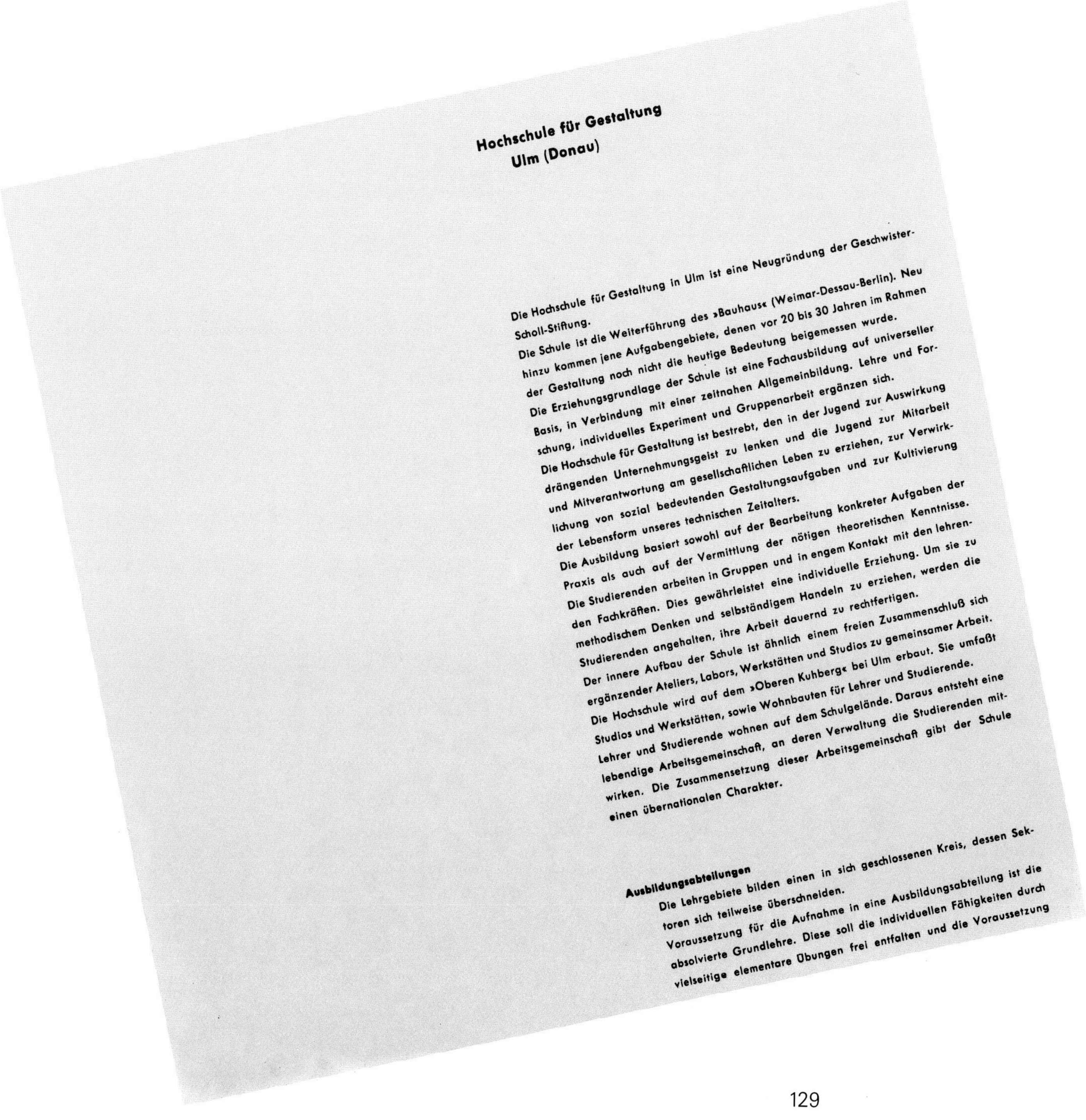

**Hochschule für Gestaltung**
**Ulm (Donau)**

Die Hochschule für Gestaltung in Ulm ist eine Neugründung der Geschwister-Scholl-Stiftung.
Die Schule ist die Weiterführung des »Bauhaus« (Weimar-Dessau-Berlin). Neu hinzu kommen jene Aufgabengebiete, denen vor 20 bis 30 Jahren im Rahmen der Gestaltung noch nicht die heutige Bedeutung beigemessen wurde.
Die Erziehungsgrundlage der Schule ist eine Fachausbildung auf universeller Basis, in Verbindung mit einer zeitnahen Allgemeinbildung. Lehre und Forschung, individuelles Experiment und Gruppenarbeit ergänzen sich.
Die Hochschule für Gestaltung ist bestrebt, den in der Jugend zur Auswirkung drängenden Unternehmungsgeist zu lenken und die Jugend zur Mitarbeit und Mitverantwortung am gesellschaftlichen Leben zu erziehen, zur Verwirklichung von sozial bedeutenden Gestaltungsaufgaben und zur Kultivierung der Lebensform unseres technischen Zeitalters.
Die Ausbildung basiert sowohl auf der Bearbeitung konkreter Aufgaben der Praxis als auch auf der Vermittlung der nötigen theoretischen Kenntnisse.
Die Studierenden arbeiten in Gruppen und in engem Kontakt mit den lehrenden Fachkräften. Dies gewährleistet eine individuelle Erziehung. Um sie zu methodischem Denken und selbständigem Handeln zu erziehen, werden die Studierenden angehalten, ihre Arbeit dauernd zu rechtfertigen.
Der innere Aufbau der Schule ist ähnlich einem freien Zusammenschluß sich ergänzender Ateliers, Labors, Werkstätten und Studios zu gemeinsamer Arbeit.
Die Hochschule wird auf dem »Oberen Kuhberg« bei Ulm erbaut. Sie umfaßt Studios und Werkstätten, sowie Wohnbauten für Lehrer und Studierende.
Lehrer und Studierende wohnen auf dem Schulgelände. Daraus entsteht eine lebendige Arbeitsgemeinschaft, an deren Verwaltung die Studierenden mitwirken. Die Zusammensetzung dieser Arbeitsgemeinschaft gibt der Schule einen übernationalen Charakter.

**Ausbildungsabteilungen**
Die Lehrgebiete bilden einen in sich geschlossenen Kreis, dessen Sektoren sich teilweise überschneiden.
Voraussetzung für die Aufnahme in eine Ausbildungsabteilung ist die absolvierte Grundlehre. Diese soll die individuellen Fähigkeiten durch vielseitige elementare Übungen frei entfalten und die Voraussetzung

any thesis that gets nailed to the wall carries an inherent risk of becoming ossified and one day being an obstacle to evolution. it is, however, not very likely that the so-called asymmetrical or organically formed typographical pattern will be more speedily overtaken by developments than was the axial layout, which is preponderantly a response to decorative rather than functional considerations. happily, we have freed ourselves from the renaissance formula and, far from wanting to go back to it, we want to exploit our freedom to the full. the old formula has been conclusively shown to be unviable, a demonstration far more convincing than its revival, and experience shows that the modern typography of the 1930s was on the right lines . . .

the major change that separates the functional typography of today from the "new typography" of the 1930s is the elimination of the heavy bold rules and bars, outsize periods, giant folio numbers, and other devices that were once regarded as the hallmarks of typographical fashion. they clung on for a while, in the guise of narrower rules that served to marshal and emphasize the type. all these elements are inessential and are now superfluous, if the type panel itself is properly organized, and if the groups of words are rightly proportioned to each other. This does not mean that such devices are to be condemned in principle; but in general they are as unnecessary as any other ornament, and when they are omitted the typography gains in simple spatial tension and quiet inevitability.

Max Bill on typography, in *Schweizer graphische Mitteilungen,* 4, 1946

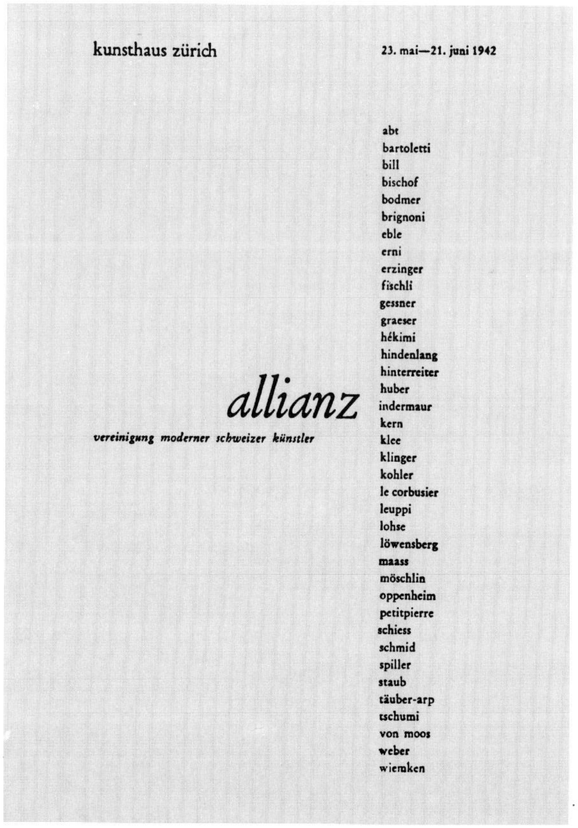


Catalogue cover for *allianz* art exhibition
Max Bill
1942

# The Ideology of a Curriculum

Kenneth Frampton

There is little doubt but that the Hochschule für Gestaltung Ulm has been the most significant school of design to come into existence since the end of World War II, not so much for what it achieved in terms of actual production, nor for the large number of designers it effectively educated, but finally for the extraordinarily high level of critical consciousness that it managed to sustain in its daily work. In many respects the Hochschule was a pioneer, not only for its evolution of design methods and for the quality of the designs it achieved with these methods, but also for the crisis of identity it suffered as a consequence of its dialectical rationality.

This much is evident from the accounts of the development of the Hochschule given in the journal *Ulm, 1 to 21.* It is clear even from a most cursory reading of this publication that, contrary to popular myth, there was never any monolithic position obtaining at the Hochschule, for the discourse that was carried on in its journal came into being solely through interchange of individual opinion.

It is not generally realized that the Hochschule survived as an institution for a comparatively long time, at least as long as the Bauhaus, if one counts the terminal date for that institution as being 1932; and although the Hochschule was not as convoluted in its history as the Bauhaus, with its cryptic men of genius and enforced migrations, it nonetheless had its own complexity and richness, which only an

Vordemberge-Gildewart did not only teach us typography and the history of painting. He himself was part of his subject; he had lived through it and helped to shape it.

After his slide lectures one always had the feeling that one had met Tatlin and Mondrian in person and had been right there when Schwitters recited his "Anna Blume."

"And if you want to see the original of this El Lissitzky," said VG (as we used to call him), "then come over to see me after the class. Over there, the last faculty house." I went. A house full of Arps, Delaunays, Van Doesburgs, Lohses, Moholy-Nagys, and Schwitterses. And his own forms of color.

Walter Müller, 1987

Friedrich Vordemberge-Gildewart in his studio

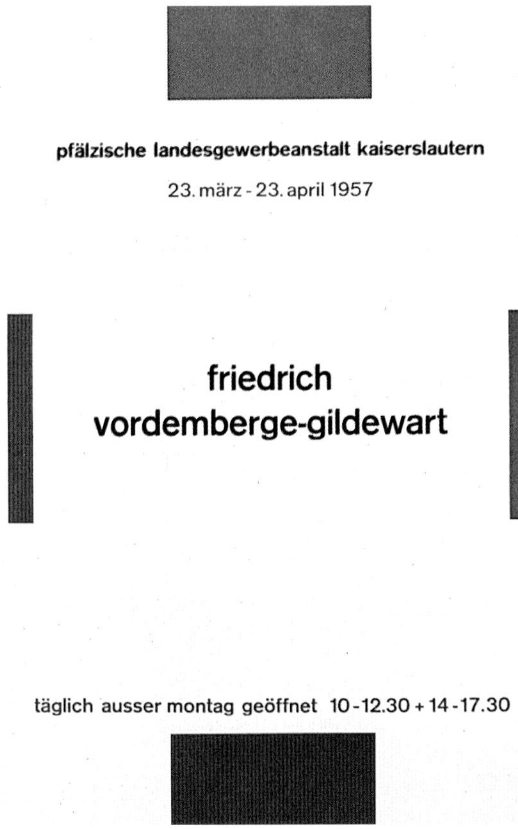

pfälzische landesgewerbeanstalt kaiserslautern
23. märz - 23. april 1957
friedrich
vordemberge-gildewart
täglich ausser montag geöffnet 10-12.30 + 14-17.30


Poster for an art exhibition
Friedrich Vordemberge-Gildewart
1957

archivist, with ample time and space at his disposal, will ever be able to master.

In an indirect way, the Hochschule für Gestaltung was a product of German resistance to the Nazi regime, for the Hochschule was created, in principle, in 1950 by the Geschwister-Scholl Foundation, in memory of two young members of the Scholl family, Hans and Sophie Scholl, who had been executed by the Nazis some seven years before. It was the purpose of the foundation to establish a school that would, in the words of the constitution, combine "as one entity professional ability, cultural design, and political responsibility."

According to Konrad Wachsmann, the Hochschule had its origins in a move on the part of the American High Commissioner for Germany, John J. McCloy, to sponsor, with considerable American aid, the foundation of a school of social research and political science as part of the American program for the postwar reconstruction of Germany. It would seem that, if Wachsmann is correct, this initiative was officially channeled through the Geschwister-Scholl Foundation and that it was Max Bill who, on being commissioned to design the building, persuaded Inge Aicher-Scholl, and presumably the American High Commission, to found not a school of politics, but a school of design. Nonetheless, a vestige of this initial political intent remained in the curriculum of the school, and this element contributed significantly to the shaping of its destiny.

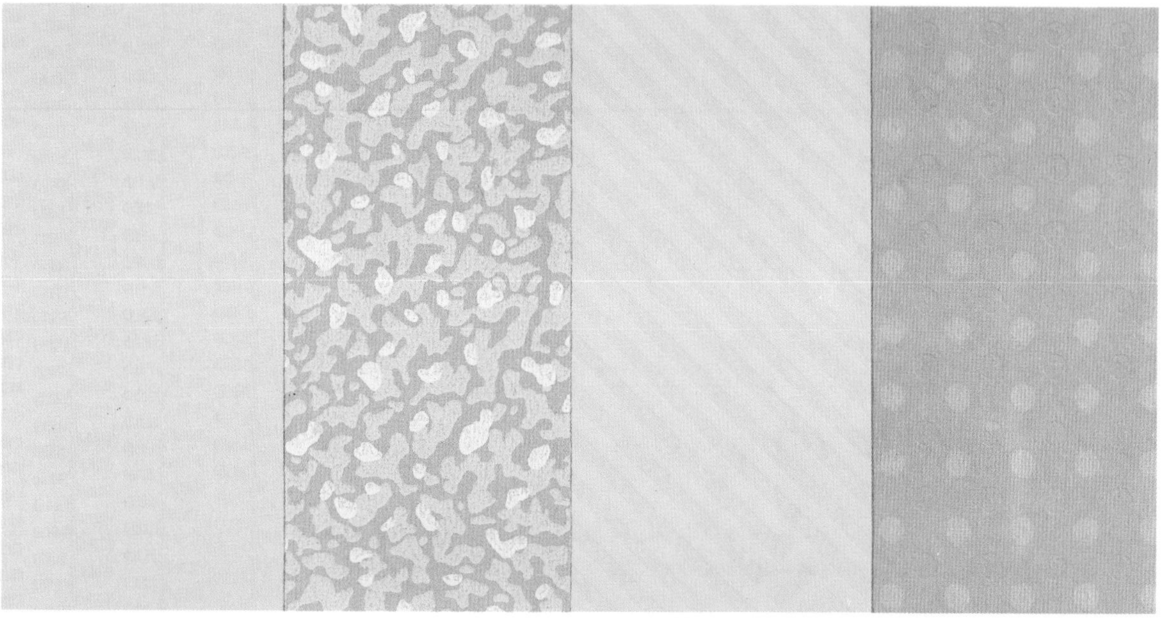

Apart from this political legacy, the Hochschule was a conscious continuation of the German Applied Art School reform movement, begun in the last decade of the nineteenth century, out of which the Bauhaus emerged in all its various incarnations. Even the name *Hochschule für Gestaltung* derives directly from the Bauhaus, since this was already a subtitle for the Dessau Bauhaus, before Walter Gropius's resignation in 1928. In any event the connection was made explicit in Max Bill's first public statement as the director of the Hochschule in 1953, wherein he wrote:

*The founders of the Ulm School believe art to be the highest expression of human life and their aim is therefore to help in turning life into a work of art. In the words of that memorable challenge thrown down by Henry Van de Velde over 50 years ago, we mean "to wage war on ugliness," and ugliness can only be combated with what is intrinsically good ... "good" because at once seemly and practical. As the direct heir to Van de Velde's school at Weimar, the Dessau Bauhaus had set itself precisely the same objects. If we intend to go further at Ulm than they did at Dessau this is because postwar requirements clearly postulate the necessity for certain additions to the curriculum. For instance, we mean to give still greater prominence to the design of ordinary things in everyday use; to foster the widest possible development of town and regional planning; and to bring visual design up*

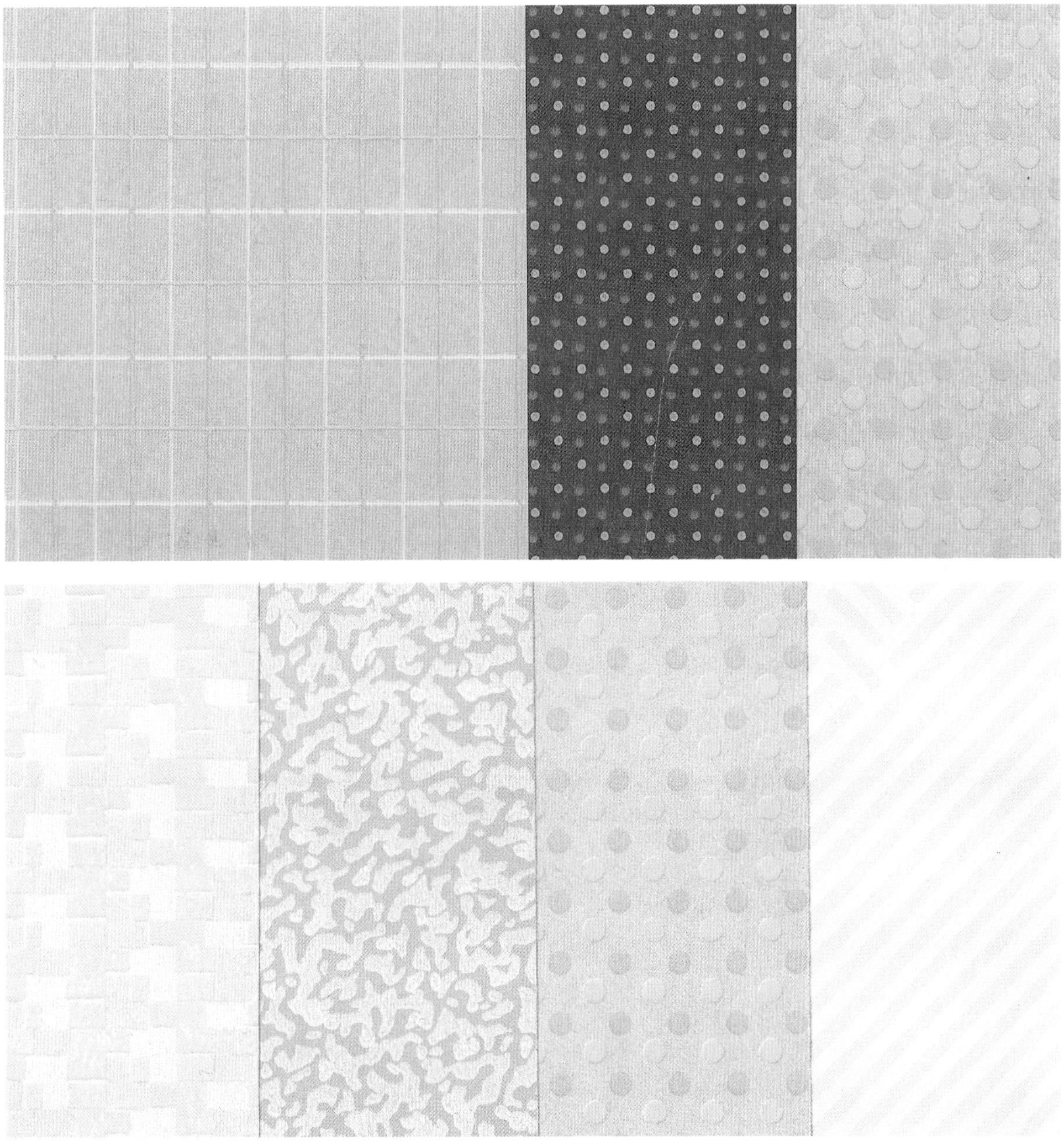

In a wallpaper design competition at the HfG, a jury consisting of Max Bill, Hans Gugelot, and Tomàs Maldonado picked out the twelve entries most deserving of development. The prizewinners were: Gui Bonsiepe, Angela Hackelsberger, Meret Mitscherlich, Herbert Lindinger, Peter Hoffmann, Helmut Müller-Kühn, Eva Pfeil, Wilhelm Ritz, Hermann Roth, and Theodor Zeitler. A working party consisting of the prizewinners under the direction of Max Bill prepared the winning designs for production. Most of the designers showed a preoccupation with using colors of approximately equal intensity to create an illusion of dematerialization of the surface. Manufactured by Salubra, Switzerland, 1956

*to the standards that the latest technical advances have made possible. There will also be an entirely new department for the collection and dissemination of useful information.*

So much for Bill's rather idealistic initial statement of intent. But was this formulation still the intent by 1955, at the time of the formal opening, or by 1958 when the first issue of the Hochschule quarterly journal, *Ulm, 1,* was published? There is a decided shift in both the language and the emphasis of this journal, as we may clearly appreciate from the following opening statement.

*The Hochschule für Gestaltung educates specialists for two different tasks of our technical civilization: the design of industrial products (industrial design and building departments); the design of visual and verbal means of communication (visual communication and information departments). . . . The school thus educates designers for the production and consumer goods industries as well as for present-day means of communication: press, films, broadcasting, television, and advertising. These designers must have at their disposal the technological and scientific knowledge necessary for collaboration in industry today. At the same time they must grasp and bear in mind the cultural and sociological consequences of their work.*

The resignation of Bill in 1956 and his replacement by a triumvirate found its reflection in these discreetly formulated goals from which any reference to city and regional planning had been eliminated. It

A major part of human communication today takes place visually, through such media as photographs, posters, and signs. To design these forms of communication in accordance with their function, and to establish methods of doing this that match the need of our time, is the task of the department of Visual Communication. Graphic design, photography, typography, and exhibition design are accordingly treated as a unity, and this will later be completed by the addition of motion pictures and television.

The department is organized like a graphic studio and undertakes practical commissions. In working with corporations, the aim is to give the corporation an appropriate external image through the design of its forms of communication, from the letterhead and the logo to the stand at a trade fair.

Research within the department is directed toward fitting visual statements as closely as possible to what they have to say and giving precision to their meaning. Use is made of recent scientific advances in perceptual and semantic theory.

The students within the department participate in the practical and theoretical work in relation to the state of their knowledge.

Necessary prior qualifications are a completed professional training and practical experience in one of the specialties, together with a basic knowledge of the whole field.

From *HfG-Info*, 1955

Advertisement
Otl Aicher, for Braun

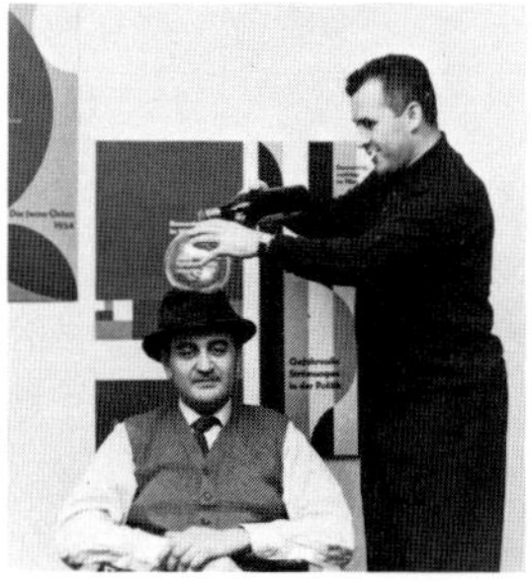

Otl Aicher and
Tomàs Maldonado

also found reflection in the four-year curriculum outline that followed, above all in the foundation course that was mandatory for all first-year students. This course, which was established as a *Grundlehre* by the Argentinian painter/designer Tomàs Maldonado, ostensibly comprised the following subjects: visual method, workshop practice, presentation methods, design methodology, sociology, perception theory, twentieth-century cultural history, and a remedial course in mathematics, physics, and chemistry. Judging from the highly schematic exposition given in *Ulm, 1,* this course attempted to place a distinct and unusual emphasis on mathematics; first, on the creative and manipulative use of mathematical constructs in pragmatic design training, and second, on mathematical logic as the conceptual basis of design method. At the same time, the sociological and cultural aspects of the course emphasized western superstructural transformations since the industrial revolution. One should note in passing that the workshop practice was markedly different from that of the Bauhaus; its emphasis being entirely away from any kind of craft production and toward the photo-reproduction of material and the making of prototypes. In fact, training was only given in wood, metal, plaster, and photography.

That there had been a major shift in orientation between Bill's brief tenure and the triumvirate rule of 1958 is also reflected in the curricula of the four departmental courses of

*middle left*
Window display
Otl Aicher, Hans Conrad, for Braun
1958

*above left*
Flexible exhibition system for trade fair stand
Otl Aicher, Hans Gugelot, for Braun
1955

*above right*
*Das gute Spielzeug*
A toy exhibition that toured many countries over a period of 30 years
Otl Aicher
1956

*below*
Braun pavilion at radio exhibition, Frankfurt Fair
Otl Aicher, Hans Gugelot, for Braun, 1959
At Ulm, between 1955 and 1960, Otl Aicher and Hans Gugelot, with their associates, laid down the outlines of the new corporate identity of the Frankfurt firm of Braun. Together with Erwin and Arthur Braun and Dr. F. Eichler, they forged a new philosophy for the company's attitude to society, to its customers, and to the conception and design of its products, exhibitions, photography, advertising, use of language, and architecture

Industrial Design, Building, Visual Communication, and Information. If the heritage of the Bauhaus, initially acclaimed by Bill, still manifested itself in the recreation of a common foundation course and in the importance attatched to some form of workshop practice, the departure from the Bauhaus tradition found clear expression in three sets of academic courses that were common to all four departments. First, in the return to *socio-cultural history,* a subject that had never been regarded as having any kind of validity within the millennial perspective of the Bauhaus; second, in a course known as *operational research,* comprising group theory, set theory, statistics, and linear programming; and finally in courses dealing with the theory and *epistemology of science,* branching out into behavior theory and the theory of machines.

Irrespective of the level initially attained in this ambitious program, there is no reason to doubt but that this curriculum served not only to structure the pedagogic program, but also to publicly proclaim the ideology of the school; and, lest there should be any doubt as to the changed nature of the school, this schematic statement of intent was followed in the same month by the second issue of the journal *Ulm, 2,* which was largely devoted to a transcript of Tomàs Maldonado's address to the Brussels World's Fair, given in September 1958, under the title, "New Developments in Industry and the Training of the Designer." This, as far as I

*above* Experimental demonstration at Laboratory for Perceptual Theory
Mervyn W. Perrine
1959 – 60

*below left* Advertisement
Hans G. Conrad, for Knoll International
1953 – 55

*below center* Poster for M 125 system furnishing
Herbert Lindinger, for Bofinger, Stuttgart
1957 – 58

*below right* Anthony Fröshaug with students

The teaching has been divided into departmental work, theory of drawing, communications media, and technology. The departmental work is directed by my colleagues, Mr. Otl Aicher and Mr. Friedrich Vordemberge-Gildewart, together with the newly appointed instructor, Professor Anthony Fröshaug from London. I shall be working with the students of Visual Communication and Information in a course on the theory of drawing. The teaching of the subject of communications media (typography) (second year) is in charge of my colleague Otl Aicher. The instructor for the subject of communications media (photography) will be the American photographer, Thomas Rago, from Callahan's school at the Institute of Design in Chicago. The historical seminar for Visual Communication and Information is conducted by my colleague, Mr. Vordemberge-Gildewart. Professor Franzen continues to conduct the sociological seminar under the title of "Problems of Communication."

Tomàs Maldonado, address at the beginning of the academic year 1957 – 58

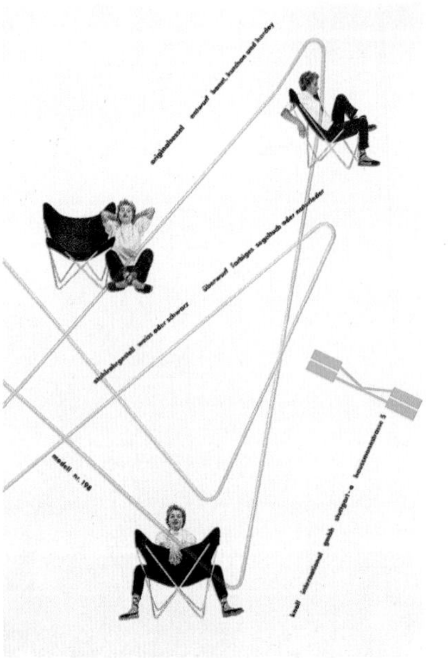

know, was Maldonado's first public declaration; the range and complexity of the argument warrants a brief analysis of its salient points, particularly as this discourse clearly exerted a major influence on the formation of school policy. In retrospect, one cannot see that either of the other members of the triumvirate – neither the graphic designer Otl Aicher, who in any event betrayed little taste for intellectual discourse at this time; nor the sociologist Hanno Kesting, whose critical sociological position could not have been very far removed from that of Maldonado, and whose studies into the nature of industrial society had been published in the previous year – would have much cause for disagreement with Maldonado's position.

Given the inescapable, almost fatal orientation of the Hochschule toward an updating of the Bauhaus, the initial point of interest in Maldonado's 1958 address lies in his measured critique of that legendary institution, and in particular for the distance he took in placing this legend in its proper historical context. Thus, of its dependence on the arts and crafts heritage, Maldonado observed:

*The inaugural manifesto of the Bauhaus in 1919 at Weimar announced – not without declamatory élan – the union of the arts and the crafts and their future integration in a higher entity: architecture. It is a typical "arts and crafts" manifesto, which Ruskin and Morris could have signed without contradicting themselves.*

Visual Methodology

The first assignments given in the Basic Course of the HfG within the field of Visual Methodology have to take into account the fact that the students differ in their previous training and that it is not possible to assume any prior methodological knowledge. They also differ widely in their command of techniques of presentation.

Initial assignments are consequently given that not only serve to develop the ability to represent but are impossible to undertake successfully without very specific methodological premises. The students have to be induced not simply to approach these tasks intuitively but, so far as is possible, to deal with them systematically, and thus to acquire a modicum of methodological knowledge. Finally, these assignments must stand in some kind of relationship to the later work in the departments.

It is advisable to set problems that have a manageable number of possible solutions. One such problem would be the representation of communications systems through two- or three-dimensional graphs. A graph is understood to mean a system of points and of lines linking these points.

At the same time it is possible to conduct an experimental investigation to determine whether principles can be found that will enable carriers of information to be designed in such a way as to optimize communication: a question that is equally important for all the departments of the HfG.

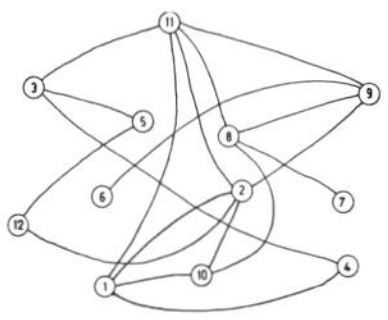

Working hypotheses. The variety of graph configurations is limited by the following working hypotheses:

the lines joining the points must be straight;
all the linking straights must be of the same length;
linking straights must not intersect;
the points must fit into a grid;
the presentation must be in as few dimensions as possible. . . .

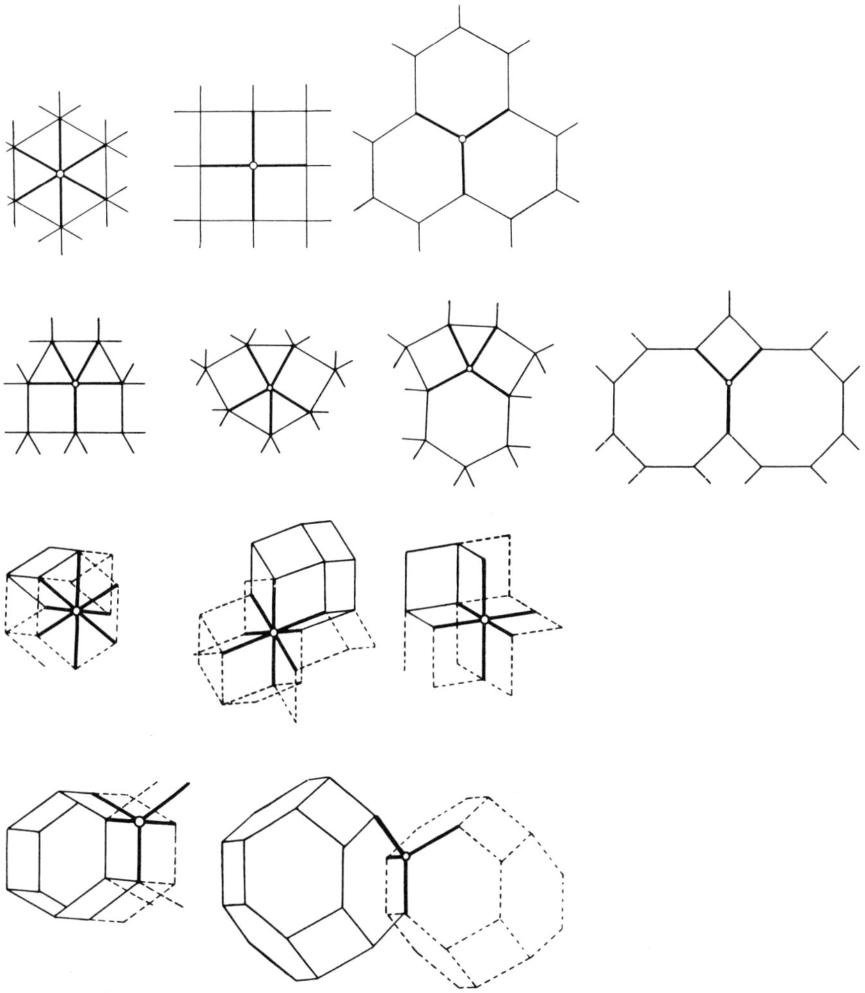

While of its problematic influence, in the period that followed its demise, he argued, with unusual insight into the ambiguities of its achievement:

*The American economic crisis of 1930 gave the day to styling — a new variation of industrial design whose influence has in fact extended up to the present day. The Bauhaus, its followers, and its sympathizers denounced from the start the commercial opportunism of styling, its indifference to artistic and cultural values. But the problem was no easy one: from time to time the stylist created products that could not but have been approved by the partisans of the Bauhaus. Stylists such as Henry Dreyfuss and Walter Dorwin Teague were sometimes damned, at other times, deified.*

That American styling since the thirties had stolen much of the Bauhaus thunder and that, in any event, a good deal of Bauhaus and post-Bauhaus design could be regarded justly as some form of neoacademic formalism, that is, as the substitution of one aesthetic formula for another, was a position equally shared at that time by both Maldonado and Reyner Banham. At the same time, Maldonado could not accept Banham's acclaim of fifties industrial styling as the manifestation of legitimate folk culture. In refutation of such "consensus" populism he stated: "I am not convinced that the aerodynamic fantasies of vice-president Virgil Exner, responsible for the design of Chrysler automobiles, coincides with the artistic needs of the man in the

Example for a street network: in 1955 – 58 it was planned to redesign the network of main streets in central Zürich.

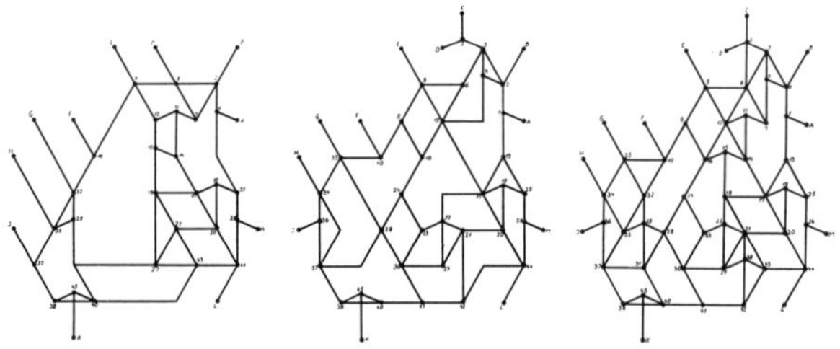

The first two graphs enable the old and the new networks to be compared with ease. Corresponding points are in the same place in the grid. This provides a basis for critical assessment. . . .

The techniques described here are used in the Basic Course exclusively for the representation and analysis of existing situations. Their true significance, on the other hand, lies in their operative character, in their constructive use in solving design problems. In designing a ground plan, for example, it is desirable to visualize the desired circulatory relationships between the constituent spaces within a matrix, thus clarifying the topological relationships. This makes it possible to draw up a complete catalogue of all the possible graphs that meet the specific requirements. These possible solutions are then tested for their compatibility with constructional requirements and with the proportions and dimensions dictated by functional considerations. In this way, one achieves both a complete overall view of the design possibilities and a rough design.

Anthony Fröshaug, in *Ulm*, 4, 1959

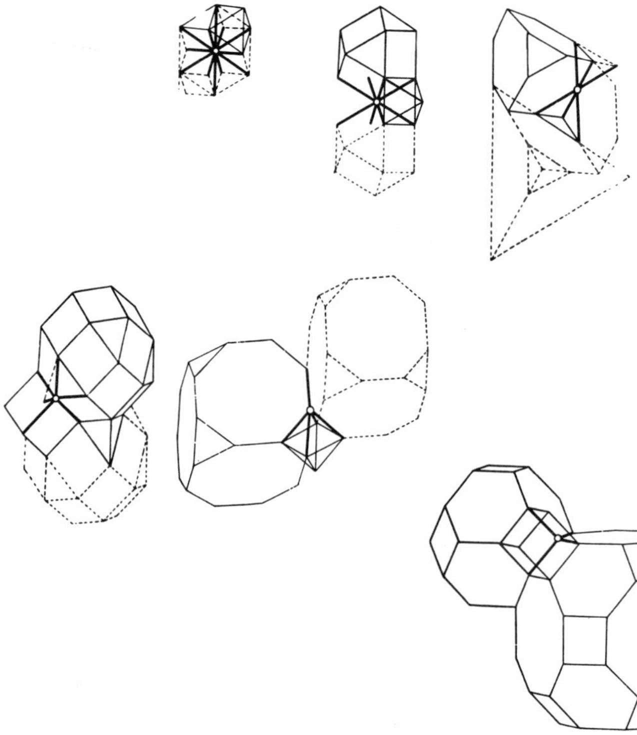

street." Maldonado was to dismiss as irrelevant Banham's analogous characterization of the difference between elite design and pop art as a distinction between "rare" and "wild" flowers.

Maldonado followed this dismissal of liberal populism with an equally critical rejection of radical idealism as represented by Gregor Paulson's position on styling of some ten years previous. Paulson had argued that the proper task for the industrial designer was one whereby the aesthetic factor becomes integrated into the *use value* of the product, rather than being assimilated, by virtue of the function of style in the marketing process, into the *exchange value* of the product. Maldonado's refusal to be persuaded either by liberalism or by this latter-day *neue Sachlichkeit* formula gave him the opportunity to pose the interrelated issues of his later criticism: firstly, under what circumstances will industrial production be capable of freeing itself from the rhetorical demands of neo-kitsch marketing? and secondly, how may we rationally determine the phenomenon of consumption in relation to need? In 1958 he was to state: "Neither the psychoanalysts nor the professional critics of our civilization can give us a comprehensive explanation of all the phenomena of the world of consumption. The Marxists themselves do not succeed. One of them, the French philosopher Henri Lefebvre, recently wrote: 'By the side of the scientific study of the productive relations that effect political economy, there

Corporate identity
Otl Aicher, for Stuttgarter Gardinenfabrik, curtain manufacturers
1959

Martin Heidegger and Reyner Banham at the HfG
1959

Poster
Otl Aicher, Herbert and Meret Lindinger
for Clima hotel group, Vienna and Innsbruck
1957 – 58

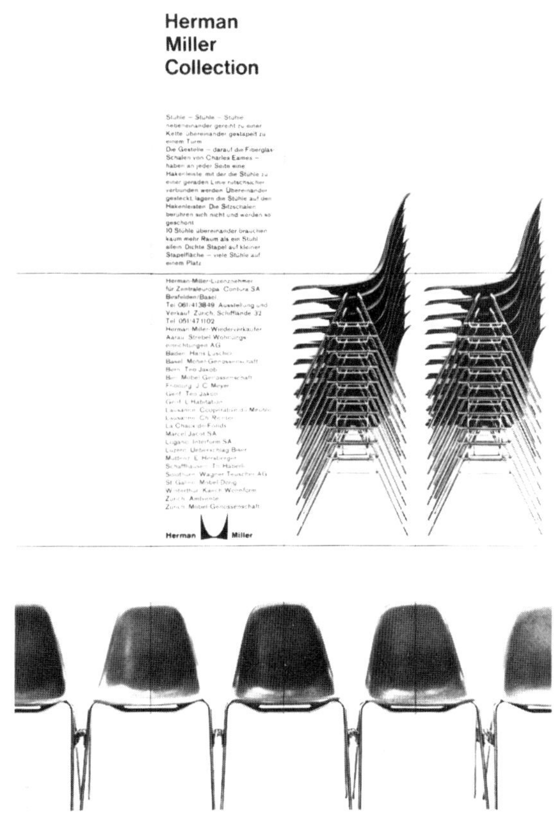


Advertisement
Otl Aicher, Tomàs Gonda, for Herman Miller Collection
1961 – 62

is ... room for a concrete study of appropriation: for a theory of needs.'"

Aside from these complex issues turning on the nature of industrial production and consumption, Maldonado praised the progressive aspects of the Bauhaus for its commitment to the "learning through doing" approach of Hildebrandt, Kerchensteiner, Montessori, and Dewey, and for its pragmatic opposition to the verbal emphasis of the humanist tradition. Nevertheless, it was clear that this particular pedagogical approach had now outlived its usefulness and that a new philosophy of *praxis* was needed. To this end, Maldonado proposed scientific operationalism, of which he remarked, "it is no longer a question ... of knowledge, but of operational, manipulable knowledge."

By "operationalism" Maldonado seems to have been referring to that philosophical system developed in the early fifties by Anatol Rapoport and published by him, in 1953, under the title *Operational Philosophy*. Given the persistence of the Bauhaus heritage, it is hardly surprising to find that Rapoport's philosophy was really a methodological updating of John Dewey's pragmatic-instrumentalism. The appeal of Rapoport's method lay in his attempt to provide a precise system for the evaluation of alternative courses of action. It is a measure of his discretion that, despite his dependence on mathematical logic, Rapoport was at pains to distinguish both his and Dewey's system from that of

In many fields of social life people are addressed, guided, or placed in contact with each other through visual images. The task of this department is to design these images in accordance with their function. This is why typography, graphic design, photography, and exhibition design are treated as a single area, which will shortly be augmented by the addition of

*right*
Poster for trotting races
Tomàs Gonda
1959 – 60

*left*
Poster: Danger
Sven Weisshardt
Second-year student,
1959 – 60
Instructor: Otl Aicher

the logical positivists, with their belief that philosophy should become a purely analytic discipline, akin to mathematics. Instead, Rapoport thought of his operationalism as being a *synthetic* action-oriented discipline. In 1953 he wrote that operationalism "is the philosophy of action-directed goals. It starts with logical analysis but transcends it by relating this analysis to society." At the time this seems to have been relatively close to Maldonado's own notion of scientific operationalism, of which he has since written: "By scientific operationalism I intended then a model of action oriented toward overcoming the dichotomy between theory and practice. Later on, following Kotarbinski, I preferred to call it 'praxiology' — and even more recently, the 'philosophy of praxis,' as seen in Gramsci."

Rejecting its arts and crafts origins and ever conscious of the perspectives of Marxist analysis, the Hochschule, bound to the service of neocapitalism, had little choice in the fifties but to look beyond the limits of these traditions for a mediatory ideology from which to develop not only a satisfactory heuristic method, but also a theory of design. A theoretical basis seems to have been proffered by Rapoport's operationalism, save for its incapacity to deal in an adequate manner with the intrinsic significance of form itself. For this the Hochschule seems to have turned first to Max Bense, whose communications approach to the determination of aesthetic need was first outlined in his book *Aesthetica,* motion pictures and television. The term "visual communication" has emerged to denote this area, in accordance with international usage.

Research within the department is directed toward fitting visual statements as closely as possible to what they have to say. To this end methods must be evolved that take account of the advances made in recent decades in the theory of perception and meaning.

From HfG curriculum document, 1958 – 59

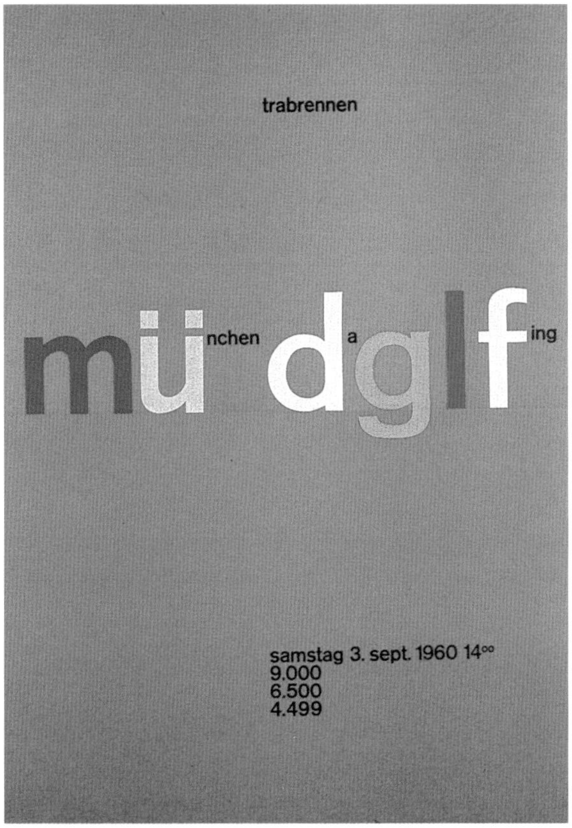

*left*
Poster for trotting races
Aicher Development Group
1960

*right*
Poster for Stan Getz
Peter Croy
1959 – 60

published in 1954; and then to the writings of Charles Morris, whose first semiotic works had appeared in the Unified Science publications of the late thirties. *Operation* and *communication,* these are the two "poles" that are to play major roles in the evolution of Hochschule theory.

One cannot complete an outline of the ideology of the Hochschule without some reference to its teaching methodology. It is evident that with the change in the directorship in 1957, the school began to take a more rigorous approach to the problem of both design and design training. That is to say, it encouraged a logical approach to the organization and generation of basic form, with the intent of applying such procedures to actual design problems. These operations varied from simple projections to three-dimensional transformations, from matrices to the manipulations of lattices, from the progressive deformation of regular grids to the rotation of ellipses centered on such grids, from the application of graph theory to topological studies, from exercises in solid geometry to the development of three-dimensional modular components that were capable of being combined in alternative sequences. Where in the visual communications department these exercises were often decisive in determining the general approach to an informational problem, such as the design of a subway map, in the building department a modular method was invariably adopted as a way to approach a variety of problems, from the

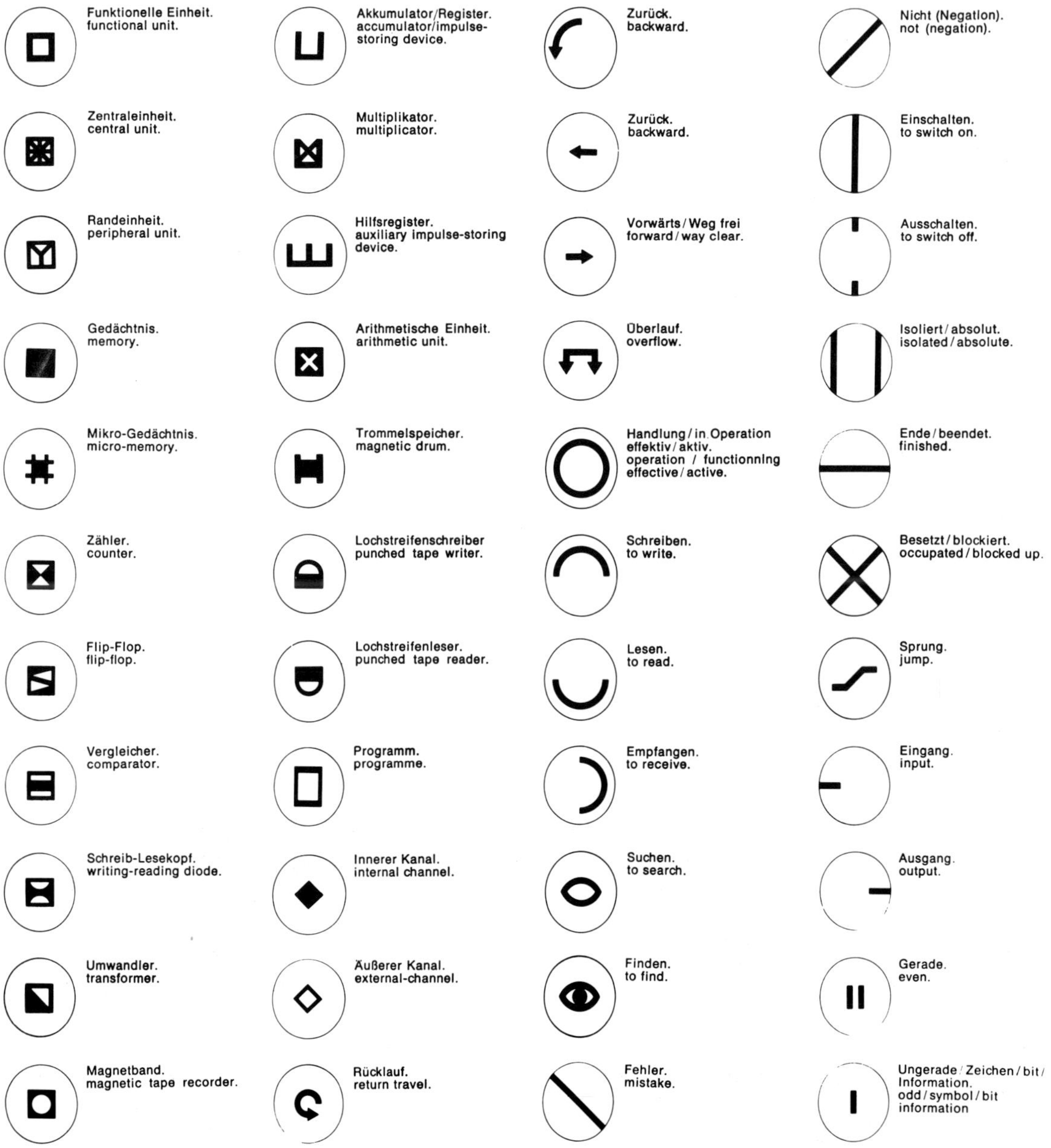


Part of a sign system
Tomàs Maldonado,
Gui Bonsiepe,
for Olivetti, Milan
1961

design of a diaphragm for a building panel to the combinatorial range of a set of prefabricated elements.

But the pursuit of a mathematical methodology did not restrict itself to this rational, yet free manipulation of formal propositions. In the field of design programming, where initial criteria and alternative solutions have first to be established and evaluated, the rigor rapidly developed into a form of heuristic determinism and into a logical positivism of design that would often be tempted to forego a solution rather than arrive at a synthesis that could not be entirely determined algorithmically. In these instances, design method rapidly degenerated into what Maldonado has since characterized as "method-idolatry." That Maldonado's own attitude to the positivistic approach was at first somewhat ambivalent seems to be reflected in the endless controversies that arose inside the Hochschule during the early sixties, which amounted to a long-drawn-out confrontation between the pragmatic designers on the one side, epitomized in such brilliant figures as the late Hans Gugelot, and the methodologists on the other; the most extreme faction, according to Maldonado, being led by a guest professor at the Hochschule, the Swiss political economist and art historian, Lucius Burckhardt.

Although little mention of this conflict emerged in the quarterly journal, it is quite clear that it nonetheless took place. This much is evident from Gunter Schmitz's paper

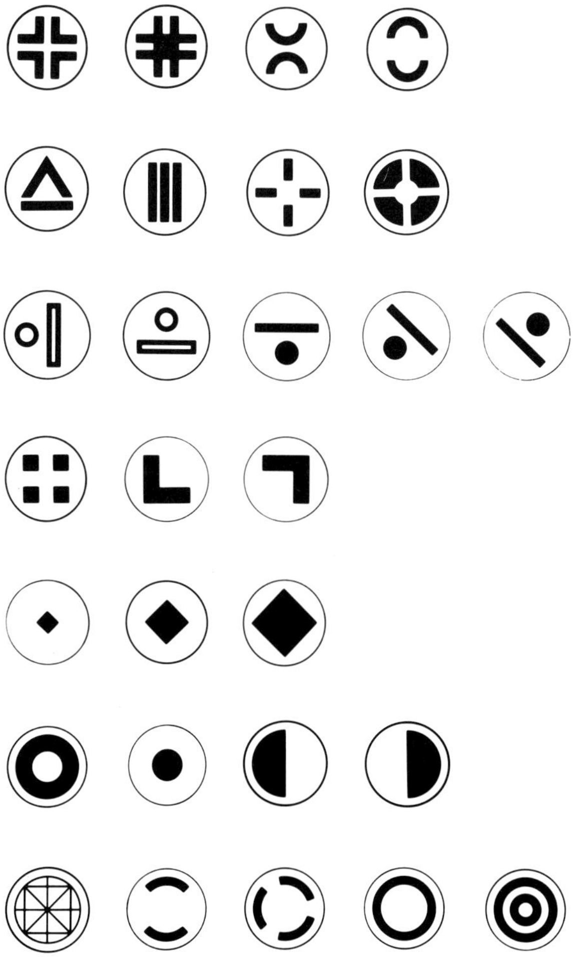

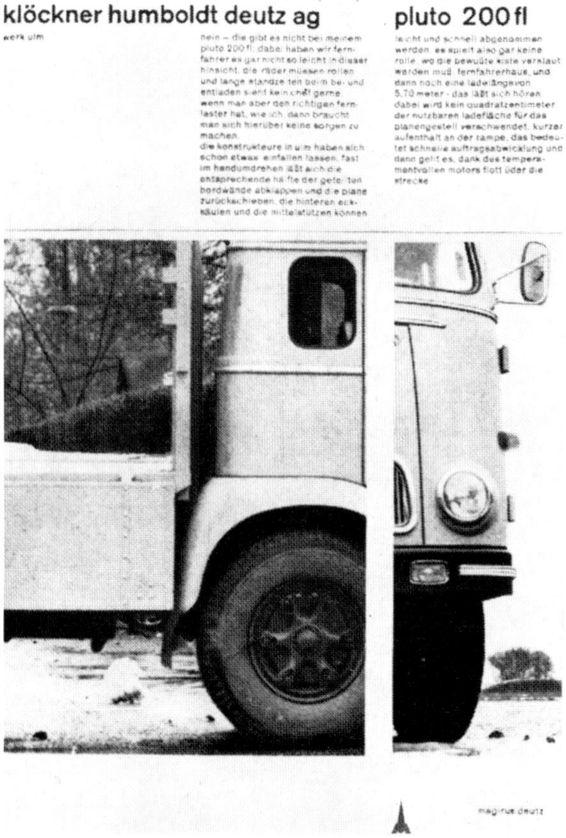


*right*
Typographic house style
Tonci Pelikan,
for Klöckner-Humbold-Deutz
Instructor:
Josef Müller-Brockmann
1962

*left*
Design of a sign system
Karl H. Remy
Second-year student,
1962 – 63
Instructor: Gui Bonsiepe

on the Hochschule, read before the AIA/ACSA Teachers' Seminar of 1968, wherein he stated:

*From 1960 to 1962 a controversy over the evaluation of the theoretical courses in relation to practical design work engaged the school. At the base of this lay the question of the exact role that analytical methods should play in the design process. The tendency towards an objectification of creative activities had nourished the dangerous yearning for a methodology which would automatically lead to original and perfect results.*

Schmitz then went on to briefly characterize the development of design teaching and practice within the school, from the beginning of Otl Aicher's directorship to its closure in 1968.

*Since 1962 the Hochschule has tried to balance the results and methods of the different scientific disciplines with the practical requirements of the design process, or, to put it another way, the Hochschule tries to avoid a mere accumulation of theoretical courses indigestible to the student. The concept since that time has involved an accentuation of the instrumental character of theory and the performance of practical design work on an experimental basis. As a consequence the number of general theoretical courses is reduced in favor of lectures which are more directly connected with design problems. On the other side the design problems stimulate prospective investigations, where theory plays an important role.*

Exhibition installations generally have a restricted life span in use. Their spatial layout should be adaptable to changing exhibition requirements. The volume occupied should be extensible and reducible. Finally, they should be reusable without visible wear and tear. Exhibition structures have a dynamic character that is shared by no other structures whatever.

The look of these exhibition structures is the result of the use of a universal, flexible, and demountable system, best calculated to adapt to changing requirements.

Exhibition structures should be composed of the smallest possible number of nonbulky elements. In volume and weight they must be adapted to the requirements of transportation, storage, and assembly. Their mechanical structure should permit a rapid and uncomplicated assembly process.

From a brochure produced by Development Group 5, 1959

Stand at the Hanover Industrial Fair
Otl Aicher, for BASF
1962

An idea of the Hochschule curriculum and ecucational policy at the time of its stability, so to speak, that is, just prior to its enforced closure in February 1968, may be best gained from the tables that Schmitz made available to the Montreal seminar, wherein the number of hours spent in various subjects and tasks during a four-year course in the building department was made quite explicit. Since this breakdown reveals the pedagogical emphasis in its prime, it seems fitting here to make a brief summation of the academic balance achieved at the Hochschule prior to its closure. By then it should be noted that the foundation course, or *Grundlehre,* had been discontinued, after Maldonado had been appointed as head of the industrial design department with the reorganization of 1962. After this year, students were channeled into one of three departments from the very beginning; that is, from the first year they entered directly into their chosen specialty, be it *building, product design,* or *visual communication.* The department of information had been eliminated in Aicher's reorganization. A certain communality of approach was now assured by common theoretical courses taken primarily over the first two years by students of all three departments.

It is clear from this that the highly ambitious educational program projected in *Ulm, 1,* in 1958 had been heavily trimmed in the Aicher reorganization of 1962, yet something of the unusual initial orientation

Exhibition stands are built like stage sets, as a rule, and their use of expressive resources increases in line with the versatility of the materials used. The design excesses that result from these short-lived sham structures turn many exhibitions and trade fairs into oppressive surfeits of form and color that overwhelm both the objects themselves and the information that is meant to be conveyed. The visitor is subjected to a visual overload that impairs his capacity to absorb information. The lack of credibility that adheres to exhibition stands in such circumstances tends to be transferred to the firms, the products, and the information.

What is wanted is exhibition stands whose constructional, functional, and aesthetic standard matches the confidence that a firm seeks to inspire for its products.

Economically speaking, it is advisable, instead of laying out individual sums repeatedly, to invest more money in a structure that can be used longer in the maximum number of variants.

Otl Aicher, in an HfG seminar, 1962

Otl Aicher and Peter Cornelius

*left*
Claude Schnaidt, Hans Roericht, Otl Aicher, and Peter Cornelius at a photographic exhibition of Cornelius's work

*right*
Roland Fürst and Hans Roericht working on a commission for Lufthansa 1965

remained in the curtailed curriculum, for all that the term and even the topic "operational research" had been discreetly dropped. Thus we find the first-year theory courses, taken by students of all three departments, with roughly 27 percent of their time devoted to mathematical techniques applicable to design, some 37 percent of their time to sociology, economics, political economy, psychology, and ergonomics, and the remainder of their time to a very brief survey of twentieth-century cultural history, including architecture, industrial design, film, literature, painting, music, and visual communications. In other words, the mathematical, political, and socio-cultural emphases of 1958 remained largely intact, although by the end of the first year, cultural history was abruptly concluded, with the remainder of the courses then being broadly divided between systems theory on the one hand and political economy on the other.

Corporate identity means that all the objects, the services, and the installations of a single undertaking identify themselves in a single consistent way.

The corporate identity is the visual counterpart of the firm's conceptual image: just as a positive image corresponds to a positive reality, an appropriate corporate identity conveys desirable qualitative characteristics. . . .

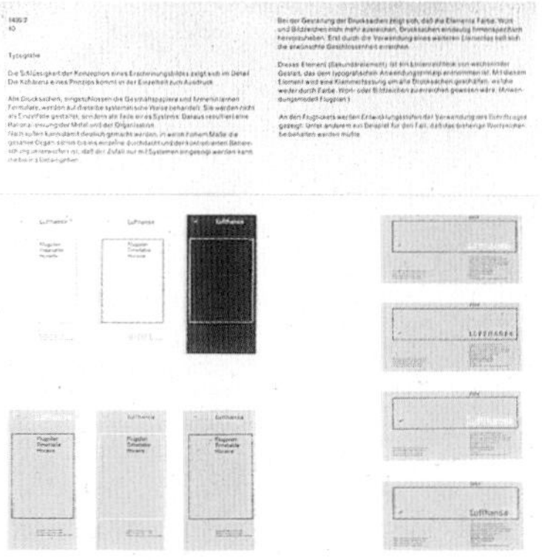

A modern business needs quantitative differentiation not only in the products or services that it offers, but also as a social entity in its own right, above and beyond its immediate commercial objectives. The business must try to find a profile, in order to create a clear and recognizable public image for itself. Anonymity does not inspire confidence, any more than excessive self-praise does. . . .

Increasing awareness of the efficacy of a strong corporate identity, and the special organizational circumstances of a worldwide airline company, suggest the advisability of formulating rules that will bring the visual design work into line with the ideas and intentions of the management of the company. . . .

The conceptual validity of a corporate identity shows itself in detail. The coherence of a principle is expressed in little things.

All printed matter, including business papers and forms for internal use, is designed in the same systematic way. Nothing is designed as an individual item but as part of a system. From this there springs a rationalization of methods and of organization.

Outsiders can thus be made clearly aware that the whole organization has been thought through and placed under tight control; and that, where as little as possible is to be left to chance, systems have to be worked out down to the last detail.

A misconceived systematization leads to schematization and sterility!

From a brochure by Development Group 5, 1962

Corporate identity for Lufthansa
Otl Aicher with Hans Roericht, Tomàs Gonda, Fritz Querngässer
and in association with Hans G. Conrad, an HfG alumnus working for Lufthansa
1962–63
The corporate identity covered the following elements: a house color scheme, pictorial and typographic logos, typeface, formats, graphic and typographic layouts and standards, style of photography, quality of materials, packaging, exhibition systems, characteristics of architecture, forms *(Gestalt)* of interior design, style of working clothes and uniforms. The Ulm designs have subsequently been realized and further developed, partly at Ulm itself and partly at Lufthansa, by HfG alumni (Conrad, Wille, Roth) and others

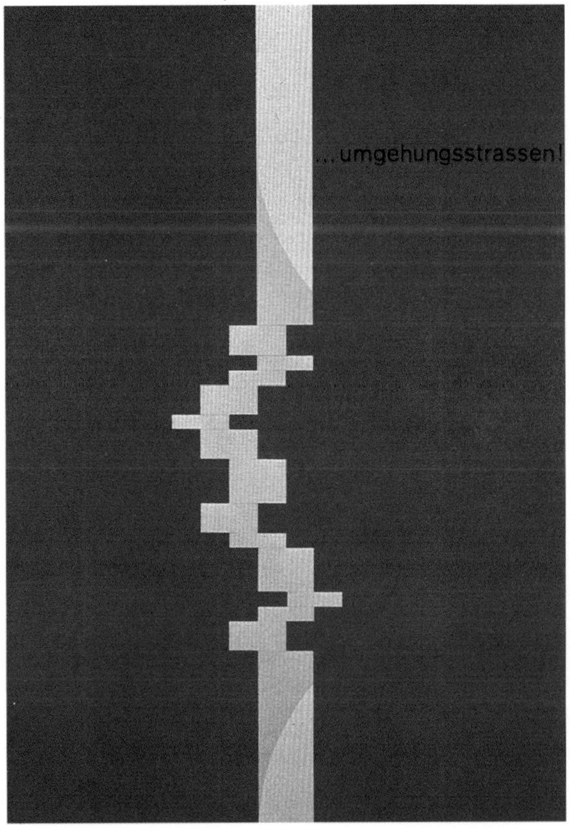

*right*
Poster: Design congress
Dieter Wagner
Second-year student,
1963 – 64
Instructor: Herbert Lindinger

*left*
Poster: Bypass roads
Rolf Müller
Second-year student,
1961 – 62
Instructor: Otl Aicher

# The Development of a Critical Theory

Kenneth Frampton

Despite the reorientation of the school under Aicher's stewardship, the socio-cultural criticism emanating from the Hochschule continued to grow over the next five years, most particularly through the contribution of Maldonado, Gui Bonsiepe, and Claude Schnaidt. These three happened to announce their common critical attitude in the review section of *Ulm, 7,* published in 1963, wherein their notes were respectively addressed first to a criticism of the intrusion of neo-Dada into the field of industrial design; second, to an appraisal of Leonardo Benevolo's *History of Modern Architecture,* and finally to a review of Georg Klaus's critique of Norbert Wiener's information theory, which had then just appeared in Klaus's book *Cybernetics in the Light of Philosophy.* The aim of the Klaus study was to refute the Wiener reduction of information to a mere quantifiable assessment of its relative density and predictability. Rejecting the implicit Wiener split of *sign* from *import* and his classification of information as a mere quantum, akin to energy, Klaus argued that "all information must rather have a definite meaning, must be a carrier of some significance."

This apparently banal but nonetheless antipositivist statement had of course been the basic assumption behind the Maldonado seminars in semiotics, given as a regular course in the Hochschule from 1957 to 1960, the first fruits of which were the Maldonado essay "Communication and Semiotics" that appeared in *Ulm, 5* in 1959, and the Bon-

In the field of graphic design there is a similar evolution toward a new position. The designer's activity has extended beyond the traditional tasks, such as posters, book design, and exhibition design; the criteria for choosing a specialty are no longer primarily artistic ones. The challenge of pictorial communication has been taken up in just those areas where it is truly of decisive significance in our civilization. Traffic signs have been the subject of thought and design work to the same degree as sign systems for electronic installations or for machines. We have pressed forward into a whole world of sign languages, sign information systems, and have involved ourselves in the constantly expanding process of creating languages with the help of visual media.

Otl Aicher, from his opening speech at the HfG exhibition at the Landesgewerbeamt, Stuttgart 1963

Alexander Mitscherlich visits the HfG exhibition in Ulm, 1963
Herbert Lindinger, Alexander Mitscherlich, and Inge Aicher-Scholl

Poster for the day of commemoration for Hans and Sophie Scholl
Herbert Lindinger, 1962 – 63

siepe unpublished text "Über formale und informale Sprachanalyse: Carnap und Ryle" that was written in 1960. Strangely enough the only adequate publication of the work of these seminars did not appear in the journal *Ulm* but in a little known publication entitled *Uppercase,* edited by Theo Crosby. Thus, *Uppercase, 5* of 1963, dedicated in the main to work of the Hochschule, featured texts by Maldonado and Bonsiepe, a design case study by Walter Muller, and a semiotic glossary.

In retrospect the most significant aspect of this whole publication was the distance it implicitly took from a positivistic design approach and the corresponding stress it placed on form as a necessary communicative element. In his "Notes on Communication" Maldonado refused the positivistic split of operation from communication in a text that is remarkable for its perception of *function* as being an integral part of *culture* and vice versa.

*The classification of the products of a culture — operative world of the artifacts and communicative world of the signs — becomes less and less convincing. In reality, all these products of a culture belong to one common system. The artifacts are operable in the extension — and only in the extension — that they are capable of communicating a definite meaning unit to the operator; the signs for their part are communicative in the extension — and only in the extension — that they can directly or indirectly influence a behavior in an operative way.*

The Communications Industry

Especially in relation to visual design, the HfG has formulated a program that is not only distinct from similar Bauhaus initiatives but radically new: the Bauhaus could never have conceived such a program, for objective historical reasons. For the industry that we now know as the communications industry, or even the "consciousness industry," motion pictures, television, radio, and press, began to take shape only in the, 1920s, and some parts of it even after the Bauhaus was closed. And it is in the consciousness industry that the dramas and farces of mass communication now take place. In addition, with the transition from an economy of dearth to an economy of plenty, advertising has moved to a central position in visual design, as a new institution of social control. These technological and economic changes have made it unthinkable to transplant the Bauhaus *en bloc*.

HfG touring exhibit at the Kornhaus, Ulm
Herbert Lindinger, Claude Schnaidt
1963
Subsequently shown in Stuttgart, Munich, and Amsterdam

Maldonado extended this argument to embrace the field of ergonomics and in particular the province of machine design, where the *operative* and *communicative* aspects become dramatically intermeshed and where the critical "man machine" relationship of advanced industrialization acquires an undeniably concrete dimension. In this respect, the advanced ergonomic theories of Chapanis, Fitts, and Taylor were welcomed by Maldonado for the stress they placed on the redesign of the machine and for their mutual intent to resolve the "man-machine" couple in such a way as to liberate man as much as possible from the tyranny of the machine.

This measured critique was extended in the Maldonado/Bonsiepe essay "Science and Design" that appeared in 1964, in *Ulm, 10/11*. Broadly speaking, this paper was an attack on the simplistic borrowing of design methods from the field of "human engineering," beginning with a critique of the established methods of experimental psychology for their untenable lineality of approach and going on to upbraid that aspect of ergonomics that grounded itself in a servomechanical model of the human being — a schema whereby the complexity of man becomes reduced (usually under conditions of *extremis*) to the so-called H-factor. The authors concluded their survey of heuristic methods, derived from the margins of applied science, with a highly skeptical appraisal of the procedures of market and motivational research and

But it is not the new parts of the HfG's program that are disliked by the majority of institutions concerned with design education. The tension and the hostility spring from the fact that the HfG pays more attention to the question of how design relates to the sciences than to the question of how design relates to the arts.

Gui Bonsiepe, in *Ulm,* 12/13, 1965

Donau Zeitung

Diplomatische Beziehungen zu Israel — Bruch mit Irak

Vietnam-Politik in Sackgasse

Kurz gemeldet:

Deutschland-Erklärung des Westens zurückhaltend

Kommunique der Kompromisse

Koalition in Niedersachsen

Mehr Rente für Kriegsopfer

Deutschland - England 0:1

The typographical problems that arise in newspapers and magazines are primarily a matter of page makeup: the correct arrangement of texts one above the other, and of pictures and text. The aesthetic and compositional approach – whether conventional or modern – does not suffice to establish a way of working. For every possible makeup situation, a system has had to be devised that defines all the elements of the page – line, column, picture, headline, caption, marginals – with such precision that they fit together as interchangeable and endlessly permutable units. The grid format is the framework, and in any combination the individual elements must fill that framework. The arrangement thus defined has economic advantages (rationalization) and provides the basis for a more closely defined aesthetic formulation. First and foremost, the functional requirements of legibility, ease of orientation, and clarity must be satisfied.

Otl Aicher, in an HfG seminar, 1962

Front page of a daily newspaper
Gerd Zimmermann
Second-year student, 1964 – 65
Instructor: Herbert Kapitzki

of the propensity of such research to convert "undifferentiated needs into definite demands." The concluding paragraphs of "Science and Design" convey the irony of their attack.

Maldonado's own views on the teaching and practice of industrial design have never perhaps been more clearly articulated than in the article he wrote (presumably late in 1963) for the education volume of Gyorgy Kep's *Vision and Value.* In this text Maldonado defined the metier of industrial design in the following terms: "Industrial design is an activity whose ultimate aim is to determine the formal properties of the objects produced by industry. By 'formal properties' is not meant the external features, but rather those structural and functional relations that convert an object into a coherent unity from the point of view of both the producer and the user."

This definition, which was and still is distinguished by its precise qualification of the term "formal" and by its insistence on the need to satisfy the user as well as the producer, appeared as the fulcrum of Maldonado's argument, which, while it exposed the commercial limitations of industrial design in a competitive society, could not bring itself to condone the pathetic mimicry of neocapitalist products in noncompetitive societies such as the Soviet Union. Disturbed by the complacent claims made for Soviet design by Yuri Soloviev, at the Aspen Design Conference of 1961, Maldonado wrote:

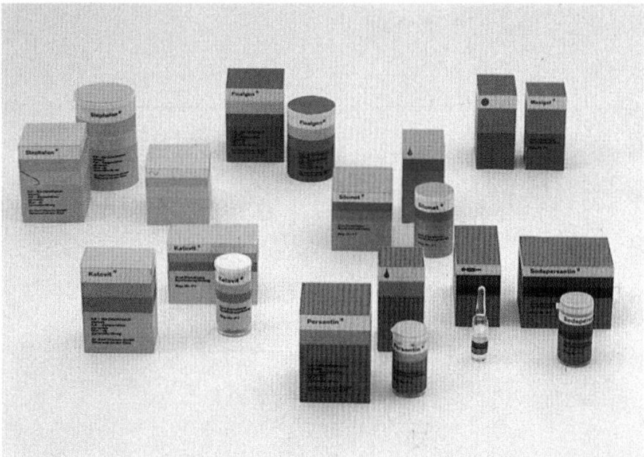

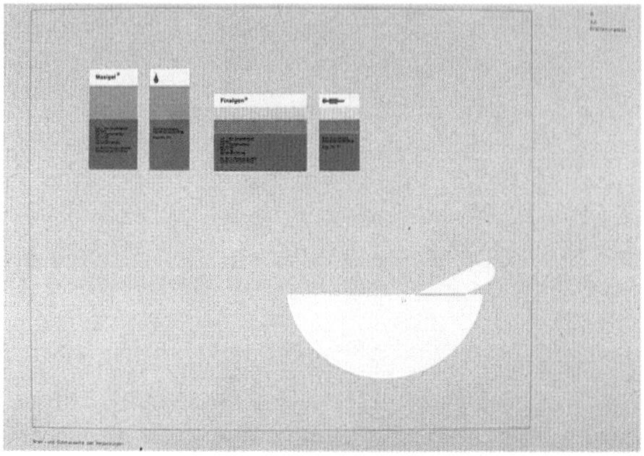

*above right and left*
Corporate identity for a pharmaceutical company
Ewald Duffner
Fourth-year student, 1964 – 65

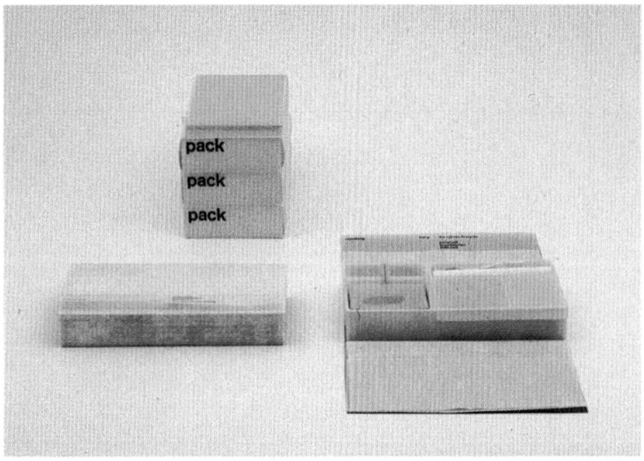


*middle left*
Packaging
Hartmut Kowalke
Third-year student, 1964 – 65
Assistant: Nick Roericht

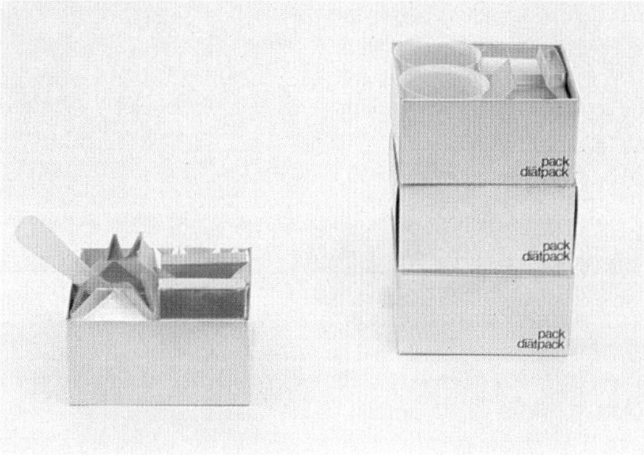


*below left*
Packaging
Josef Breuer
Third-year student, 1964 – 65
Assistant: Nick Roericht

*below right*
Packaging for tomoebilin
Shizuko Yoshikawa
Second-year student, 1962 – 63
Instructor: Josef Müller-Brockmann

*Curiously enough, the industrial designer in the Socialist countries is not fully conscious of the new possibilities that his economic and social system — at least in theory — offers to his profession. If this were not so, how can we explain that the frankly pathological manifestations of American and European industrial design are adopted by the Soviet designers as models worthy of imitation and perfection? One does not expect from the Soviet designers the imitation of our weaknesses, but rather the full exploitation of their own, specific possibilities. One expects them to tackle problems we are not allowed to tackle. For instance, technical products themselves require an urgent revision as far as their structural and functional properties are concerned, but in the framework of our competitive society, initiative in this direction cannot be imagined, because the main activity of our society is to merchandise these products. … The designers of a noncompetitive society are in a favorable position for attacking this new kind of task, but until now not very much has happened. One can only hope that this cannot be traced to the same reason that in the past caused Soviet architects and urbanists to commit such mistakes as maintaining naive confidence in a tradition in which no one any longer believes.*

Thereafter, in the same text, Maldonado added to his definition of industrial design the rider that its interpretation and application would be differentiated according to the follow-

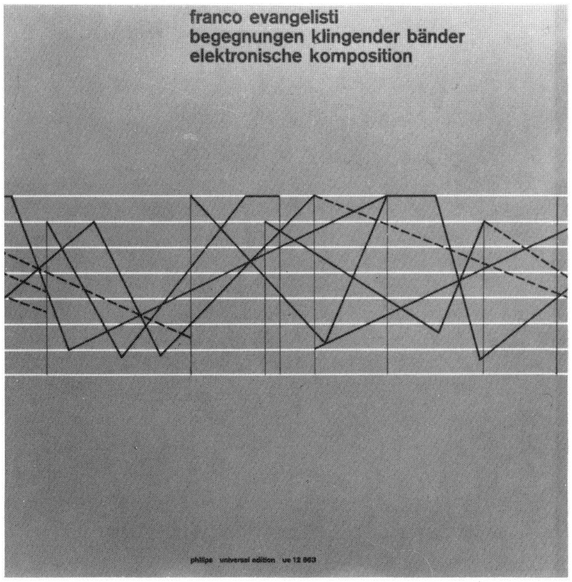


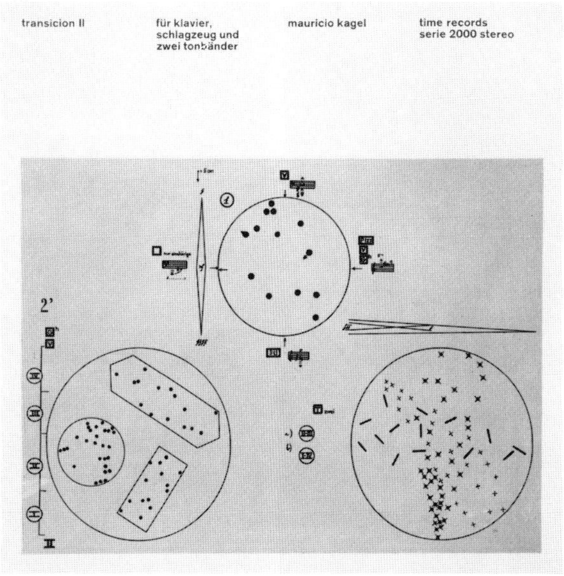


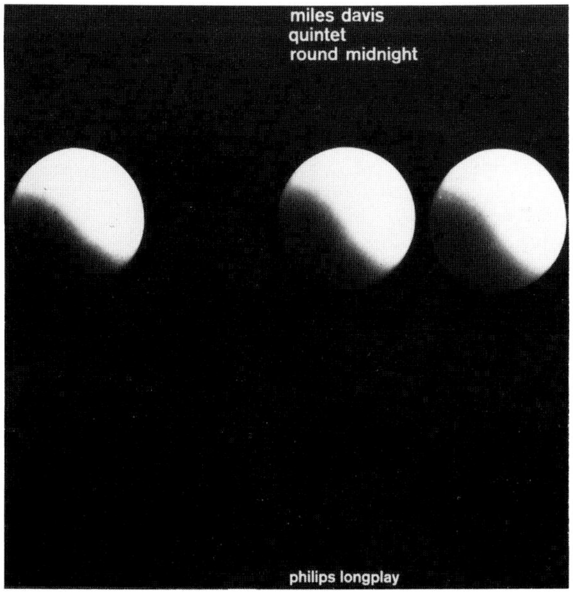


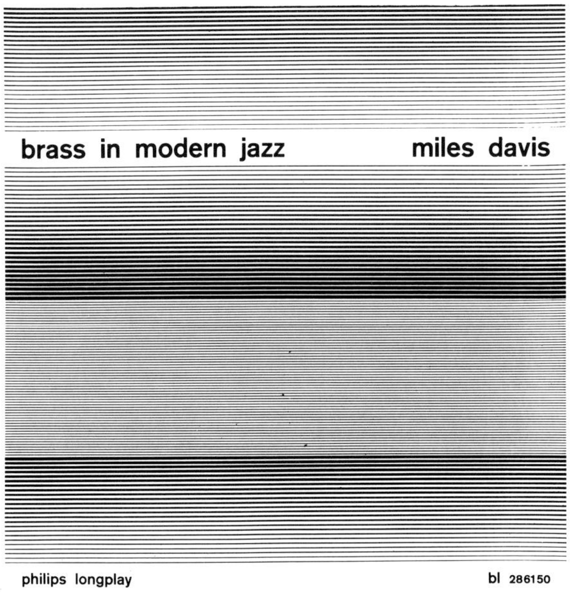


Album covers
Instructor:
Tomàs Maldonado

Begegnungen Klingender Bänder
Josef Breuer
Second-year student, 1962–63

Transicion II
Lothar Spree
Second-year student, 1962–63

Round Midnight
Hartmut Kowalke
Second-year student, 1962–63

Brass in Modern Jazz
Hans-Rudolf Buob
Second-year student, 1962–63

ing variables: "(1) the social and economic context, i. e., whether the profession is exercised in a competitive or noncompetitive society; (2) the degree of the structural and functional complexitiy of the objects to be designed; (3) the degree of dependence of the particular object to be designed on the traditions of craft and the traditions of taste."

For Maldonado, *design* in general, after a dialectical overcoming of both the "degeneracies" of admass populism and the paradoxical "alienations" of bureaucratic socialism, has to be returned to a strict distinction in practice between puristic formalism on the one hand and formal order in its broadest sense on the other. In the last analysis, for Maldonado, this distinction can only be made in the context of preserving human values, an issue with which the second half of the twentieth century has yet to come to terms. Given the economic and highly abstract imperatives of our present society, this is understandable, since the reintegration of such values ultimately presupposes a dialectical definition of "needs" that would have to transcend, without excluding them, the primary demands of production and use. Such a definition would have to assimilate these basic criteria into a perspective that takes cognizance of the fundamental limitations of human life — *eros* and *thanatos*, hedonism and mortality.

The position taken by Schnaidt and Maldonado with regard to the particular predicament of architecture, as it

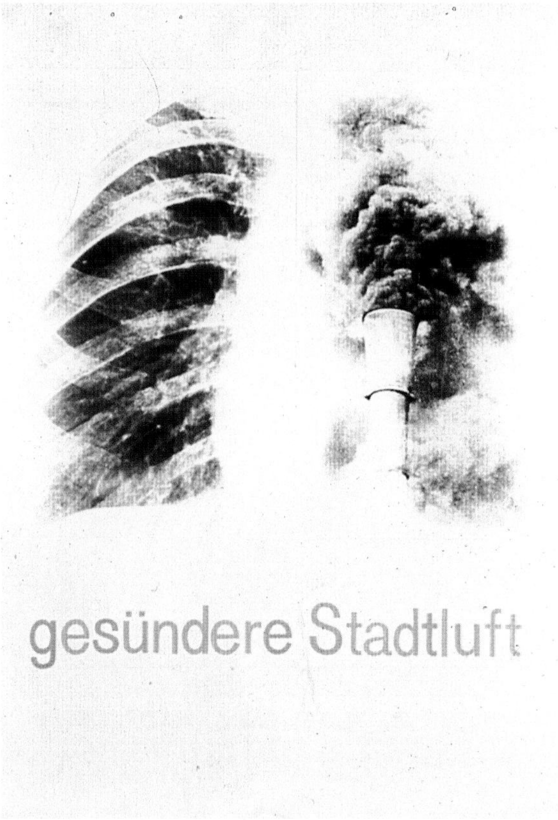

*right* Poster: Ermanno Olmi's film *Il posto*
Eckhardt Jung
Second-year student, 1964–65
Instructor: Kohei Sugiura

*left*
Poster: Cleaner Air in Cities
Giovanni Anceschi
Second-year student, 1963–64
Instructor: Herbert Lindinger

was then being practiced and taught in the sixties, remains remarkably timely. Their views, now almost ten years old and strongly influenced by the conditions of the time, retain nonetheless a certain general validity that makes them as applicable today as when they were first written. In fact, in a decade, little has changed except that the opportunities for the architect to make a significant contribution to the society are possibly even more limited now than they were in the early sixties.

In his essay "Prefabricated Hope" that appeared in *Ulm, 10/11,* Schnaidt attempted a comprehensive analysis of the failure of industrialized building. Schnaidt then argued the by-now-familiar sterility of treating this prospect from a purely technical standpoint. He wrote:

*It is difficult to apply profitably industrial production methods in the building of housing developments containing less than 500 dwellings. Given the current density of population, 500 dwellings require at least 2.5 hectares of land. ... To create such sites one must acquire many small lots, paying the owners a surplus value estimated according to the expected value of the lot after main supplies and sewerage pipes have been laid. This is where speculation is let loose. The sale and resale of building sites to the profit of the few is a course that is becoming increasingly ruinous to the community. On the outskirts of numerous major European cities, the price of real estate has increased tenfold in the last*

*left*
Poster: Ulm Bach Concerts
Peter Polland
Second-year student, 1964 – 65
Instructor: Kohei Sugiura

*right*
Poster: German Athletic Championships
Gerd Zimmermann
Second-year student, 1964 – 65
Instructor: Kohei Sugiura

*ten years; in 1950, the ground rates represented about 10 percent of the selling price of a house; by 1960, it had risen to 45 percent. The reduction in the cost of housing that can be achieved by industrializing building seems ridiculously small in comparison with the increase caused by land speculation. … The future of the industrialization of building will depend on the solution found to all these problems. This is why it is erroneous, if not dishonest, to speak solely of technical matters when evoking decisions that affect this future. The choice is not, as they would have us believe, between so-called traditional building and prefabrication. It is between a disordered, slow, and precarious development of technical progress in building as a whole and a coherent, rapid, and planned industrialization for the benefit of the community.*

One need hardly add that this critique was, in many respects, an implicit attack on the work of the Hochschule building department in which Schnaidt himself functioned as a teacher. In a parallel criticism of architectural education given as the Lethaby Lecture at The Royal College of Art, in 1965, under the title: "The Emergent World: A Challenge to Architectural and Industrial Design Training," Maldonado was to argue that the upgrading of architectural school curricula had largely resulted in a shifting of the academic scenery, in which the fundamental pedagogical orientation had remained unchanged. A primary aspect of this apparent

*right*
Poster:
*Schlachtbeschreibung*
(Battle Report), book by
Alexander Kluge
Peter von Kornatzki
Second-year student,
1964 – 65
Instructor: Kohei Sugiura

*left*
Poster: Danger — Yield
Klaus Hofmann
Second-year student,
1965 – 66
Instructor: Herbert Kapitzki

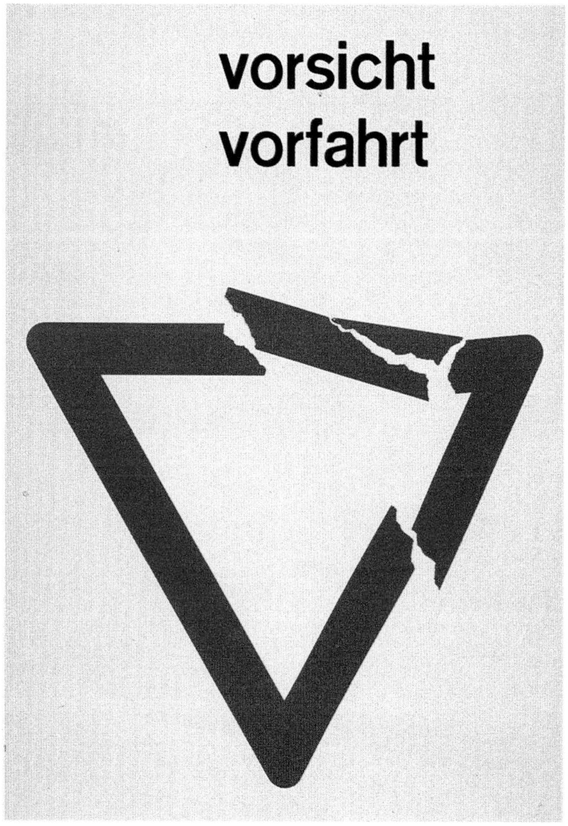


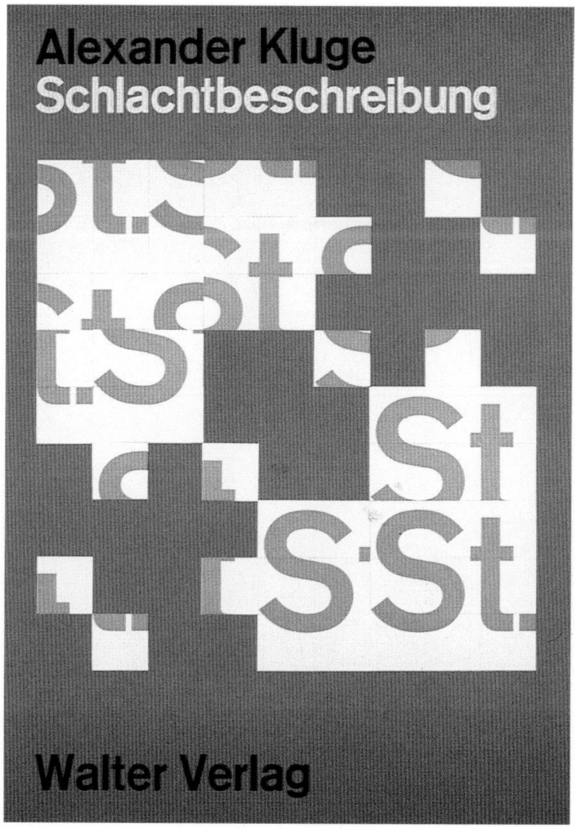


transformation had clearly been the universal adoption of basic design courses, along the lines of the Bauhaus, while the most common secondary change, largely unrelated to the first, had been the wholesale acceptance of modern architecture. Of this Maldonado remarked, "On the altar where Palladio was worshipped, Wright, Le Corbusier, Gropius, Mies van der Rohe, Fuller, Louis Kahn or Kenzo Tange are now being honoured. The idols have changed but not the doctrines." Yet for Maldonado not even those schools that had attempted to restructure their curricula along scientific lines were entirely free from criticism, for he could see all too clearly, after his own experiences at the Hochschule, how a naive worship of scientific method could lead to designs even more abstracted than before from any legitimate form of socio-cultural reality.

It would seem that by 1966 the "critical theory" of the Hochschule had already reached the threshold of disputing by implication the viability of design schools *per se,* and there is little reason to doubt but that Otl Aicher's essay "Planning All Awry?" which appeared in *Ulm, 17/18,* of that year was nothing but an oblique attempt to counter the autocriticism of his "left wing" faculty; Aicher urging all designers, not only planners, to accommodate themselves to the power constraints of neo-capitalism. It is interesting to note that the antimonumentalism of his position would have been shared by Schnaidt, but not the ultimately apolitical,

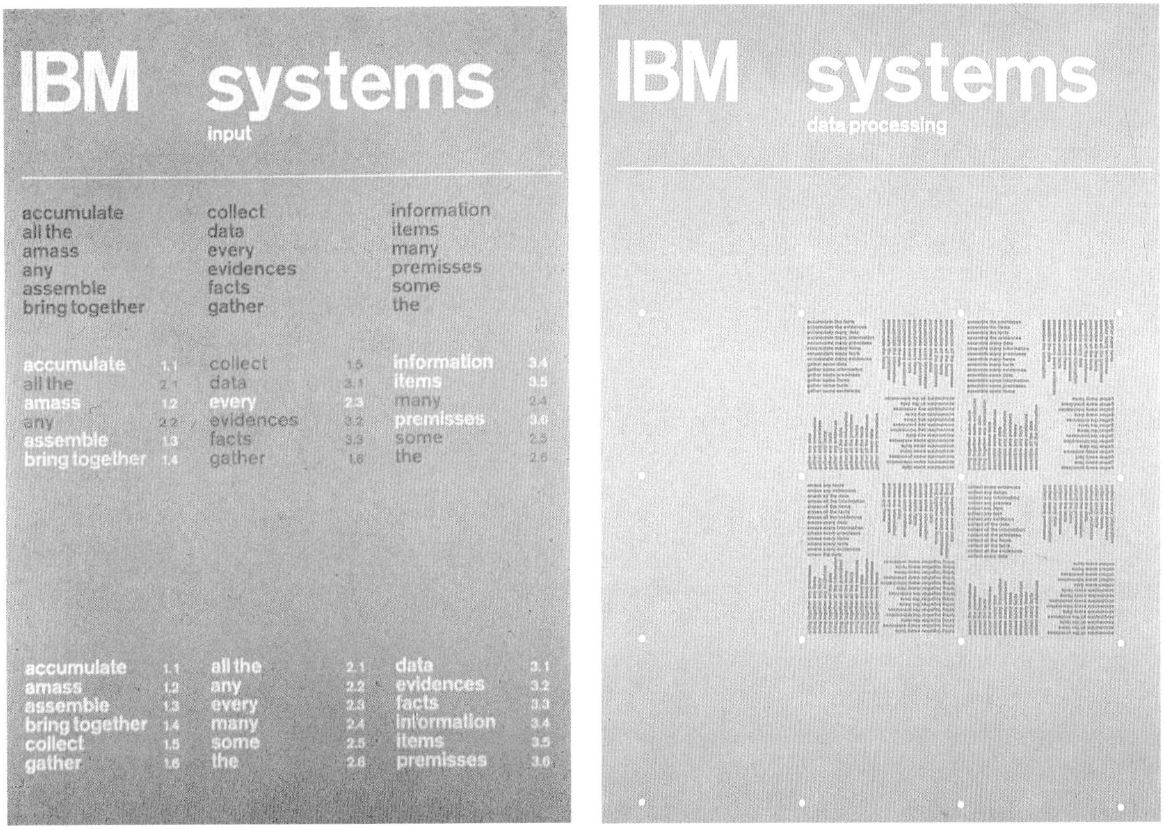


Poster: IBM Systems
Thomas Nittner
Second-year student, 1964–65
Instructor: Kohei Sugiura

mystifying scientism of his conclusions, wherein Aicher stated:

*Viewed in this way, conventional architects still form the professional category that is best suited to satisfying the demands of a planner. Only the days are past when the proximity of art shed some of the glory of genius on the architectural profession. The specific skills of the architect are today just adequate to enable him to elicit from the fast facts assembled by science a tangible plan tailored to political objectives, and then he must again leave it to the politicians to make the final decision as to its implementation. The planner loses nothing by recognizing in the development director a figure of political power who is set above him as regards both the definition of objectives and the assessment of feasibility. The planner loses nothing if he tolerates the presence of scientists who can make forecasts in respect of an applied development theory. He will undoubtedly be forfeiting his chance of a niche in cultural history because he will have surrendered his sole authority. But in return he will enjoy a special advantage: the prospect of seeing what he plans actually being realized. The prospect, that is, of gradually narrowing the gap between plan and reality that today condemns the plan to impotence.*

Aicher was apparently unable to realize the inadequacy of this position, wherein, irrespective of the "forecasts of science," the contradictions

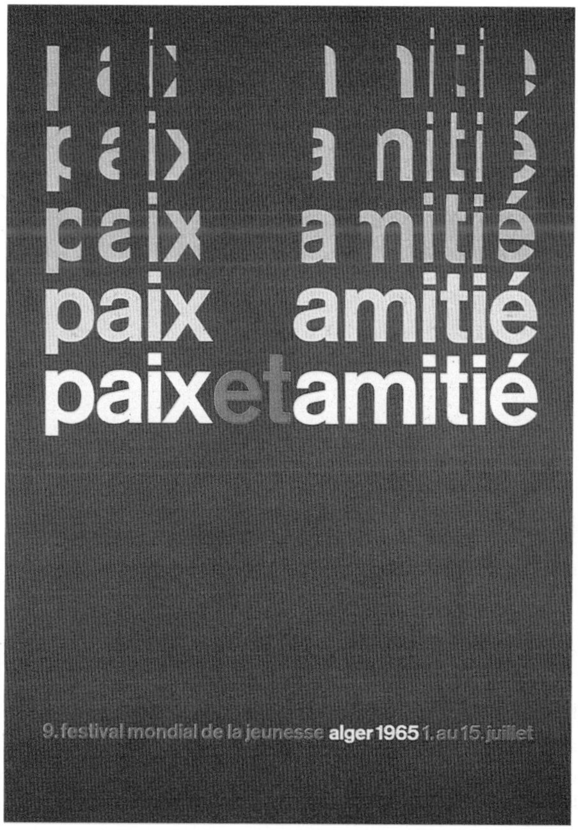


Posters: Peace and Friendship
Anne Preiss
Second-year student, 1964–65
Instructor: Kohei Sugiura

of society as they impinge on design are incapable of resolution through the "mythical" abrogation of power on the part of the designer. The *neue Sachlichkeit* architects of the Weimar Republic certainly had no taste for the "glory of genius," but this had little evident effect on the realization of their plans; particularly after 1933, when the figures of power chose to define the overall objectives in entirely different terms.

In any event Aicher did not go unanswered, first inadvertently by Maldonado, in the same issue of the journal, *Ulm, 17/18,* in a text with the provocative title, "How to Fight Complacency in Design Education," and then in the penultimate issue of the journal, *Ulm, 19/20,* in 1967, in a seminar report by Abraham Moles addressed to "Functionalism in Crisis," and in an essay by Claude Schnaidt entitled "Architecture and Political Commitment."

Where Maldonado, while pleading the case for C. S. Peirce's "university of methods," stressed conflict and disorder and the reciprocal link obtaining in the Third World between violence and necessity, Moles went straight to the raison d'être of the Hochschule and argued in effect that its basis had been overtaken by the success of the "economic miracle," since the pure functionalism it professed was no longer required by the economic system it was pledged to serve. While diplomatically evading the ultimate consequences of this argument, Moles presented his case with characteristic irreverence.

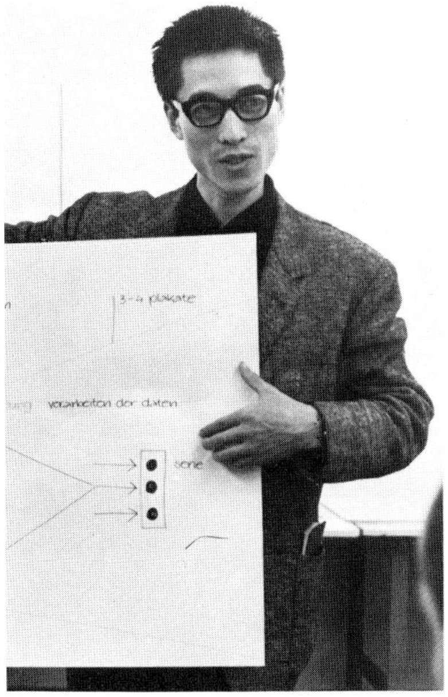

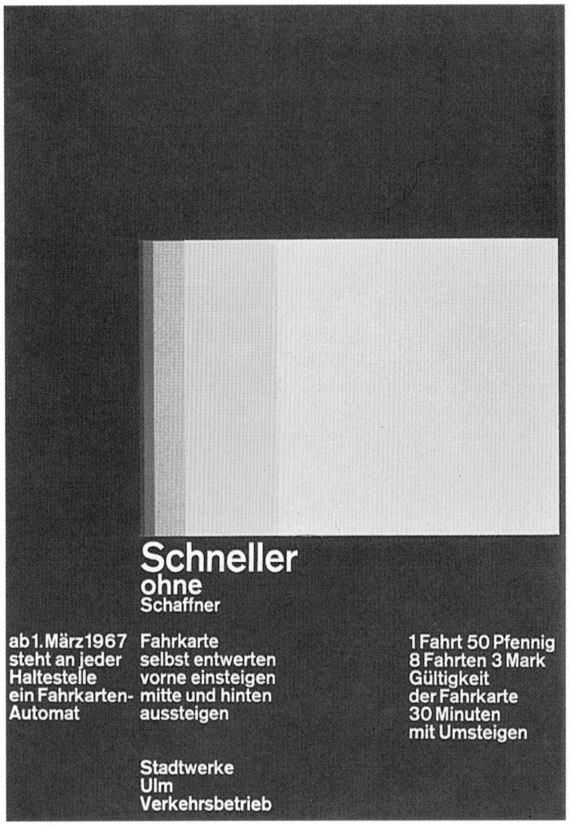


Kohei Sugiura

Herbert Kapitzki and Anne Preiss

Poster: Quicker with No Conductor
Part of corporate identity for Ulm city transportation system
Karl Gröbli
Second-year student, 1966–67
Instructor: Herbert Kapitzki

*Affluent society as an economical theory purports that the machinery of production has to run permanently; therefore the consumer has to be stimulated to consume at any price. Consumption and production are linked into a combined system that runs at an ever-increasing speed. Functionalism necessarily contradicts the doctrine of affluent society that is forced to produce and to sell relentlessly. Finally functionalism tends to reduce the number of objects and to realize an optimal fit between products and needs, whereas the production machinery of an affluent society follows the opposite direction. It creates a system of neo-kitsch by accumulating objects in the human environment. At this point the crisis of functionalism becomes manifest. It is torn between the neo-kitsch of the supermarket on the one side and ascetic fulfillment of function on the other side.*

It was left to Schnaidt to articulate in unequivocal terms the consequences of this crisis in its wider ramifications, and in many respects his text was to be the last major contribution to the critical theory of the Hochschule before its self-dissolution in February 1968. His indirect response to the Aicher model of planning reality requires little comment, save that his arguments lead him to advocate regional decentralization.

*While architects take refuge in aestheticism, fantasy, and technocracy, man's environment and everyday life are steadily deteriorating. The*

## Linie 2
## Eselsberg-Böfingen

**Montag bis Freitag**

| | | | | | | | | | | | | |
|---|---|---|---|---|---|---|---|---|---|---|---|---|
| Söflingen | 538 | alle 7 Min | 850 | alle 9 Min | 1130 | alle 6 Min | 1403 | alle 9 Min | 1610 | alle 6 Min | 1922 | alle 12 Min | 2309 |
| Gemeindeplatz | 540 | | 852 | | 1132 | | 1405 | | 1613 | | 1924 | | 2311 |
| Berliner Platz | 542 | | 854 | | 1134 | | 1407 | | 1615 | | 1926 | | 2313 |
| Hauptbahnhof | 544 | | 856 | | 1136 | | 1409 | | 1617 | | 1928 | | 2315 |
| Basteistrasse | 546 | | 858 | | 1138 | | 1400 | | 1619 | | 1930 | | 2317 |
| Georgskirche | 548 | | 900 | | 1140 | | 1422 | | 1620 | | 1932 | | 2318 |
| Königsstrasse | 550 | | 902 | | 1140 | | 1424 | | 1623 | | 1935 | | 2320 |
| Gaswerk | 552 | | 904 | | 1142 | | 1426 | | 1625 | | 1937 | | 2323 |
| Blücherstrasse | 554 | | 906 | | 1143 | | 1428 | | 1627 | | 1937 | | 2325 |
| Justizgebäude | 556 | | 908 | | 1146 | | 1420 | | 1629 | | 1939 | | 2327 |
| Schillerstrasse | 558 | | 910 | | 1248 | | 1422 | | 1631 | | 1942 | | 2329 |
| Ostplatz | 600 | | 912 | | 1252 | | 1424 | | 1633 | | 1943 | | 2330 |
| Neutorstrasse | 602 | | 914 | | 1254 | | 1426 | | 1635 | | 1945 | | 2333 |
| Donauhalle | 604 | | 916 | | 1200 | | 1428 | | 1637 | | 1947 | | 2340 |
| Westplatz | 606 | | 918 | | 1202 | | 1430 | | 1639 | | 2026 | | Ende |
| Sonnenstrasse | 608 | | 920 | | 1204 | | 1432 | | 1642 | | 2028 | | |
| Stadion | 610 | | 922 | | 1206 | | 1435 | | 1644 | | 2030 | | |

**Samstag**

| | | | | | | | |
|---|---|---|---|---|---|---|---|
| Söflingen | 558 | alle 9 Min | 755 | alle 10 Min | 1304 | alle 9 Min | 2355 |
| Berliner Platz | 620 | | 755 | | 1304 | | 2359 |
| Georgskirche | 624 | | 755 | | 1304 | | 2400 |
| Schillerstrasse | 628 | | 755 | | 1304 | | 003 |
| Donauhalle | 632 | | 755 | | 1304 | | 006 |
| Stadion | 634 | | 755 | | 1304 | | |

**Sonntag**

| | | | | | | | |
|---|---|---|---|---|---|---|---|
| Söflingen | 600 | alle 10 Min | 755 | alle 12 Min | 1304 | alle 9 Min | 2355 |
| Berliner Platz | 622 | | 755 | | 1304 | | 2359 |
| Georgskirche | 623 | | 755 | | 1309 | | 2400 |
| Schillerstrasse | 624 | | 755 | | 1304 | | 003 |
| Donauhalle | 634 | | 755 | | 2309 | | 006 |
| Stadion | 626 | | 755 | | 2304 | | |

*left*
Commuter fares announcement and timetable
Part of corporate identity for Ulm city transportation system
Karl Gröbli
Second-year student, 1966–67
Instructor: Herbert Kapitzki

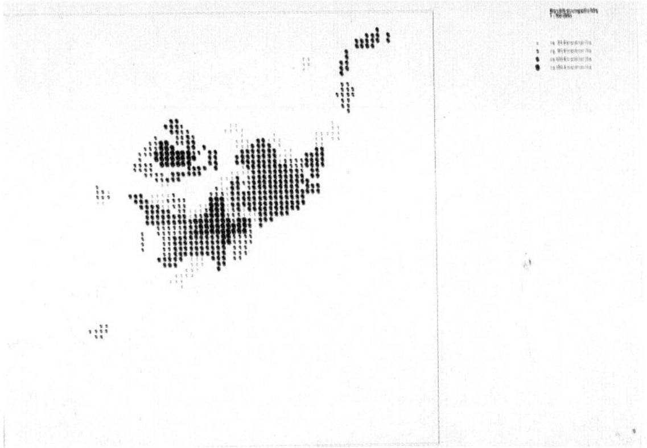

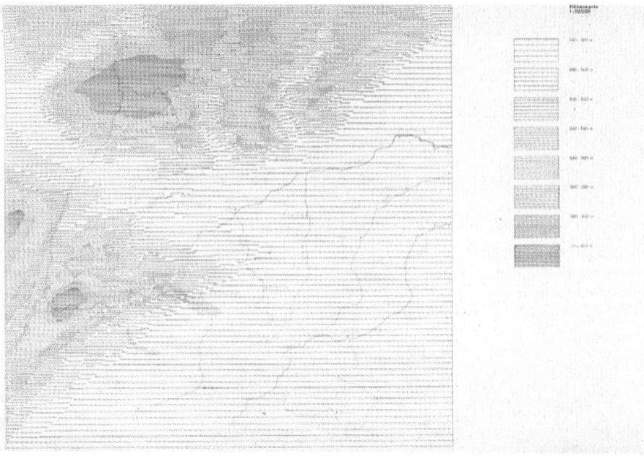

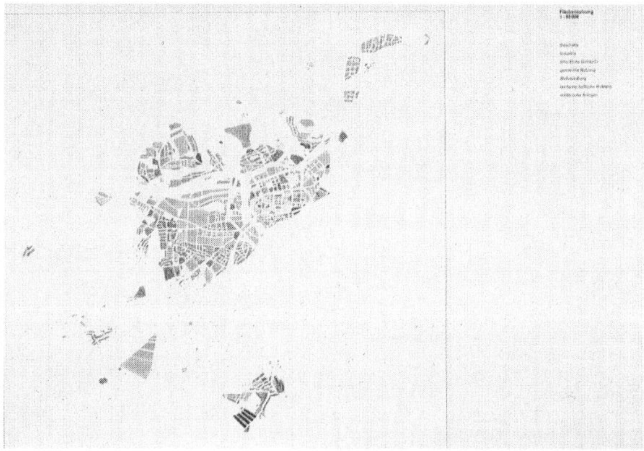

*right*
Map design
Gerd Zimmermann, Anne Preiss
Third-year students, 1965–66
Instructor: Kohei Sugiura

*megalopolises that are taking shape are stricken at the least failure of their overburdened infrastructures. They call for prodigious amounts of money to function at all. . . . The annual subsidy received by the Paris Passenger Transport Board is four times larger than all the allocations made to help the industrialization in Brittany during the past ten years. . . . The concentration of industries and their head offices in and around the metropolis and the continuous increase in rents that compels those working there to put up their homes far afield have made certain reductions in working hours a purely illusory gain. After all, a cut of 6 to 8 hours a week means very little when 2 to 3 hours a day are lost traveling to and from work. And all this lost time comes off the leisure that people are forever talking about. . . . Apart from the loss of time, money, and lives, the problem of home-to-job distance causes another kind of trouble, this time of a social nature with repercussions on both the individual citizen and the urban region. The latter has gone onto "half-time" and its inhabitants have followed suit. Thus a man sets off at dawn from his village, his suburb, his satellite town that provides the labor needed for the big city. He is away the whole day and he comes home in the evening depleted of energy and longing for nothing else but peace and quiet. And for this reason it is rare for him to contribute anything to the community in which he lives; he has no*

Schematic representation of complex data:
Norbert Kurz, 1966 – 67
Instructor: Otl Aicher

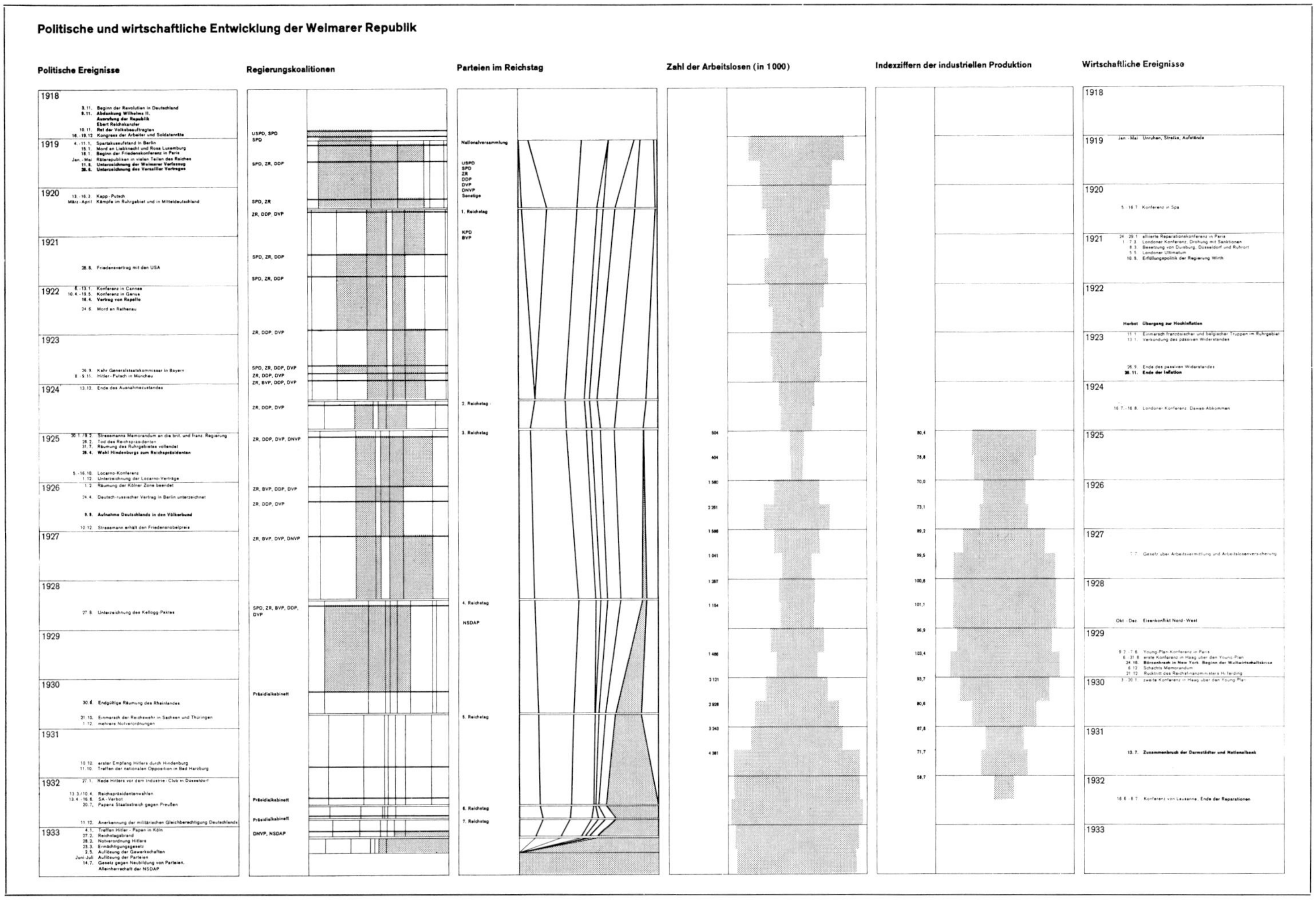


*ideas, no criticism, no impetus to give it. As far as his environment is concerned he might just as well be dead. . . . What is the basic cause of concentration? When a manufacturer sets up in a developed area he can use the existing infrastructure and equipment. And these — water, gas, electricity, telephones, sewage, communications, public transport services, public buildings — are paid for by the community. Thus the manufacturer is enabled to avoid the expenditure involved in setting up, renewing, and adapting this infrastructure. . . . He is thus able to increase his profit margin. Put differently, the community has to bear what has been called the "social cost of private enterprise." Political commitment requires one to demand that the brunt of the social cost of private enterprise should no longer be borne by the community.*

The unequivocal and sometimes simplistic remedies that Schnaidt prescribes for "Planning All Awry?" categorically reveal the radical nature of his own political affiliations, but this unfortunately in no way detracts from the general accuracy of his analysis nor from the pertinence of his revolutionary perspective.

The critical theory of Bonsiepe, Maldonado, and Schnaidt was fated to return the Hochschule to its point of departure. Having started its existence as a school of design, in lieu of a school of politics, it was paradoxically returned to its political destiny by men whose lives were dedicated to design. The vicissi-

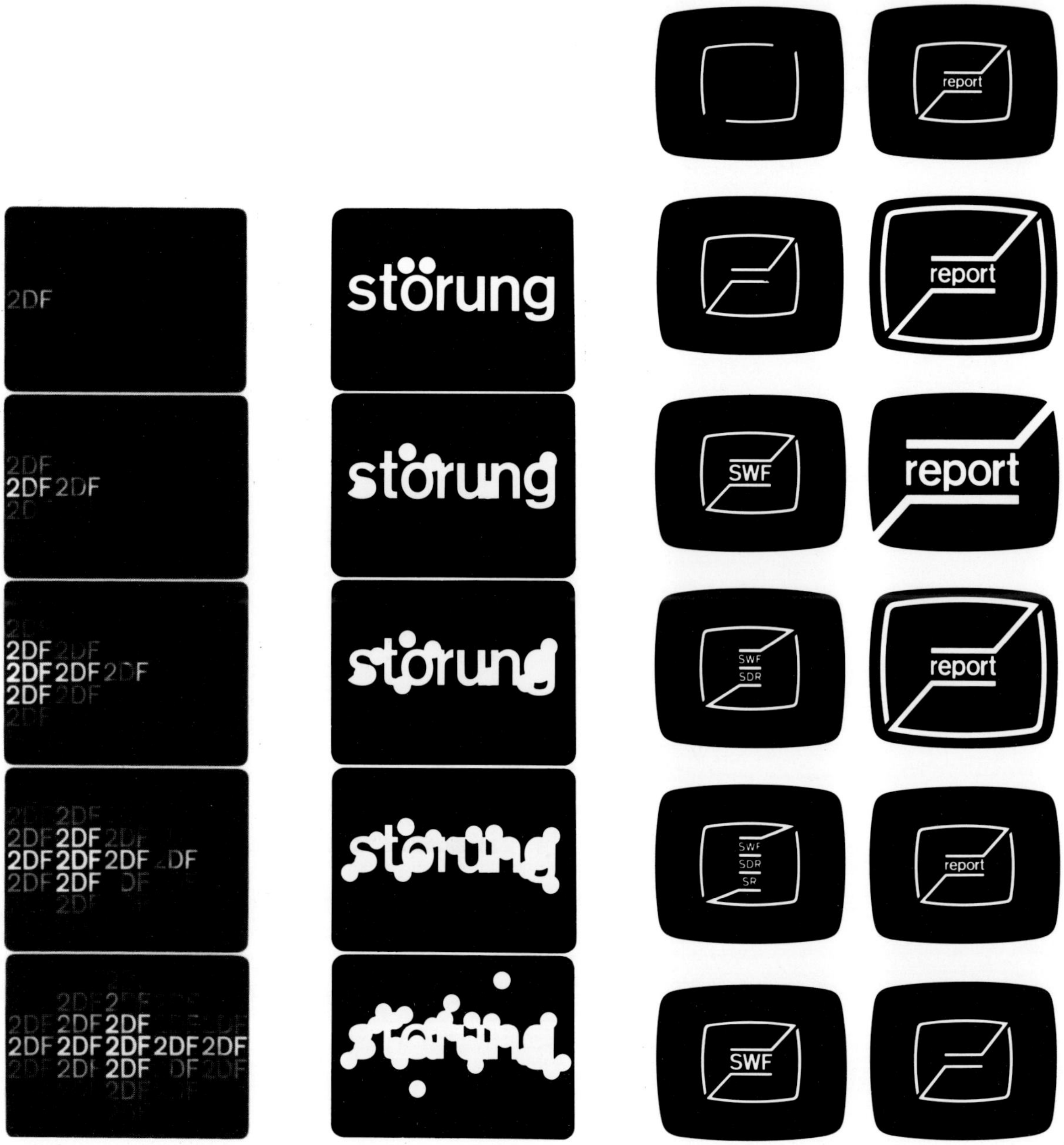


*two right columns*
On-screen identity for Southwest Broadcasting (SWF, Südwestfunk), Baden-Baden: network announcements, program identification
Herbert Kapitzki

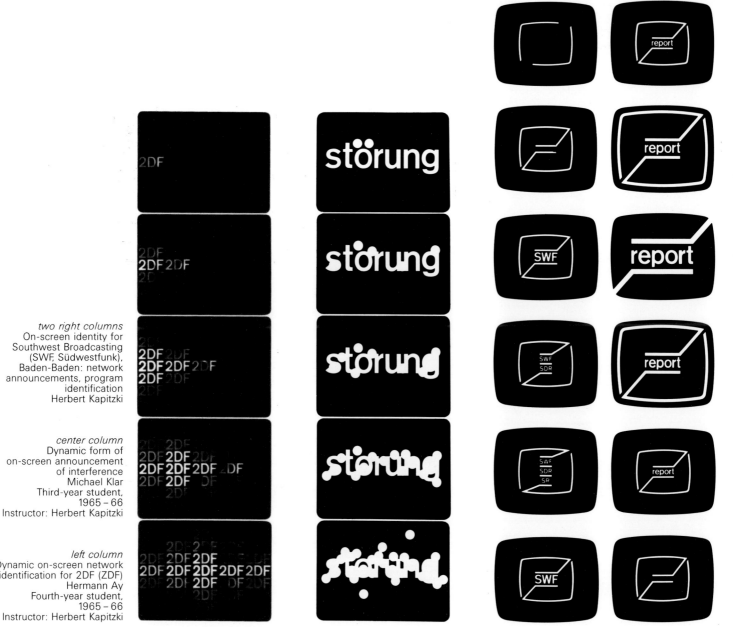

*center column*
Dynamic form of on-screen announcement of interference
Michael Klar
Third-year student, 1965–66
Instructor: Herbert Kapitzki

*left column*
Dynamic on-screen network identification for 2DF (ZDF)
Hermann Ay
Fourth-year student, 1965–66
Instructor: Herbert Kapitzki

tudes that their respective theories passed through, over a decade, tend to confirm that this development arose naturally out of adopting a certain attitude toward design. For design as the self-determination of man on earth, through the exercise of his collective *consciousness,* still remains with us as a positive legacy of the Enlightenment. Despite the admass absorption of the modern movement, the fundamental frustration of its *genuine* realization in every domain of life still testifies to the present containment of its liberating force. This much was stated by Schnaidt when he wrote of the historical co-option of the movement: "Modern architecture, which wanted to play its part in the liberation of mankind by creating a new environment to live in, was transformed into a gigantic enterprise for the degradation of the human habitat," and by Bonsiepe when he wrote in the last issue of the journal, *Ulm, 21,* the following text of resignation:

*Admittedly there is little evidence of realization in training institutions that the communications industry is a consciousness industry, whether it is concerned with the engendering of truth or untruth in consciousness, with enlightenment or ideology. The more visual designers concentrated on the aesthetic perfection of the designs, the more the communication industry was able to keep its power out of sight. The insistence on the aesthetic as one aspect of design is undoubtedly warranted and was capable of*

*left*
Corporate identity for
Southwest Broadcasting
Herbert Kapitzki, 1967

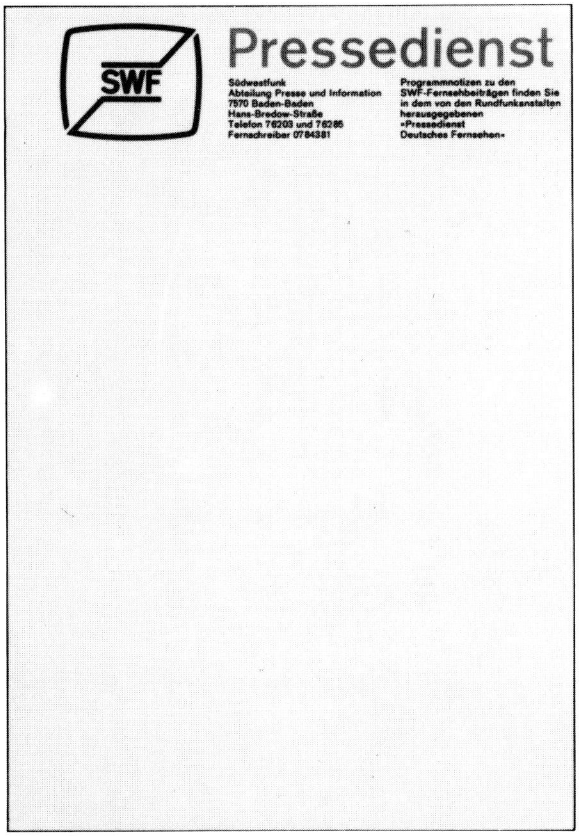

*above right*
Studies in semantics
Wolf Seeberg
First-year student, 1966 – 67
Instructor: Gui Bonsiepe

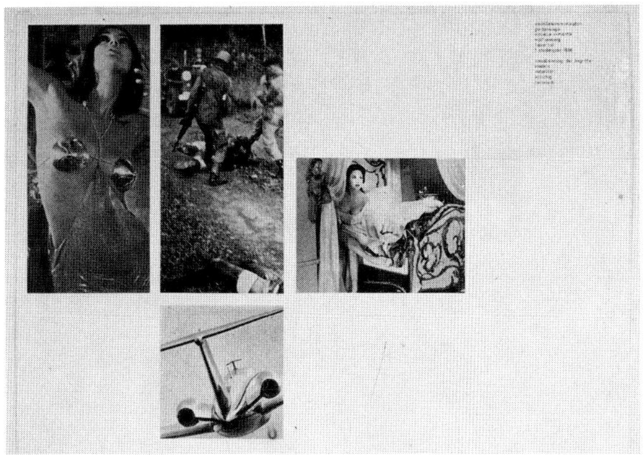

*below right*
Concept visualizations:
"modern," "authoritarian,"
"kitschy," "technical"
Harald Heimbucher
First-year student, 1966 – 67
Instructor: Gui Bonsiepe

*retaining its validity over the years. But the aesthetic cannot be maintained in unsullied and apolitical detachment from the social. Formerly, the aesthetic figured as the anticipation of a state of affairs that implied liberation from the constraints of necessity. But the aesthetic met with a fate that could not have been foreseen. It was found that it could very readily be pressed into the service of repression. The forms of power have been sublimated. In the course of this sublimation the aesthetic — which was and still is a promise of the state of liberation of mankind — has been harnessed by the agencies of power and thus used to acquire and maintain power. No consequences have as yet been drawn from this change in the role of the aesthetic insofar as it affects either the theory or practice of training in visual communication.*

From *Oppositions*, 3, May 1974

Photograph Otl Aicher

Photograph Meret Lindinger

Photograph Christian Staub

The avant-garde has grown old. What once was eccentric is now dead center; what once was abnormal is now the norm. The "new photography" of the 1920s, then unconventional and battling against a fraudulent "art photography," has now itself become a convention. But we have no new theses of equal boldness to match against those old theses that once were so passionately proclaimed. We are more cautious and less noisy; for we are understandably wary of manifestos and doctrines. We have seen what happens when they are pursued with blinkered logic until they become enmeshed in their own contradictions. In the name of the New Objectivity, the object was so provocatively handled as to lose its reality and become alienated; and photographers battled against pseudopainterly, unphotographic concepts – such as bromide oil prints and soft outlines – only to follow modern painting into abstraction through the use of solarization, negative prints, and other manipulations.

Our present standpoint is less programmatic and harder to define. Photography conveys information, and at the same time it is subject to aesthetic criteria. Under the influence of the great reportage photographers, the charm of

Photograph
Sigrid Maldonado

everyday life has been discovered. Any photographer who subscribes to the doctrine of "art for art's sake," who moves in an esoteric setting of Surrealist props, abstractions, and darkroom tricks, has forfeited his claim to our interest. The more straightforwardly the resources of the "new photography" are subordinated to a statement, the more closely they correspond to our concept. We often emphasize the materiality of a structure, or the sharply drawn outline of a detail; and yet, at the same time, we like to interpret movement through the blurred outlines that are proper to it. We recognize that formal and aesthetic experimentation needs to continue, but we remain aware that photography is not ersatz art.

The photographs shown here might all have been taken for a practical purpose. They all represent a real situation or process, and might all therefore serve to convey information in a reportage context. But do they retain their value when seen in isolation? We believe that they do. They are less ostentatiously modern, but they have become more human.

Christian Staub, in *Ulm,* 3, 1959

# Information

This department set out to prepare its students for writing careers in press, motion pictures, radio, and television. They learned to write factual texts for use in the modern mass media. The training was directed toward producing a writer who does not specialize from the outset in one particular field but who has mastered the different problems, methods, and techniques of all the mass media.

Instructors: Max Bense, Gert Kalow, Abraham Moles

The training is structured like an editorial office or the advertising department of a company. The basics of journalism and working methods are learned as they become necessary in practice. It is planned to expand the department into radio and television. Prior educational requirements: *Abitur* or equivalent high-school certificate of linguistic attainment. Good general education.

From *HfG-Info,* 1954

Max Bense teaching
1958

Experimental curriculum for Information at the HfG: Professor Max Bense

Just how important the idea of information has become for language is apparent from the teaching content. The curriculum firmly opts to look at texts – from "simple" utilitarian copy to literature – in terms of the quantity of information that they contain. In the department of Information, general and specific issues concerning texts of all kind are dealt with, and the novelty lies not only in the close collaboration with the department of Visual Communication but also in the fact that general semantics and information theory are the disciplines that underlie the work of both departments. As both teaching and research in this field are in their infancy, the present curriculum is an entirely experimental one.

You write that all obstacles and misunderstandings have now been eliminated – I will believe you; because I have had a feeling that I would not find it hard to work with you and the others.

Arno Schmidt, in a letter to Tomàs Maldonado, 1955

The department of Information is divided into two sections, concerned respectively with theory and practice. The information theory section, in turn, operates in two phases, one purely theoretical and one experimental.

"Theoretical information" covers the presentation and use of all the theoretical resources, i.e., the basic disciplines that are necessary to build up an account of the theory and practice of information. "Experimental information" is developed primarily in discursive seminar work: experiments are carried out in the use of traditional forms of information and the creation of new ones, partly with a view to finding means of quantifying information.

From *Texte und Zeichen,* 8, 1956

The social order is decisively influenced by the quality of information purveyed by press, radio, motion pictures, and television. A newly developed discipline of information and communication theory seeks to give a theoretical underpinning to this situation. The department of Information uses the results of its research to work on the development of unambiguous and comprehensible linguistic techniques for use in press statements, advertising copy, scientific texts, art criticism, and elsewhere. This work is a necessary complement to the endeavors of the department of Visual Communication; text and pictorial communication are closely connected on many levels. The thematic material mostly relates to the work areas with which the HfG deals.

From *HfG-Info,* 1955–56

Professor Max Bense and Dr. Elisabeth Walther will continue with the theory of information. Mr. Gert Kalow from Heidelberg will embark on the journalistic training of the students. In the following trimesters, screenwriting for motion pictures and television will be studied under Dr. Martin Walser and Mr. Peter Dreesen.

Tomàs Maldonado, address at the beginning of the academic year 1957–58

The Bauhaus was a response, a constructive response, to the chaotic internal situation of a defeated Germany after World War I. The Hochschule für Gestaltung in Ulm was and is a response to the situation after World War II: the destructive forces that have spent themselves in wreaking unimaginable havoc are now faced by a constructive principle that concerns itself not with prettifying the surface but with structural change.

# On Contradiction

Michael Erlhoff

One of the contradictory things about the bourgeois collective consciousness is that it constantly represses the very thing that is its essence: contradiction, conflict, antagonism. Every time someone has thought through the fundamental facts of bourgeois society — with the consequence that the conflict between executive, legislative, and judicial powers has been institutionalized, and the opposition between subject and object, or between nature and consciousness, has been formulated — there has been a rush to deny the conflict, to paper over the cracks. For the sake of peace.

The result is that the conceivable utopia of a peace that might be possible (the resolution of conflicts) is displaced onto the individual level of "a little peace and quiet." Society's dilemma is manifested and internalized in terms of individual disorientation; historic conflicts pale into insignificance by comparison with private crises. And, historically, whenever the world out there gets too hot for comfort, people choose themselves a Kaiser or a Führer, or cry out for some positive leader figure to act as a beacon. Or those inexpressible private virtues come into fashion that are conjured up for achievement's sake under the names of spontaneity, freedom from anxiety, creativity, or communicativity. And so misery acquires an ideological sparkle; conflict is defused and becomes a mere formula.

As for the HfG, however, it was indubitably one of its basic qualities that it — or the

**The task in hand is to create cells of order in the world. The present-day situation in Germany cannot profitably be viewed in isolation from the global situation. The world we live in is filled with strident conflicts. Slogans rule, on every side. The age of technology and so-called mass civilization has knocked the ground from under the feet of humankind, the majority of whose members still think in the categories of past centuries. The Hitler period, our own German trauma, was only a model of what can happen when people approach the present with mental equipment that is still firmly stuck in the Middle Ages – or even in a world of naive mythology – and are then exposed to entirely new situations. This model, with all its potential for new horrors, is not by any means restricted to Germany.**
**The world is full of hate and full of kitsch. You have heard of "civilization and its discontents." You have heard the word "homeless" applied to twentieth-century man, who has learned to subjugate the world but not how to make himself truly at home in it. What we are trying to do here relates to this situation. I might quote Gottfried Benn, who said that there is no remedy for chaos, or for the self-destruction of humanity, except *Gestalt.* To find forms for the unfixed, the flux that threatens us on all sides.**

Gert Kalow, address at the beginning of the academic year 1957–58

Originally conceived as a teaching and research department on a par with the other departments, the initial effort here did not go far beyond the experimental planning phase.

Initial plans: Hans Werner Richter and Inge Aicher, around 1948–49. Contents: political methodology / press / radio / advertising / photography / filmmaking. Education for autonomous political action, for active participation in public life, study of forms of sociopolitical organization. Department to serve as basis for other departments. Advertising to be largely education and information.

Plan, partly realized: Aicher/Bill, 1952–56. On the basis of Aicher's plans for socially committed journalism and text presentation: concentration on the main thrusts of the HfG: environmental design. Information department as a support in media terms for the objectives of the design departments. First teaching 1955. Intructors: Bill, Gomringer, Bense.

Plan, partly realized: Bense, 1956–58. Comprehensive syllabus drafted, stressing information theory. Systematic theory and practice of communication and information (logic, metrics, semiotics, theory of media and of translation), practical writing for television, radio, journalism, and advertising. Instructor: Bense. Guest instruc-

Gert Kalow

people who were active within it — faced up to genuine conflicts, recognized them as the true motive force of thought and action, and set out to work with them. As a college of design it had a responsibility to resolve conflicts and to find positive solutions; but in practice it actually courted conflicts, both internal and external, exacerbated them, and enmeshed itself in constant discord. And not only in theory, but in the objects themselves and in the College's own role as an arbiter. This went on quite deliberately, and from the very beginning.

Even in the initial stages, Inge Scholl was working with incompatibles. She brought former émigrés together with people who had stayed in Germany, and assembled a heterogeneous group of intellectuals and writers; she involved a writer whom the Americans disliked — Hans Werner Richter — while at the same moment negotiating with the Americans for money for the HfG; she neither joined in the German mourning for a lost war (how could she have?) nor tamely handed German history over to be colonized by the Americans.

And then a college of design — of *Gestaltung,* or new form — was set up in a Germany in which there was a lot of concern with "reconstruction" but none at all with new form. This was a Germany, furthermore, that was being reconstructed largely by the very same people who had presided over its previous destruction: architects from Speer's circle, politicians like Globke, and many teachers, scientists, doctors,

tors: Enzensberger, Fabri, Walter, Franzen, Kalow, Rübenach, Vesely.

Plan and realization: Kalow, 1958–60, 1962–64. Shift of emphasis to "literature as a living medium" (Kalow), language as a working material, training in established genres: radio play, essay, poems, criticism, short stories, stage plays. Attempt to find a "bridge between literary studies and the craft of literature." Installation of a sound studio with donations from radio companies. May 1960: Kalow and Rübenach organize radio drama conference of *Gruppe 47*. Department increasingly set apart from other HfG departments in terms of content. Instructors: Kalow, Kesting, Franzen, Frank.

Plans: Aicher/Kluge, 1964. Drastic financial cutbacks, together with low student numbers, lead to the integration of the department of Information and its students with the department of Visual Communication and the more recently established department of Filmmaking.

Notes of an interview between Gert Kalow and Herbert Lindinger, Frankfurt 1986

**We need a more solid training for the new generation of communicators. The rightly deplored phenomenon of "journalese," the prevalence of half-educated individuals among those who work with words, is the result of an institutionalized deficiency in our educational system. Our universities have been unpardonably slow and reluctant to catch up with the evolution of modern industrial society. Half a century ago, the Institutes of Technology separated themselves from the Alma Mater. And now the cultural commentators are horrified to see that intellectual evolution has not kept pace with technological evolution. No wonder!**

Gert Kalow, "Sprache als Fach?", *Output*, 14, 1962

**the department trains communicators for press, radio, television, and motion pictures – media that have a constantly growing impact on modern industrial society and decisively affect the ways in which it functions.**

**the department works in close conjunction with the department of visual communication. the aim of the training is to produce a writer who has not specialized from the outset in one particular medium, but who knows and has mastered the problems, methods, and techniques of the various media. the work of the department primarily emphasizes the experimental study of present-day communications techniques.**

From the HfG curriculum, 1958–59

Students in the department of Information

and others, who had previously been in the service of the Reich. A foreigner was brought in to build the HfG, and foreign instructors to teach there. The HfG became one of the first institutions to create a new image of Germany abroad; and yet there was none of the consolatory – or, latterly, boastful – pride that went with the slogan "Made in Germany."

There were, furthermore, a number of objective contradictions and conflicts that must even then have been impossible to overlook. Even a new consciousness needs a sense of history, and so there initially was a tendency to identify, perhaps rather too readily, with the Bauhaus, and specifically with a Bauhaus seen in retrospect as conflict-free. Too late, it became apparent that rationality can always be twisted into a rationale (so can irrationality); and that a craving for purity can all too easily lead to self-deception and the falsification of history.

There was one source of conflict that was (as Max Bill has made clear in conversation) entirely unforeseeable, but that turned out to have a decisive impact on the history of the HfG. The College was founded at a time when – politically, ethically, and materially – Germany lay in ruins, and *Gestaltung* seemed both historically necessary and in itself moral; but by the time the College began to operate, economic progress in West Germany was going full tilt, and it was overtaken by headlong political stabilization ("No Experiments"), market saturation, and political conservatism.

The department trains communicators for press, radio, television, and motion pictures. It works in close conjunction with the department of Visual Communication. The aim of the training is to produce an information practitioner who has not specialized from the outset in one particular medium, but who knows

Bernhard Rübenach and Elke Koch-Weser

and has mastered the problems, methods, and techniques of the various media. The work of the department primarily emphasizes the experimental study of present-day communications techniques.

Alongside the departmental work, which is intended to provide practical training in writing for press, radio, motion pictures, and television, the following subjects are taught in the department: information theory, semiotics, linguistics, mathematical operational analysis, history and operational theory of communications media, history of modern literature, photography/filmmaking/sound, typography, sociology, theory of science.

This is not only the department with the smallest number of new students; not only the one showing the clearest departures from the work of the Bauhaus: it is also the department most often discussed in public, the *enfant terrible* of the HfG, and almost a school within a school. Its guest instructors have included Albrecht Fabri, Erich Franzen, Hans Magnus Enzensberger, Käthe Hamburger, and above all Max Bense, who left the Hochschule für Gestaltung in March of 1958. Bense, who said of himself that he had spent "four years pumping intellectual substance into the College," and who was thereafter called by many students the *umpmeister,* first formulated the idea of this information school in a program that was printed in Alfred Andersch's magazine *Texte und Zeichen.* He has also given an account of what he taught at the HfG in his three volumes of *Aesthetica*.

The Information department, originally conceived by Bill as a sort of public relations group for the HfG, has since evolved into a training school for writers that is as novel as it is unique. The writer is no longer to be educated as he used to be educated – rather in passing – at the universities. He is no longer to teach himself by doing the job, or in other words to learn to conform. A new understanding of language, of sociological orientations, and of mathematical aesthetics underlies the training at Ulm – or has done so up to now. Mathematical information theories, theories of signs and of communication developed from the science of logic, linguistic studies, statistical aesthetics: a writer with such a training is not easy to integrate into the mass media as they are today.

To keep up with all this while hanging onto its own wiser and more socially responsible priorities, the HfG had to change, take a dive into the tide of progress – and be carried along with it.

This led to inner conflicts, which were often bitterly fought out. There was the rising against Bill, ostensibly directed against his style as Rector but in fact much more against his interpretation of *Gestaltung* and his view of the Bauhaus. This was a rebellion among the younger instructors, certainly; but its result was that for the first time – as can be seen from the publications that then emerged from Ulm – the Bauhaus became controversial: Early against Late, Gropius versus Meyer.

This change went along with new attitudes, including an increased interest in industrial design and a change in the Basic Course model (a close reading of the present book will reveal the course of these shifts and disagreements).

The HfG was now in the thick of the contradictions of German industrial society – between labor and capital, design and packaging, public commissions and commercial marketing, wisdom and profanity. Objects were constructed, but they at once became commodities. And to try to democratize a commodity-based society, using commodities, was not an option. This particular contradiction was accepted and worked up into voluminous methodological theories. There were more internal quarrels, and there was discord on a theoretical and also –

He represents a new kind of "information practitioner": a type who takes as his point of departure an image of the technical functioning of public opinion and journalism. Here, as in all the other classes at Ulm, the model that is envisaged is a totally artificial world – that of technology.

In the Information department, the schooling concerns itself with method, rationality, the scientifically justifiable form of a product, rather than with "literature" or "how to write" in the sense of "self-expression." There are no "elegant stylists" here, any more than Product Design produces "the handmaidens of the sales department," or Industrial Building produces "designers of architecture," or Visual Communication produces "commercial artists." The information practitioner must be able to construct a product – even if it is a dog-bites-man story – to which he has given a *Gestalt* that he can justify by reference to the ancillary disciplines, principally information and communication theory and sociology.

The relationship to language, as it becomes evident in the Information class – language no longer as a construct that has grown in an irrational way under the influence of history and national identity, but as something that can be traced back to logical symbols, syntactic and semiotic processes, "grids," "integrating elements," and "prefabricated elements": language as an operative medium. These are ideas that indicate something of what counts at Ulm, both pedagogically and ideologically. What counts is no longer a humanistic education; nor is it art – applied in any form whatever. What counts is the mastering of technological methods in an industrial age marked by the phenomena of cybernetics, automation, and function, as applied not only to the mechanical but to the human sphere, as in the areas of planning and organization.

The mental landscape that begins to emerge on all sides, defying all the deperate efforts that are made to avert its coming, is the one that Oswald Spengler and Ernst Jünger discerned as a threat, and described with horror, decades ago. A situation that may well be lamented, but a development that seems impossible to halt: the human being as a manipulated and functioning element in a technological cosmos. . . .

The pieces of work produced in one of Gert Kalow's classes are intended as exercises in writing, exercises in description. The assignment is "Fog: Objective and Verifiable." One of the taboo words at Ulm is "art." In practice, however, the strictly objective may very well turn into a new kind of art: and this applies to language, as to other media.

"Fog, and sick eyes, delay the takeoffs and landings of airplanes, upset the navigation of ships, delude automobiles. Recommended remedies: pig's gall, feeling your way, calling out, direction-finding equipment, bugle, motor horn. If unsure or nervous, be doubly cautious; half-speed ahead only. The gods conceal themselves on Olympus. They send a mist to protect their favorites in battle. A thief vanished in the murk. Murk and mist agree. They weaken the familiar and strengthen the unfamiliar. Murk and mist are originally the same word." (Cornelia Vargas.)

"Fog. Billions of tiny suspended droplets of water. They appear when the air is cooled. They spread, enshroud, penetrate. If a moving body displaces them, they swirl back in to occupy the empty space that it has just vacated. Their objective was equal distribution.

"And yet I have an island in the fog. Circular, varying in size, like the islands that other people have. Sometimes the circles intersect, but they never entirely coincide. The extent of the shared space depends on the distance between the people in the fog. My island looks different from all other islands, because its position is unique, defined by its center, which is me. A round, grayish-white wall, which always turns the same side to face me, encloses it. In my island space there is no fog; it is only dim, cool and damp, and I can assess colors, forms, and distances as usual. Where a wall begins for me, there may be

although it's hard to distinguish – on a personal level within the faculty.

A college of design could not restrict itself to theorizing. It had to produce something, and was thus inextricably bound up with the system of production and exploitation (which is inevitably hard on theory). On top of this, there was the HfG's permanent financial crisis: it was not adequately subsidized, and was thrown back on its own productive resources. And so, with great effort and labor, and – we are told – amid bitter political controversy, the HfG set up its own development departments, in which the idea was that projects would be carried through to the final preproduction stage and then offered to industry. By this time, a number of upwardly mobile industrialists had started to get interested in the HfG; it was becoming clear that design was something that could stimulate consumption and thus production.

The Basic Course changed once again. In response to external pressures – and thus not without conflict – the teaching became more vocational, more directed toward practical objectives. And once again there were internal wrangles, some of them between those departments that were heavily involved in commerce and those others that lacked the opportunity and the desire to be so.

All this, however, was hammered out and – theoretically – resolved. A magic word was found that offered release from the bonds of the marketplace:

island space for others, and where a wall begins for others, there is island space for me. An imaginary labyrinth of perambulating walls of fog and intersecting spaces, which no one can see as a whole." (Elke Koch-Weser.)

"Fog: a mixture of air and many small particles of water. One cannot see far. White-light faded colors. Things cast no shadows. Droplets combine into drops, course down the window, float and settle and slide and fall.

"Fog: clouds on the ground. Contrasts shrink. Flattened volumes. Surroundings vanish. Fog: hazy-moist points to clear-dry; fog points to nonfog." (Gui Bonsiepe)

"Fog: naturally autumnal – transitory – scary – clammy – apprehensive – blinded. Somebody somewhere is indulging his solitude. Fog: image of all possibilities. Symbol of beginning. Midgard – Asgard – Genesis. Gods of all sorts flitted, invisibly powerful. Faith moves mountains of fog. Number disposed. Fog: now no more than an illusory confinement. A pillow for the one who hammers out his dreams on islands at his ease. He feeds the mist. In there a television tower is a mother oak. The leavings of abuse: an exhausted image. We must send the word on vacation." (Ilse Grubrich.)

Bernhard Rübenach, "Der rechte Winkel von Ulm," radio documentary, 1959

Conference of *Gruppe 47*
Topic: radio drama
Those present included Günter Eich, Günter Grass, Wolfgang Hildesheimer, and Dieter Wellershoff

Abraham Moles teaching

Gui Bonsiepe

The more the visual designers concentrated on the aesthetic perfection of their designs, the more effectively the dominance relationship intrinsic to the communications industry could be concealed. It is undoubtedly legitimate to insist on the importance of aesthetic considerations as an aspect of design; and for years this was regarded as the crucial factor. But aesthetics does not hover aloft, somewhere above society, intact and apolitical. At one time it was seen as an anticipation of a hypothetical liberation from the bonds of causality. But aesthetics then suffered an unexpected

the word was "systems." Systems (rather as in the present-day trend toward "environments") force the merchandizer to adopt not just individual products but whole concepts. In addition, the HfG now moved into designing for the public sector, designing for society. By the end there was a clear effort to replace the private client with an apparently more respectable, or even more moral, source of money: the public service (although this provider, too, exacted a price).

Ultimately, though, none of this could resolve the HfG's inherent contradictions, and especially not the structural conflict between design and social morality: there was an inevitable gulf between, on one side, the vital social function of design as a cultural phenomenon and directing force, and, on the other side, its need to maintain itself in the market.

Neither the public, as represented by the media, nor the politicians were impressed by the spectacle of Ulm working out its inner contradictions.

Bourgeois society – as has been said – misunderstands itself.

And so it came to pass as the proverb says: He who lives by contradictions shall perish by contradictions. The HfG was torn between its own internal demands and those imposed on it from outside, and by the conflicts implicit in the idea of setting up a "Hochschule für Gestaltung" at all. But because the story of the HfG remains to this day a chronicle of contradictions, and because it so insistently exemplified them in fate. It became apparent that it is perfectly possible to apply it to repressive ends. The forms of dominance had become sublimated. In the wake of this sublimation, the aesthetic – which was and is a promise of human liberation – was taken over by power interests and thereby put into use for the acquisition and maintenance of dominance.

Gui Bonsiepe, "Kommunikation und Kunst," *Ulm, 21,* 1968

the students in the department of information are trained to impart design to the language of the modern communications media. they will work for example as writers in press, radio, television, and motion pictures. the department works in close contact with the department of visual communication. the department of information is presently engaged in a restructuring of content and

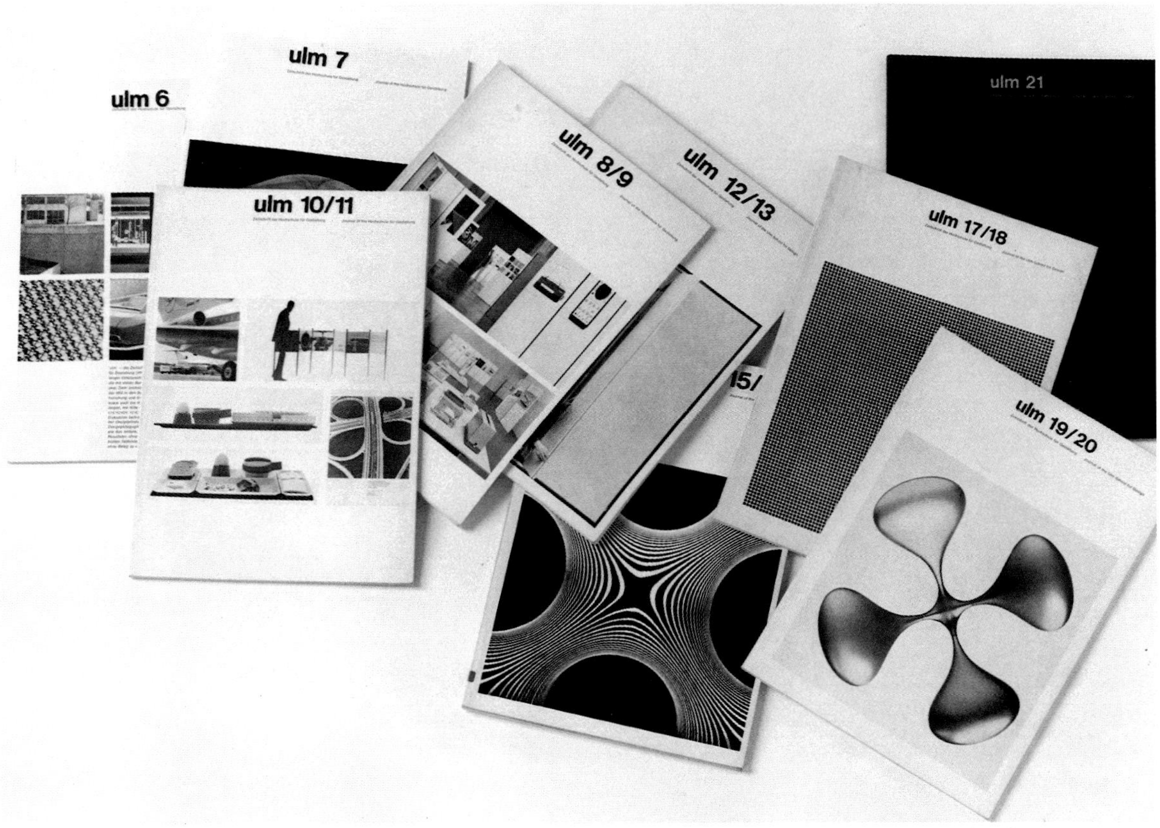


Reporting on the work of the HfG: the quarterly periodical *Ulm*

practice, it can now be recognized — alongside its equally contradictory contemporary, the Social Research Institute in Frankfurt — as a historically significant institution.

organization. this year the students of the department of information are participating in the work of the filmmaking sector, and are also receiving assignments specific to their own department.

From HfG prospectus, 1964–65

the students in the department of information are trained to impart design to the language of the modern communications media. they will work for example as writers in press, radio, television, and motion pictures. the department works in close contact with the department of visual communication. it is intended that the department of information shall undergo a restructuring of content and organization. in the current academic year no lectures or seminars will be held.

From HfG prospectus, 1965–66

# Filmmaking

The filmmaking sector was set up as a separate department of the HfG in 1961. The aim was to train a filmmaker of a new type, who would be an *auteur,* in charge of every aspect of his picture, and take responsibility for the development of the art form. To this end there had to be a combination of development, research, and training. The main points of the training were: the acquisition of well-tried craft techniques, an approach to contemporary issues, the extension of the means of formal creation, the involvement of other disciplines, the releasing of filmmaking from its mystique, the relationship to industrial process, and the making of model motion pictures.

Instructors: Alexander Kluge, Edgar Reitz, Christian Staub.

Filmmaking at the HfG

1951 – From a projected curriculum: "Filmmaking and photography are of great importance for present-day life. Their influence, partly in conjunction with typography and graphic design, is still growing in importance. The Studio for Visual Design works both on the creation of autonomous works (films, photographs, books, magazines, exhibitions, etc.) and in connection with the needs and problems of the other departments of the College."

1952 – Prospectus: "Work with filmmaking and television forms part of the forward planning. As a continuation of the Visual Design department a studio for artistic design is planned."

1955 – Prospectus: "The department of Visual Design embraces all the areas that use visual means either in isolation or conjointly; that is, commercial graphics, photography, typography, exhibition design, and filmmaking and television."

1956 – Detten Schleiermacher (architect), Enno Patalas (journalist), and Martin Krampen (graphic designer and cameraman) submit to the HfG authorities a program for "Filmmaking and Television at the HfG."

1958 – *Ulm, 1:* "Typography, graphics, photography, and exhibition technique are accordingly dealt with as a single area. which in the near future will be extended to cover filmmaking and television."

1958 – For the first trimester of the academic year 1958–59 Herbert Vesely is appointed a visiting instructor in the department of Information. His work in the department: a synopsis, treatment, and script outline for a television film. Topic: Big cities in Germany since 1945.

1960 – Christian Staub writes a "Memorandum on the founding of a Filmmaking and Television department at the HfG," of which the following is a shortened version:

"In the prospectuses of the HfG, 'motion pictures' and 'television' are announced as future areas of teaching within the departments of Visual Communication and Information. At present, however, these areas do not exist as part of the curriculum but only as vague intentions. This fact causes the students to ask awkward questions, and in some cases even to leave the College. (To my personal knowledge, for instance, Miss Peters and Mr. Weidmann have left our college for just these reasons.)

"In the Federal Republic there is still no training establishment in which filmmaking and television are systematically taught. Some experiments to this end have begun at the universities of Münster, Munich, Frankfurt, and Bonn – without, so far, very much to show for them. The professional training of those who make motion pictures today is largely haphazard and in most cases inadequate; and this is reflected in the low standard of German documentary and feature films.

"In contrast to the situation in the Federal Republic, in other countries serious efforts are made to offer the new generation of filmmakers a solid and worthwhile training. Paris, for example, has not one but two film schools, namely the Institut des hautes études cinématographiques (IDHEC) and the Ecole technique de la photographie et de la cinématographie. In addition, future directors and cameramen have the opportunity, alongside extremely rich commercial programming, to see three different motion picture classics every day at the Cinémathèque Française. These opportunities, as well as sociological factors about which I have less knowledge, have led to the much-discussed renaissance of the French cinema.

"In West Germany there is now a latent crisis of filmmaking, largely caused by television. If this crisis is to be resolved, there is, in the long-term view, only one possible solution: better films. This would also undoubtedly bring with it a better standard of television programs. This interaction will sooner or later force the motion picture industry or the television industry to provide money for development and laboratory work that will open up new paths through experiment. I therefore believe that, given a serious effort, it will

## Ulm Freedom

Editorial Discussion

Erlhoff:
When the Hochschule für Gestaltung was founded, one very important consideration was that it should be conducted as a genuinely private institution. Nowadays the idea of ensuring freedom through private administration seems absurd; but then, in a much tougher social situation, it seemed possible.

Lindinger:
Ultimately, wasn't it an unreal idea even then? Especially as there is no tradition of private colleges in Germany.

Riemann:
I think there was a dependent relationship even then. The money had to come from somewhere, after all, and promises had to be made in return. The autonomy of the HfG was already compromised by the way it was set up.

Staubach:
Yes, there was repression right from the start: for example, when the German steel industry agreed to finance the construction and then backed out because of an allegation from an ex-Gestapo man that those involved in the HfG were all Communists. And they for their part made no attempt to confront the conflict in political terms; all they did was point out that no Communists were in fact involved. So instead of having a court case as to whether Communist involvement was legitimate or not, all they did was prove that they weren't Communists themselves.

prove possible to interest the motion picture and television industries in projects of this kind.

"I therefore propose that the following immediate action be taken: (1) set up a new Filmmaking and Television Development Office; (2) appoint a film director as guest instructor and head of the new office; (3) contact other suitable experts for classes in visual communications and information; (4) contact institutions and organizations for advice and help (ask for curricula of motion picture academies); (5) contact possible sponsors with a view to enabling the Filmmaking and Television Development Office to become a documentary film production agency in its own right and to finance the continued motion picture and television teaching program out of the proceeds; (6) contact possible donors of technical equipment and money; (7) install and equip an editing room, and possibly a small archive room for research; (8) set aside one half or whole day of teaching time in every week for motion picture subjects in the department of Visual Communication and Information; (9) look for students about to take diplomas who would be suitable and willing to work in the new Filmmaking and Television Development Office from the fall onward.

"If this new Development Office were to prove its worth and succeed financially, it would then be easy for the College to open a new department of Filmmaking and Television. The career objectives of students in this new department would be to train as scriptwriters, documentary film directors, cameramen, editors, television reporters, and producers. The opening of this new department of Filmmaking and Television would be necessary because a young person who wants to enter upon any of these careers has no desire to spend four years on visual communication and information in general before being able to work in his own specialty."

1960 – Christian Staub submits his memorandum to the faculty conference and asks the HfG to start a Filmmaking and Television department. The faculty conference agrees to his proposal.

1960 – The following lectures, seminars, and exercises connected with filmmaking are incorporated in the program of the department of Visual Communication for the academic year 1960–61:

Dr. K. J. Fischer, Heidelberg: Techniques of Scriptwriting. Alain Tanner, Free Cinema, London: The English Documentary Film. Christian Staub, Ulm: History of the English Documentary Film. Dr. W. Berghahn, Munich: Television Technology; Ideal Images in the American Film. Sukopp, Ulm: Motion Picture Technology. Christian Staub, Ulm: Practical exercises with a 16 mm Bolex: Focusing Attachments.

1961 – The faculty conference decides to divide the department of Visual Communication into one sector centered on typography, graphics, exhibition techniques, and photography, and another sector based on filmmaking, television, and photography. Christian Staub is entrusted with the planning and organization of the filmmaking, television, and photography sector.

1961 – On October 2, Christian Staub submits a status report:

"Department of Filmmaking and Television at the Hochschule für Gestaltung. With the inception of the academic year 1961–62, which begins today, the department of Film-

Christian Staub

Erlhoff:
That's symptomatic of the whole history of the HfG, because conflicts of that kind were never confronted as an issue of principle. Everyone must have known the parameters that governed his own work at the HfG, but they simply kept going and learned to avoid bringing matters to a head.

Lindinger:
That was one of the great contradictions at Ulm. The independence of the HfG was supposed to be guaranteed by the private Scholl Foundation. But the Foundation didn't have enough money, and there was never enough from private donations. In the end, three-quarters of the budget was met by the Baden-Württemberg and Federal governments, and that money had to be begged for and voted through, every year. The unacknowledged price for this was a policy of operating with great caution, to keep Stuttgart quiet and Bonn in line. No sooner did it become known that the students had taken part in Easter marches, or that an instructor had collected signatures against the Vietnam War, than there were protests from Bonn or Stuttgart, which Inge Aicher had to mollify.

Staubach:
There are other examples of this tactical caution. For instance, signatures were withdrawn the next day, and Vietnam collections were disowned on the next day, as soon as it became known that an appar-

making and Television embarks on a full teaching program. It is organizationally linked with the department of Visual Communication.

"Objective: It is intended to train specialists for documentary filmmaking and television, in particular assistant directors, cameramen, and editors. The course lasts four years, and just under half of it consists of classes in theory and basics; the major part of the work is original practical filmmaking and photographic work under supervision. In addition, during the College vacations, practical work in the motion picture industry and in television will provide further professional experience. The fourth year is set aside for the diploma work, which is a major undertaking carried out entirely independently.

"For the first trimester of the academic year 1961–62, a number of directors, a cameraman, and an editor were employed for the practical filmmaking work (the so-called departmental work): these were Ferdinand Khittl, Raimond Ruehl, Haro Senft, Franz Josef Spieker, Schwennicke, and Frau Dr. Henrici. Each student was assigned one sequence of an instructional film. In addition, Dr. Berghahn continues his course on television technology. At the moment eight students have enrolled for the department of Filmmaking. They are joined by a number of students from the department of Information, who are engaged mainly in script work.

"Equipment: for the moment we have only 16 mm amateur equipment and a few spotlights. From the end of October onward a sound studio donated by the radio companies will be operational. The UFA company has provided us with a 35 mm sound projector with all its accessories. There is general agreement that for teaching purposes, in particular, it would be best to work in 35 mm. For this purpose we still need the following equipment: a 35 mm camera, a 35 mm editing table, and a special effects table. The necessary film stock cannot be provided from the teaching materials budget.

"It is envisaged that the topics for student exercises will be chosen in close consultation with the motion picture industry, so that the training will be as close to reality as possible."

From *Output,* 14, 1962

Operation "Oberhausen Group," Ulm

Making films is the personal achievement of a person or a group. This personal, i.e., organic principle of filmmaking stands in an apparent contradiction to the industrial methods of the motion picture industry. In fact, however, that industry has never really developed industrial methods at all, but works on its products individually, one after the other. The contempt for the individual imagination in motion picture practice hitherto has led to the subjectivization of the individual specialties within the industry: the star, the director, the camera operator, the noncom technician, the scriptwriter. This additive combination of ingredients is not exactly calculated to lead to a new principle of filmmaking.

The training for filmmaking in the HfG does not, therefore, take the individual specialties as its point of departure, but produces a filmmaker who is a generalist. This filmmaker has mastered the whole process of making films. In addition there is a wider perspective to be kept in view, which goes beyond filmmaking to embrace a general restructuring of the whole motion picture business, from the shoot, by way of production, distribution, and the movie house, to the public. In this respect the Filmmaking department of the HfG is aiming at a comprehensive motion picture education. It is up to the individual student to find the focal points of his own future involvement. Basic to the whole is technical proficiency and organized thinking in every field.

Alongside the technical sector there is a formal training. The following are the characteristics that the educational process is particularly intended to develop:

Imagination, aesthetic experience, ability to turn ideas into reality, precision, readiness to experiment, and advanced awareness.

ently left-wing organization was behind the appeal. From all sides the HfG was regarded as a left-wing, progressive school; but it never made any official political utterance on its own behalf. For instance, the department of Visual Communication, which had all the technical equipment — printing machines and so on — never produced anything that might have been called a political statement. German rearmament, emergency laws, etc., seem to have passed the HfG by without leaving a trace.

Lindinger:
Some things did get printed for demonstrations. But it was never an official topic for student work.

Erlhoff:
I can see that for the HfG, as a private school, the public was an important factor; so why did the College never succeed in getting that public on its side? Or the media; or at least the critics?

Lindinger:
Nowadays people find it hard to understand just why press reports were such a problem for us in those days — the article in *Der Spiegel,* for instance, which did us an enormous amount of harm. This was probably because the HfG was on a knife-edge more or less from the day it opened. The instructors had no long-term contracts: just annual contracts with three months' notice of termination, no compensation, no pension, no interim social security payments, nothing. For that they were paid the ludicrous salary of 700 marks a month, which was less than the assistants

The Filmmaking department of the HfG consists of three institutions:

The film school undertakes the teaching, which is carried on in accordance with the didactic guidelines worked out in the other departments of the HfG.

The development studio works on problems that arise in the field of filmmaking and film organization. The work of the development studio will take place in development groups. Work of this kind is very closely linked to the conceptual guidelines that apply to the development studios of the other departments of the HfG.

Commissioned research on the theory, technique, formal design, and history of motion pictures is to become a constant feature of the Filmmaking department and will find a regular place in a separate institution.

The teaching will be planned and carried out by members of the "Oberhausen Group" (Bernhard Dörries, Edgar Reitz, Detten Schleiermacher, Haro Senft, Dr. Alexander Kluge) and by staff critics of the magazine *Filmkritik* (Ulrich Gregor, Theodor Kotulla, Enno Patalas). In the first two years of the course the students will be introduced to all the potentialities of form through small-scale exercises.

By the end of the second year the following resources must have been mastered: concepts and departure points, extension of the technical resources of film, expansion of the rhetorical resources of film, planning of the formal aspects, organization, art direction, sound, camera, montage, editing, print techniques, special effects, studio technique, direction of performers, and reportage techniques. In addition, in the second year polemical studies of specific technical and formal problems, and historical, political, formal, and technical investigations, are undertaken. An important component of the teaching program is the history and analysis of motion pictures.

In the third year the theoretical topics are further explored. The technical training will be very closely linked to the developmental work of the development groups. The fourth year, as with the other departments, is devoted to diploma work. This may for instance consist of a theoretical dissertation and a short film. Students of the HfG who have earned their diploma in the Filmmaking department will have their next short feature film financed by the Junger Deutscher Film foundation.

Detten Schleiermacher, in *Output,* 14, 1962

What Does the Oberhausen Group Want?
Filmmaking in the Federal Republic is in crisis: its intellectual status has never been lower, and now even its economic existence is threatened. This is happening at a time when in France, Italy, Poland, Czechoslovakia, and many other countries, motion pictures have attained an unprecedented artistic and political status. Films like Italy's *Salvatore Giuliano,* which has just been shown in Berlin, or *Ashes and Diamonds* from Poland, show that the movies have attained equal status with the various genres of literature and the other arts and have found access to the political consciousness in particular. We have had numerous conversations with members of *Gruppe 47* and have become aware that interest in filmmaking is at an extraordinary pitch among those active in the other arts.

It is therefore necessary to release filmmaking in the Federal Republic from its present intellectual isolation, to reduce the force of the commercial *Diktat,* and to place filmmakers in a position to become aware of their responsibility to the public and to seek their own subject matter in accordance with that responsibility: they must be free to deal with social facts, with political issues, with educational issues, and with new developments in the art of the film itself: all of which have been possible only to a very limited degree under the working conditions that have hitherto prevailed.

Toward a mental revolution. In the motion picture industry there are two groups at pres-

were getting at the universities. The financial support went only from year to year and was always uncertain. One unfavorable newspaper article – which was often just sloppily phrased – would lead someone to say in one conservative quarter or another: "All that disturbance and infighting at Ulm – let's put an end to it once for all."

Staubach:
What still isn't clear is why, for example, *Der Spiegel* never had anything positive to say about Ulm, or even any sympathetic, constructive criticism.

Chemaitis:
And it wasn't just *Der Spiegel.* If you look at the media landscape of the time, there really was such a thing as intelligent and sophisticated reporting – from stations like the old Southwest Radio, for instance, or North German Radio. It isn't clear at all why the HfG never succeeded in getting the media on its side. It's the same with newspapers and magazines: not all were right-wing. There were some periodicals, newspapers, and broadcasting organizations that were democratically minded.

Staubach:
Especially since Ulm had its own department of Information to analyze the press and broadcasting media, and actually had alumni working in those areas.

Chemaitis:
Which would have made the training a really practical one, with the journalists and broadcasters putting the message across

Lindinger:
When you look through the press comment on the HfG, you notice that in the early

ent: the producers who are members of the dominant organization, together with their specific distributors; and the new generation, most of them members of the Oberhausen Group. The group is no longer restricted to the signatories of the Oberhausen Declaration of last February, but embraces all those working in the field who seek not only a reform of the German film but a restructuring of the whole way in which the German motion picture industry is organized and a mental revolution in filmmaking. This group includes directors, producers, and writers. The group seeks to work closely with *Gruppe 47,* with the so-called Cologne School (Stockhausen), and with other intellectuals outside the motion picture industry.

The program put forward by the Oberhausen Group is ultimately self-financing. In the first period, in which new models of filmmaking must be developed in a relatively short time (in France and Italy the last six years have been used for this, and in the Eastern Bloc the state bears the cost of this development work), the scheme will remain temporarily dependent on public money. This means that we need to maintain a certain noncommercial area within the framework of a free-enterprise motion picture industry.

Three objectives. In the spring of this year, the Oberhausen Group adopted three objectives:

1. The promotion of the free-form short. This is the natural focus of experimentation in filmmaking, just as in education the private school is the vehicle of the free impulses and the reforming work that cannot be handled within the ponderous apparatus of the public schools. This form of film is now seriously threatened by the abolition of the entertainment tax. The distributors have hitherto been prepared to pay, not for the shorts as such, but for the quality rating that went with them. The disappearance of the entertainment tax in a number of the West German states means that these quality ratings are no longer worth so much to the distributors. The proceeds from sales of the shorts with the "Special Merit" rating have thus sunk far below production costs. We are concerned that this financial base for the making of shorts not be lost.

2. The founding of a nonprofit foundation under the title Junger Deutscher Film (see Enno Patalas's article "Die Oberhausener Gruppe").

3. The founding of an intellectual center in which a new generation can be trained and in which there will be a place for theory and for development studies, such as every industry needs.

The department of Filmmaking at the Hochschule für Gestaltung in Ulm combines the developing of new film models with its instruction in filmmaking. This department is also a development studio, in which practical development work takes place alongside the training of students who have completed two years of basic studies. The Filmmaking department and the development studio at the HfG fit into the overall framework of the Motion Picture and Television Academy that is to be founded with branches in Munich, Berlin, and Ulm.

Alexander Kluge, in *Output,* 14, 1962

## The Filmmaker as *Auteur*

The term *cinéma des auteurs* is a French one; and its German equivalent, *Kino der Autoren,* has been in use for about two years now. It was adopted by the Oberhausen Group as one of the points in the program of their Junger Deutscher Film foundation. The term describes a form of practice in production and creation. The roles of the producer and the filmmaker are changed.

*Cinéma des auteurs* is based on the premise that film is an artistic medium that stands alongside other disciplines such as music, architecture, and literature. According to the definition given by Hans Rolf Strobel, filmmaking in this sense is "another form of writ-

years some strongly positive articles were written, and that it all changed with the crisis that surrounded Bill's departure. A large number of the feature section editors took Bill's side at that time. And that fateful report in *Der Spiegel,* with all its consequences, sprang from *Der Spiegel's* partisan support for the method people and the positivists within the HfG, whom they preferred because they sounded more scientific. What is more, the media often had very little idea of what was actually going on up there on the Kuhberg; we were mysterious, and therefore we were suspect.

Chemaitis:
It was also hard for a lot of people to understand something like Inge Aicher's continued total commitment to the College. That went along with the failure to understand anything about the College at all.

Riemann:
At the very end, when the HfG was due to be closed, many of the papers came out very strongly on its side. But then it was too late.

ing." To make this understandable, it must be explained what an *auteur* is: he is, first and foremost, no more and no less than the "originator" of his film as a whole. The writer and maker (realizer) of the film bears responsibility for every aspect of the work. This signifies not only that he is in total control of all the technical processes of filmmaking, but also that his imagination operates from the outset in the realm of his chosen medium. Adherents of the *auteur* theory like to refer to a statement by the Russian *auteur*, Pudovkin, who wrote that the art of the motion picture could only come into being on the basis of those methods that are specific to it.

The *auteur* theory embodies, firstly, the recognition that every medium of communication is a filter that blocks out particular contents (information) and favors others, and, secondly, the conviction that motion pictures are a particularly well adapted means of communicating those ideas and forms that are characteristic of the present day.

The *auteur* principle addresses the issue of "originality" in a new way. The *auteur*'s identification with his material tends to be so complete as to be autobiographical. The film is always based on "original" material. If a literary original is used (as is often done by the younger French directors), the picture is not a "filming" of the material in the traditional sense – filming with ingredients, as it were – but a new approach that uses the resources of the motion picture medium and is inspired by the filmmaker's conviction.

Through the film medium, the theme then undergoes a new interpretation: its information content changes. (Example: *ZAZIE* by Louis Malle and *ZAZIE* by Raymond Queneau.) In this case the movie and the novel are completely separate and autonomous works, which share a theme and a number of narrative events. The best analogy is a musical composition in which themes from other compositions are developed (Bach).

As a rule, one person signs the film as its *auteur*. If the authorship is shared by a team,

*left*
Edgar Reitz

*right*
Bernhard Dörries

*left*
Haro Senft

*right*
Detten Schleiermacher

Alexander Kluge

the roles of the individual members of the team are not rigidly distinguished. All have mastered the medium as a whole; they see the collaboration primarily as a way to objectify problems through discussion and through the highly complicated production process, in which one individual is often unable to exercise total control. Teams of this kind can be composed of persons whose principal interest is in one or other of the specific areas of camera work, direction, motion picture theory (and literature).

All those who work with an *auteur*, and who do not share the authorial responsibility, are

Camera exercises and filming for the first film "miniatures" and sequences, experimental work, reportage technique, and documentary

fulfilling a purely subordinate, technical function (actors, writers, cameramen, sound operators, editors, lighting operators, assistants, production managers, property men, architects, etc.). The way in which *cinéma des auteurs* works is not at present universally understood in Germany. The film papers interpret it as meaning that writers ("authors") play the decisive part in forming the film by supplying their novels as material. Nor has the law of intellectual property yet found a place for the *auteur*.

The role of the producer in *cinéma des auteurs* is comparable with that of a publisher. He finances and organizes the making and distribution of the picture made by the *auteur*. He does not try to influence his *auteurs*, or to force his own opinions on them. The contract between the producer and the *auteur* is simply a commitment to observe a budget agreed between them. The producer's role consists above all in the choice of the *auteurs* whom he commissions (just as in publishing). In fiscal terms, the difference between *cinéma des auteurs* and traditional filmmaking resides in the fact that the *auteur* is responsible for his own tax bill, while the film director is taxed as an employee.

The distribution organizations have no influence on the *auteur's* work. In the credit sequences of *cinéma des auteurs* pictures, these words increasingly often appear: "A Film By . . ."

Edgar Reitz, in *Output*, 15, 1963

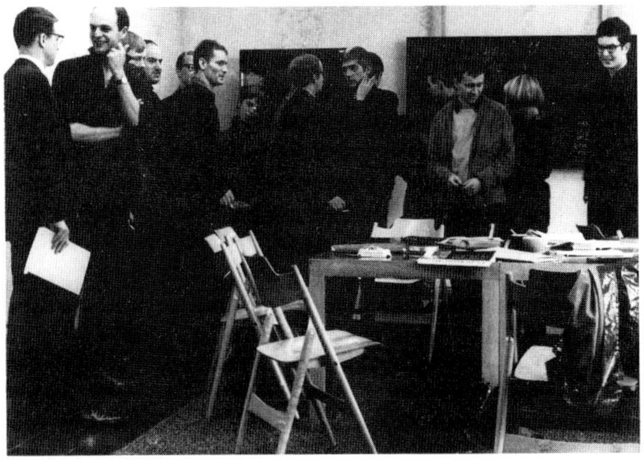

Theoretical exercises and lectures as a complement to the practical work

the department of filmgestaltung is an intellectual center for young filmmaking. it is engaged in bringing on a new generation, and its work is complementary to the work of the film- und fernsehakademie in berlin and the kuratorium junger deutscher film in munich.

development and education: the department carries out research and training productions. instructors and advanced students work together on theoretical and practical models that serve the future evolution of filmmaking (development). the department pays particular attention to the fundamentals for which it is hard to find time in the course of practical work. filmmaking is treated as a medium of communication based on a general intellectual approach to reality, whether the subject is fictional or documentary. through the concentration on this kind of work there emerges a type of auteur who is distinguished from the specialist by a greater range of responsibility. the aim is not to improve the work of the motion picture industry from its own point of view but to develop the motion picture as a general means of expression for intellectual and human experience.

mode of working: the department of filmgestaltung works like a production unit. the educational program is so organized that one trimester in each year consists of formalized instruction in the form of seminars and a system combining classes and core subjects; the other two trimesters are used for the long-term realization of working objectives, both practical and theoretical, especially in the form of teaching productions. in addition the students of the filmgestaltung department are given the opportunity to attend the lectures of the theory weeks at the hochschule für gestaltung.

prior qualifications and course content: the number of student places is limited. a precondition of acceptance for the filmgestaltung department is a completed university or equivalent education. the training in the department lasts a minimum of two or a maximum of five years. the basic unit is not a class but a system of working groups. the course concludes with a diploma, that is awarded on the basis of an independently completed film production, a theoretical study, and a colloquium.

From the syllabus of the Filmgestaltung department of the HfG, 1966–67

Films on the student movement, 1967 – 69
Brochure of the Institut für Filmgestaltung

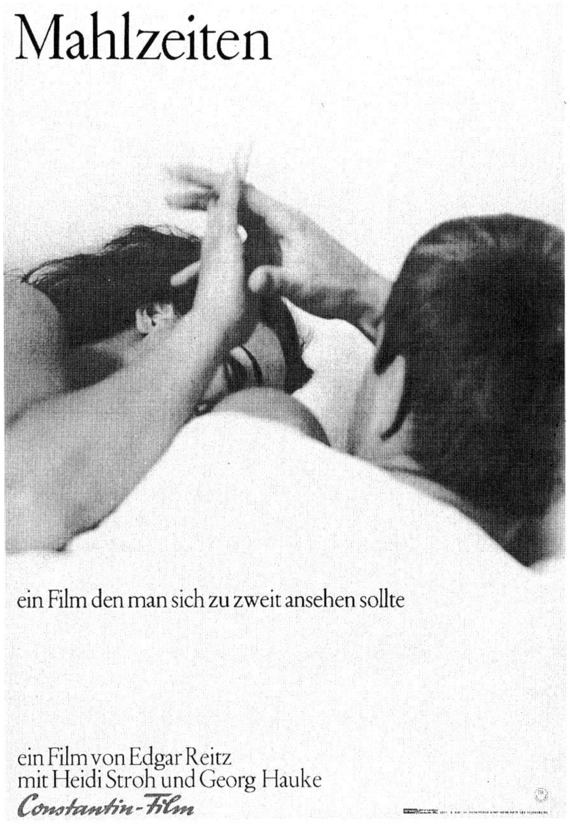


*Artisten in der Zirkuskuppel: ratlos*
Poster for film
by Alexander Kluge

*Mahlzeiten*
Poster for film
by Edgar Reitz

Experience shows that in an area deprived of cultural values you cannot achieve culture through subsidies alone – by spending money – but that it is a question of the mind. I do not believe that there is any industry – with the possible exception of the timber industry – in which the general educational level is as low as it is in the motion picture industry. The fact is that most culturally significant filmmaking projects are already financially feasible within our motion picture business as it stands at present, but that they are not carried out as a result of sheer mental incapacity. This explains why there is a demand for the creation of a focal point for the brain workers of the motion picture industry and for a raising of the intellectual level through an Academy of Motion Pictures and Television.

I would like to say something briefly about the so-called Oberhausen Group. This group is often misjudged, or overestimated. It is an association of twenty-six young producers and directors. Young in this case means between thirty and forty years old. These people came together about a year and a half ago at the Oberhausen film festival, under the impact of the UFA crisis, and raised a number of demands relevant to the politics of culture. They addressed the following three demands to the standing conference of state ministers of education and culture, and to the education and culture committee of the Federal Legislature: (1) support for quality films, and in particular for good short films; (2) a Junger Deutscher Film foundation to foster the next generation of talents on a noncommercial basis; (3) one or more intellectual centers in the form of an overriding academy.

Demand (1) is currently being discussed within the conference of education ministers; demand (2) has been agreed in principle by the Federal Department of the Interior. As for (3), the question of an academy, a number of representatives of the Oberhausen Group, including myself, have turned to the HfG. This college had already embarked on the beginnings of a training program. Discussions with members of the Oberhausen Group date back to 1958. These very fragmentary beginnings have in the past year been brought to fruition through the initiative of members of the Oberhausen Group. These efforts have met with the approval of the Martin Committee and of the prospective partners, who have been working on their own academy plans: Berlin and Bavaria.

This year, as our Constitution permits, the plans for an academy have been transferred from the Federal Government, which initially dealt with them, to the Länder. Within the framework of the education ministers' conference an umbrella organization has been set up, based in Bonn, that will oversee the three branches that have grown up – organically, as it were – in Berlin, Munich, and Ulm. The idea is that Munich becomes the television center, Berlin the center for training in specialized motion picture areas (and film criticism in particular), and Ulm will have a comparatively small setup concerned with practical training in all disciplines. The Berlin course would thus produce the specialist or the critic; Munich would produce the television man, and Ulm would produce the all-round filmmaker, as required by the doctrines of *cinéma des auteurs*. In this context Berlin in particular, but also the representative of the Bavarian education ministry, laid particular stress on collaboration with the Filmmaking department in Ulm.

Alexander Kluge, in *Ulm*, 10/11, 1964

# Industrialized Building

The students in this department were trained to solve, as architects, the problems arising from the industrialization of construction.

To this end the department trained the students to identify the tendencies embodied in this process of industrialization, and to make them the basis of their work: the growing volume and also increasing anonymity of the demand; the structural changes occurring within the organizations that project and construct buildings; the mechanization of construction work; the use of new materials and processes; the ever more frequent recourse to scientific research and systematic experimentation.

The educational program presupposed a good prior vocational training on the students' part. The aim was to prepare the architect for the increased responsibility that results from the industrialization of building, rather than to train a specialist.

Instructors: Max Bill, Rudolf Doernach, Herbert Ohl, Claude Schnaidt, Werner Wirsing

Architecture
The training takes place through collaborative work in the drawing office on practical assignments, without the usual breakdown into semesters on specific topics. The theoretical subjects are related to the student's educational level and taught in close conjunction with the practical assignments. Students are expected to have completed an architectural training, or an apprenticeship in architectural draftsmanship, or a course at a school of construction, or to have practical experience in a skilled trade.
From *HfG-Info,* 1952

Rough sketch
for the HfG building
Max Bill, 1952

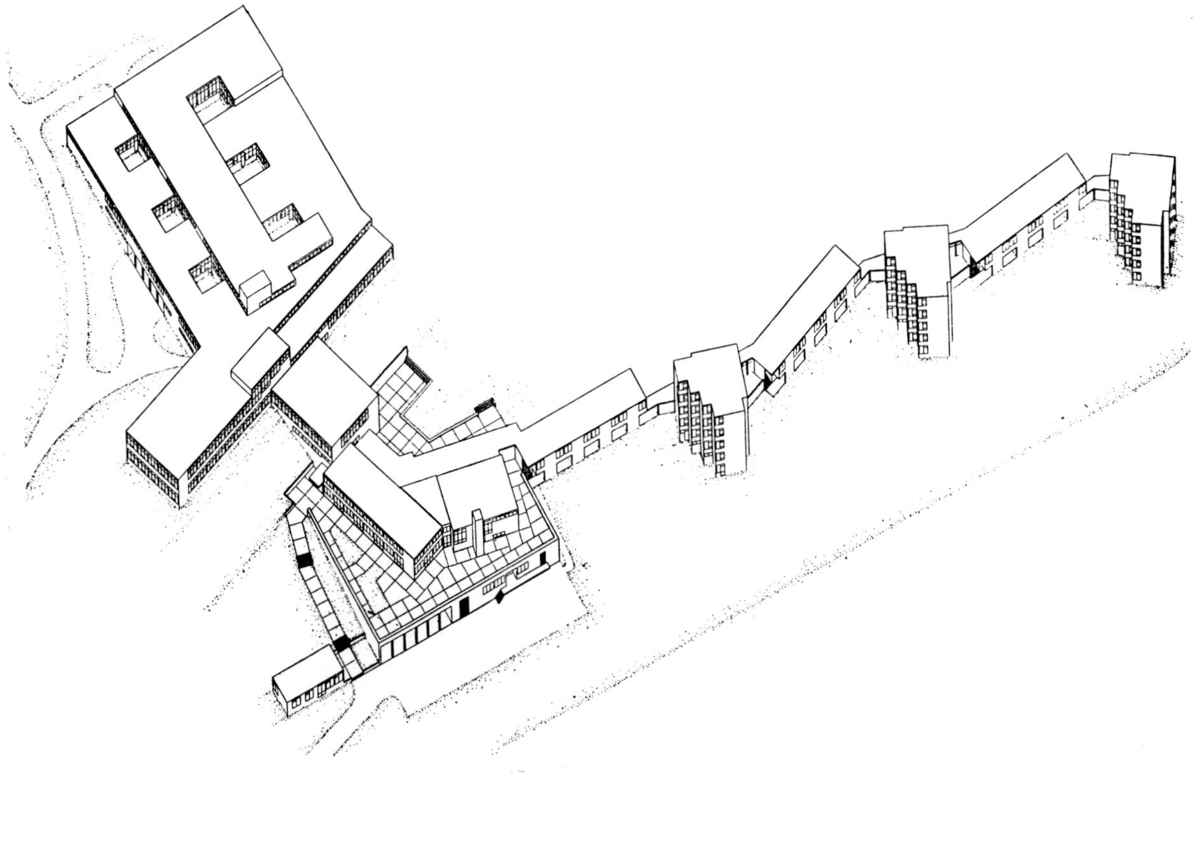

Isometric view of the HfG
Max Bill, 1952

Urban Planning
Training takes place in the drawing office with reference to practical town and country planning assignments. The theoretical topics are taught in close conjunction with the assignments. Essential prior qualifications are a completed architectural training, or equivalent.

From *HfG-Info,* 1952

*left*
Surveying the site

*center*
Building work starts on the Kuhberg
September 8, 1953

*right*
Topping out
July 5, 1954

The buildings of the Hochschule für Gestaltung are so laid out on the site that each level has direct entrances and exits. The landscape and existing trees are incorporated in the plan. The architect, Max Bill, designed the reinforced concrete construction system that is used throughout. Exposed concrete, cast in smooth formwork, has been employed for all constructional elements, all outer walls, and many internal walls. This is the first time it has been used on so large a scale in Ulm.

Margit Staber, *Süddeutsche Zeitung,* October 2, 1955

Students on the terrace

And no one view is identical with any other. To walk round the College is to set it in motion. There is no vantage point that commands a central facade. From every angle the masses seem organized in a different way, but the view is always of an evidently closed, definitive shape. There is no concentration of volume anywhere, and nowhere does a center become visible.

Apart from the four-story dormitory block, all the buildings have two floors at most. They climb the slope in a terraced arrangement. The dominant impression: grace, movement, multiplicity.

Bernhard Rübenach, "Der rechte Winkel von Ulm," radio documentary, 1959

*left*
Group photograph
Summer 1954

City of Ulm pavilion, Baden-Württemburg state fair
Max Bill, with Otl Aicher and Friedrich Vordemberge-Gildewart
1955

# Industrialized Building at Ulm

Herbert Ohl

Max Bill and Tomàs Maldonado visit Henry van de Velde at Oberägi, Switzerland

Konrad Wachsmann

Ulm was a focus. It was an institution that had adopted as its goal the overall *Gestaltung* of society and all its products: the design of a new culture. Its starting point in this enterprise was not an image of the society that it saw around it, but a process of extrapolation: its own ability to visualize the potential that still lay in existing industrial technology, the needs that resided within the mass of still unsatisfied users, and the latent possibility of a recombination of existing products and phenomena to create a new *Gestalt*.

The result was the image of a society that did not yet exist, but that could be glimpsed through hints, possibilities, faint beginnings, unattained objectives: an open, liberal, shatterproof, harmonious, constantly evolving, humane society. A society free from aggression; a society open in all directions except that of nationalism: a world society. The members of the HfG, students and faculty, always formed an international community, although the roots of the College had emerged from the soil of the Swabian city of Ulm. Those local roots had been supplemented by many participants who belonged to the history of the Bauhaus, and by many others who had been drawn to Ulm from abroad by closely related ideas.

"Industrialized building" was what we called the activity of redesigning architecture in the spirit of a new and modern society. Perhaps "industrialized architecture" would have been an even better term. Industrialization appeared to us – and still appears to me – to be a necessity, if mass needs are to be met in a technologically adequate and scientifically responsible way.

Industrialized building was only a part of the generalized Ulm endeavor to apply *Gestaltung* to the whole environment and its products. With this in mind, very soon after the institution was first set up, architecture was redefined in an openmindedly functional, but still rather classical, spirit that matched the other objectives of the College: industrial design and mass communications. The carrying over of many insights from the technological and also the human sciences – insights concerning people, constructional techniques, work in allied areas such as aircraft building, shipbuilding, and machine construction – gave the College the ability to apply the whole potential of modern industrial production to the design of building as a product.

However, as conditions in the construction industry and in architecture were by no means as favorable as they were in capital or consumer goods, it was more necessary here than in other areas of work at Ulm to concentrate on technology: it was there that the key to the new forms and design possibilities of industrialized architecture was to be found. The objective of all these efforts was – of course – the improvement of the quality of life of the people who were going to live or work in the resulting buildings, complexes, or townscapes.

The work that took place at Ulm considerably expanded the scope and range of industrialized building and systematized and developed much knowledge that had only been touched on by earlier architects and engineers. On the one hand there were studies that led to the methodical assessment of users' needs: surveys and interviews were carried out, and behavior and lifestyles were analyzed, in order to define ways in which the living and working environment might be improved. On the other hand, there was research into planning techniques, and in particular into modular rationalization, in order to make buildings transparent, producible, modifiable, to make them more economical, and to exploit the designability, the formal freedom, of the architecture that can be created through industrialized building.

Construction was to be modular, but not only in the sense of rectangular or of simple two- or three-dimensional triangular and hexagonal networks: much attention was paid to modularization using spherical constructions, with a view to elaborating functional building types rather than merely types of construction. Whether in the development of a spherical movie theater, pursued concurrently with the examination of spherical lightweight modular constructions in general, or whether in modular formulas for housing elements that could be combined into wider urban planning units: in every case the concern with modularization, based on measurements, based on elements and interconnections, was a key factor in the attainment of a modern, versatile, inspiring, industrialized architecture.

At an early stage there came into existence, alongside the teaching department, an Institut für Industrialisiertes Bauen: it carried out practical, privately financed research and development projects that provided a stimulus to new ideas. New component and cladding systems took full advantage of the properties of new materials, as with the thermal expansivity and ease of forming afforded by impact-resistant thermoplastic resins.

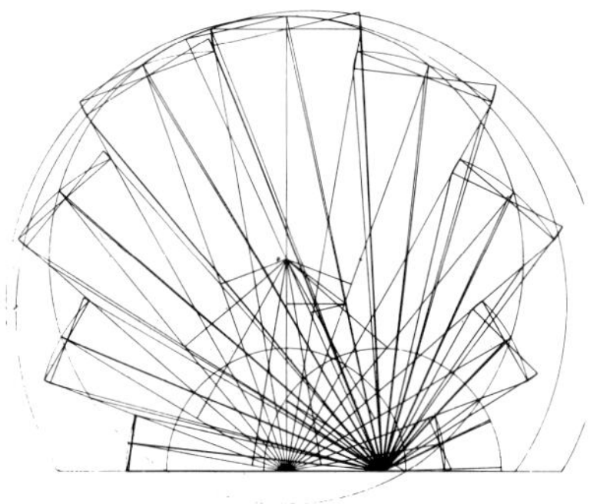

Spherical movie theater
Herbert Ohl, 1956
Creation of a projection system analogous to the human visual field

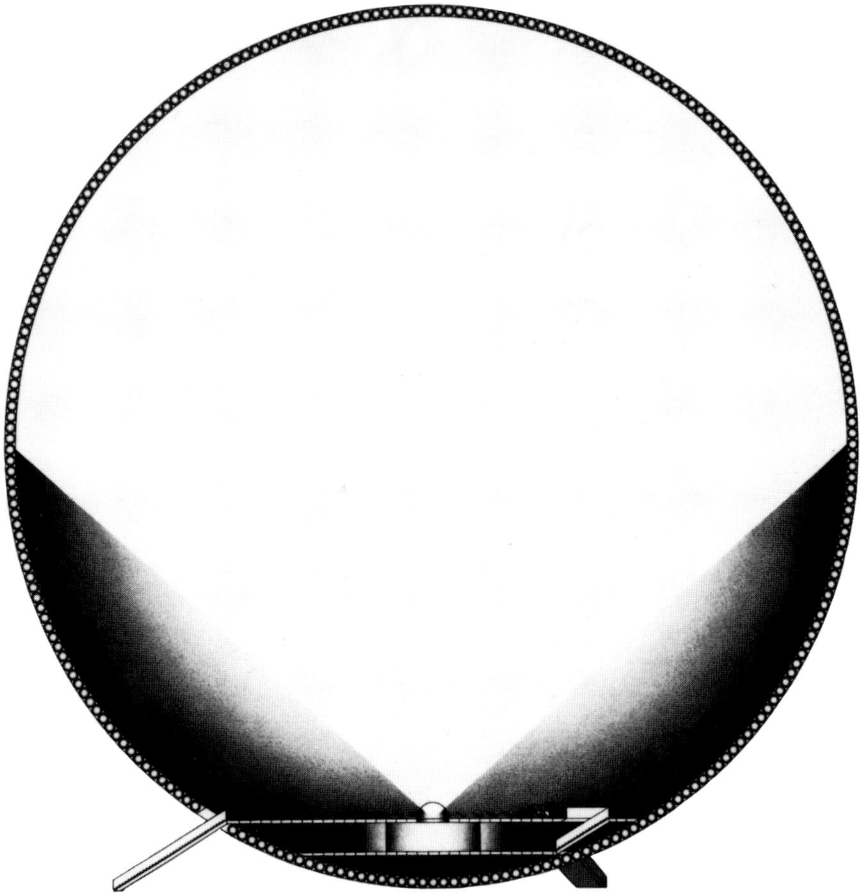

The program was completed by window and extension systems, together with lightweight building systems — using light materials assembled with new techniques, such as integral construction, that had previously been used only in vehicle and aircraft manufacture. All this formed an exemplary program of research and interpretation, leading to the development of new modular systems, modular forms, and techniques of assembly that offered formal, creative freedom as well as solutions to many technical problems that traditional methods failed to solve. The objective was to make the end product of the building process — working or living space — both adequate to its purpose and new in form, analogous to the products of manufacturing industry. After the experience of this first phase of using lightweight construction, attention was increasingly directed to the use of the classic modern material of concrete, now traditional and supposedly well understood, in order to combine it with the principles of lightweight construction to produce open constructions with wide spans that would provide a high degree of freedom and flexibility in use. Problems of all-weather construction work, of building in difficult terrain, of overcoming subsidence, of resistance to structural damage, and of harmony with natural landscape were worked out to the stage of pilot projects virtually ready for execution.

Darmstadt, 1987

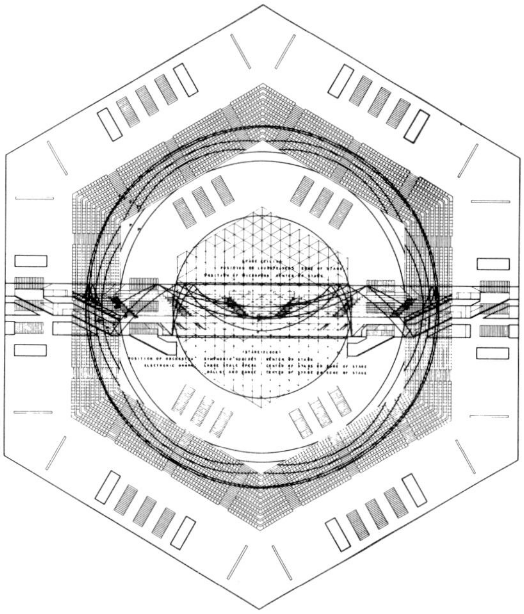

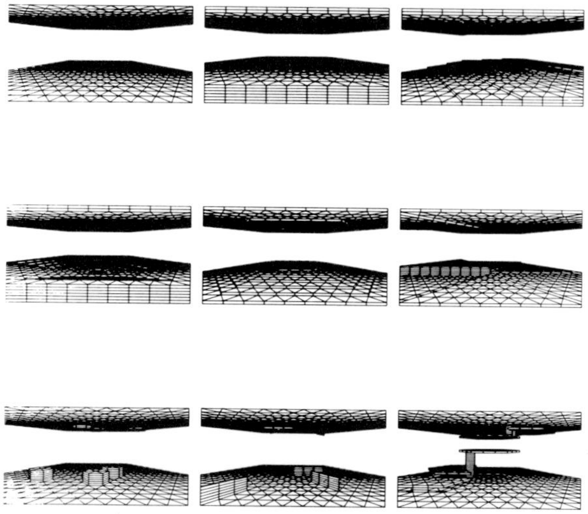

Competition entry for
Sydney Opera House
Herbert Ohl,
Maurice Goldring,
Klaus Franck
1957
Stage adaptable for concert,
ballet, and theater use

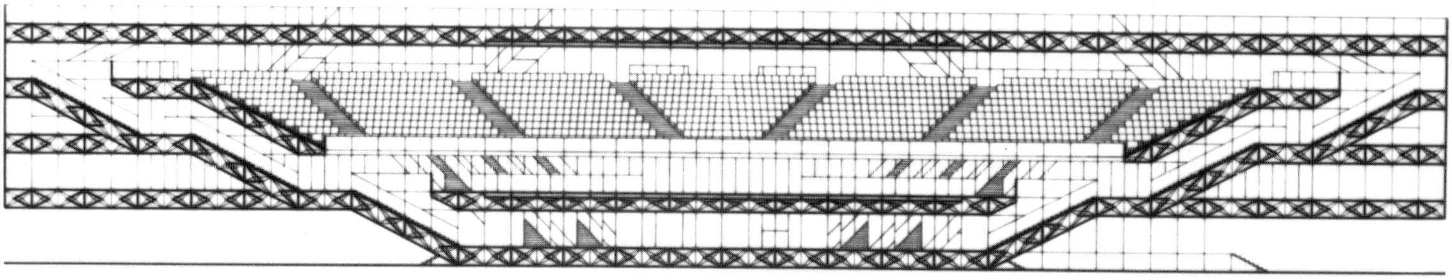

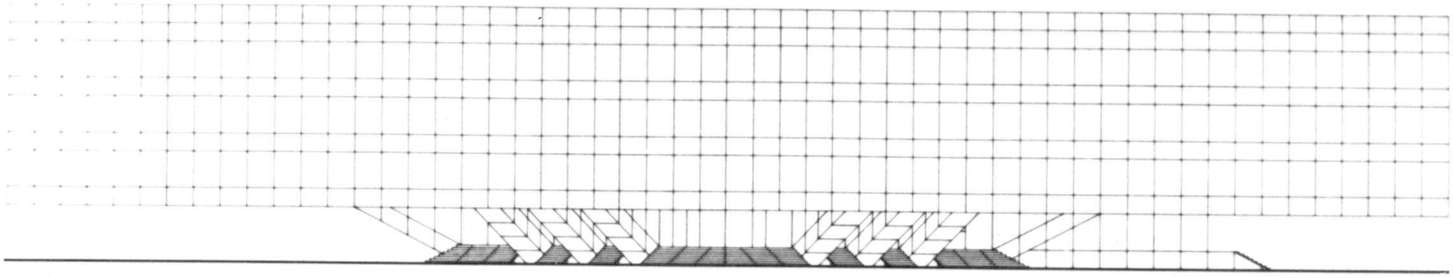

Giulio Pizzetti

The Building department has set itself a comprehensive but clearly delimited task: the industrialization of building, i.e., the application of modern production methods to building technology.

The Building department trains specialized architects who are capable of undertaking this task.

It is a fact that traditional building methods are no longer suited to the demands of the present day; and this applies in particular to the growing need for housing units. It is therefore essential to apply technological and industrial methods to building.

From *HfG-Info,* 1960

Giuseppe Ciribini with students

Bruce Martin and Claude Schnaidt

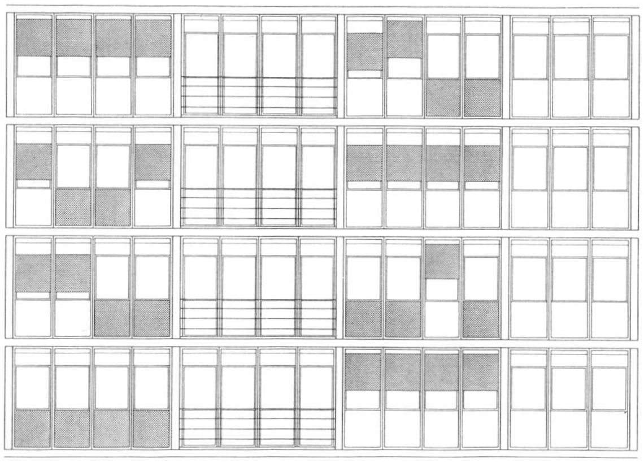

Facade system
Bernd Meurer
First-year student, 1957 – 58
Instructor: Bruce Martin

the pernicious nature of a conceptual model that takes automobiles, refrigerators, and shoes as its most favored models can be seen in our apparent powerlessness in the face of the neighborhood development that is sweeping over us like an uncontrollable force of nature. not only were we not in a position to reorganize the great expanses of ruins in our cities according to overriding viewpoints; the chances offered by the rebuilding of a devastated city have long been frittered away. now

Integrated construction system
Students of the department of Building and Herbert Ohl, Günter Schmitz, Rupert Urban
1957 – 63

We don't use the word "architecture" very much; we say "building." By this we mean that architecture is nothing more than the sum total of the activity entailed in construction, all the thoughts and decisions, all the skills and the product of those skills, all taken together. We don't just say "building"; we say "industrialized building." As a description of architecture and thus of building, in a strictly contemporary context, this seems to me to be right and necessary. The architect ought to have realized long ago that buildings can best be produced by industry, with its factories, machines, and assembly processes, and with the materials and material forms that go with them; and that, whatever happens, he is a part of this construction industry.

Herbert Ohl, from a lecture in New York, 1961

*center right*
Herbert Ohl

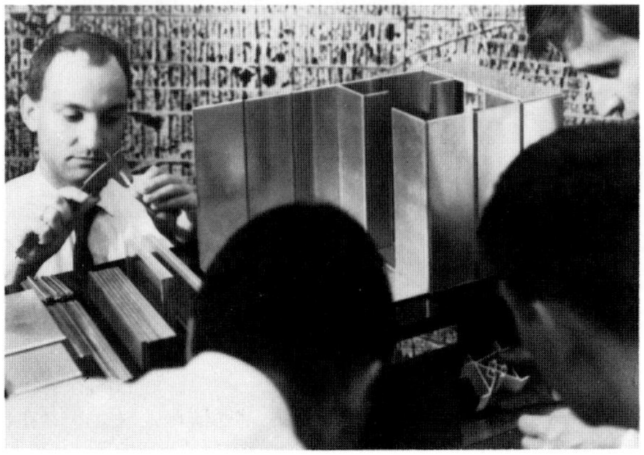

*bottom right*
Bernd Meurer and workers at the Industrialized Building Institute

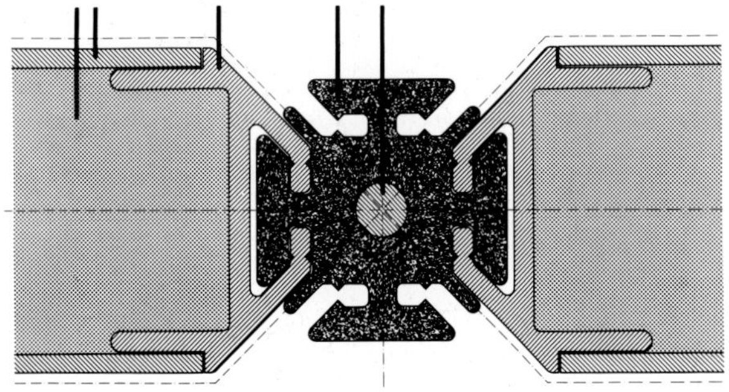

*bottom left*
Section of aluminum sandwich panel and neoprene jointing profile
Günter Schmitz, Rupert Urban
Instructor: Herbert Ohl

## Designed Spaces

Michael Erlhoff

Or: "Vermin is not neat and clean; in glass houses it's never seen." (Paul Scheerbart.)

One obvious thing about architecture is that it tends — by constantly opening up spaces and rebuilding them, erecting signs, and piling up monuments — toward the creation of a unified whole under dictatorial rule. As — historically speaking — the lowest of the arts, it always aspires to the heights; and it does so unbidden, in sheer uncritical exuberance, as if all this were the nature of architecture.

There can really be no question that, first and foremost, architecture is order. It sorts masses, leads us by the hand, structures sequences, coordinates our outer life. By definition, it resists chaos, anarchic proliferation, and, ultimately, the "building blocks of nature." Architecture is intervention, and this, however conditioned and molded by time, is its spatial essence. There is one factor that architecture is reluctant to acknowledge, although it constantly exploits it: that of playing tricks with time, spiriting it away or translating it into a metaphorical, spatial perspective. Architecture surrounds time with space (communication: walking about within the walls); brings it to an end; or presents intervals of time in we are helpless in face of the instantaneous urges that spring from the general acquisition of mobility through the motor car. the lamentations that are heard about the laying waste of the countryside are rather an assertion of generalized impotence than an attempt to stem the flood. events have caught us totally unprepared. an appeal for the saving of what is left is not in itself an adequate basis for controlled development in the future.

in an almost exclusively consumption-oriented world, we concentrate on the automobile for its own sake: we force up its production without building the social organization that alone would make it into a rational extension of the human body. increased road building could never be enough. the road system would have to reach a point of sophistication that would guarantee an organized flow of traffic right into the constipated centers of our cities. we concentrate on the automobile, reduce the design dimension to the viewpoint

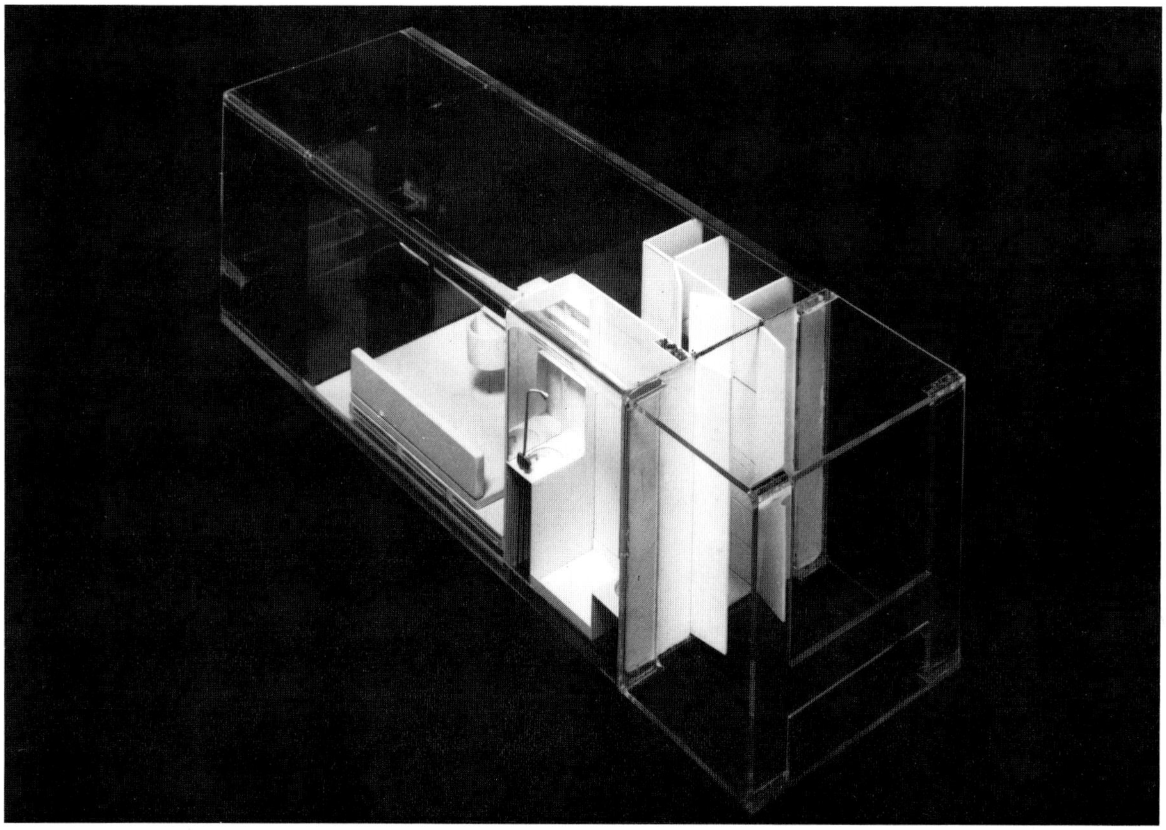

Modular utilities for student dormitories in cellular construction system
Herbert Ohl, Bernd Meurer
1960

the guise of shifts from one space to another.

All of which endows architecture with its tendency to grow and tend toward totality. It makes no difference whether this takes place to the greater glory of God or as part of the dynamics of bourgeois society, which launched into architecture very early on (Boullée, Ledoux, and industrial landscapes of all kinds) and visualized progress in terms of spatial expansion or ascension in a scale.

"We shall know the highest that can be built – can be built, if human beings have the will to do it. We architects must inspire them with this will," wrote Bruno Taut; and he designed his cathedrals of upward spatial expansion – and his cathedrals of light, of which the mystics had long ago dreamed, and which, in a very different guise, the Fascists were later to turn into reality. There is, clearly, an inner contradiction within architecture that leads it to aspire to domination, to the achievement of the utterly real, and thus to get embroiled with authoritarian systems. This is how Fascism came to be so attractive to architects (Terragni, Nervi, and some people from the Bauhaus), and how Le Corbusier came to be so fascinated by Stalinist city of a pipin farina, and forget that the reality of the automobile is changing the places where people live and where they work. dwelling, working, and living are changing as a result of the fact that everyone, not only the privileged, is acquiring a greater radius of activity. society is passing into a state of constant, ever-widening movement. the transportation component in products from food to building materials is increasing. it is the social consequences of the automobile that are changing our environment and demanding that resources be made available to bring them under control.

i would like to use the example i have just given in order to bring up another issue of principle. besides our consumer orientation, there is another block that serves to explain why we are so overwhelmed by developments in our civilization. it may also explain why the various attempts at regional planning only touch the fringes of the complex, and why the ugliness of our environment is increasing rather than diminishing.

this further block has its basis in a specific cultural attitude. attitude in the sense in which we speak of the enlightenment, or of romanticism: an epistemological system for the interpretation of reality and for the designing of a possible future. the attitude i want to discuss here has pathological features. it is based on a blind faith in science that has produced an exaggerated urge to analyze accompanied by a diminishing capacity to act. for ten years now, as a result of significant scientific advances, especially in cybernetics, we have been prey to a methodological optimism that postulates that the results that guarantee further progress will appear of their own accord as soon as we have the data to program the problem. progress based on initiative is replaced by automatic progress.

Otl Aicher, address at the beginning of the academic year 1963–64

the aim of the department of building is to train architects capable of solving the problems that arise from the industrialization of construction. to this end the department trains the student to identify the tendencies embodied in this process of industrialization and to make them the basis of his work: the growing volume and also the increasing anonymity of the demand; the structural changes occurring within the organizations that project and construct buildings; the closer collaboration between all those involved in construction; the greater demands for precision and organization; the mechanization of construction work; the application of new materials and processes; the growth of standardization; the increasingly continuous nature of design work; and the ever more frequent recourse to scientific research and systematic experimentation. the educational program presupposes a good prior vocational training on the students' part. The aim is to prepare the architect for the increased responsibility that results from the industrialization of building, rather than to train a specialist.

From the teaching program of the HfG, 1964 – 65

Cellular reinforced concrete construction system: assembly of the cellular units
Herbert Ohl, Bernd Meurer, Willi Ramstein
1959 – 60

plans; it was the motivation for German town planners after 1945, and also for the systematized building methods that were worked out at Ulm and elsewhere, which could have been used to make everything, everywhere, uniform.

These giants within society regarded themselves as meta-artists, all-unifying and all-embracing. If there was to be a total synthesis of the arts, a *Gesamtkunstwerk,* then let it be an architectural one. Gropius exclaimed in 1919: "Painters and sculptors, break through into architecture and become fellow-builders, fellow-strivers for the ultimate goal of art: the creative conception of the cathedral of the future, in which all will be in one *Gestalt.*"

All will be assembled, that is, by the architect — master builder, artisan, artist, politician, supremely gifted individual — who, when he builds, leaves monuments for eternity and destroys other monuments to that end. For architecture (although those involved are hardly likely to say so) is an unremitting and competitive campaign of psychological repression, translated into very solid terms: new buildings devour old ones, big ones devour little ones, complex ones devour single ones. under the name of order, what archi-

*left*
Suspended construction: competition entry
Willi Ramstein
Third-year student, 1960 – 61

*right*
Student dormitory in cellular construction system
Herbert Ohl, Bernd Meurer
1962

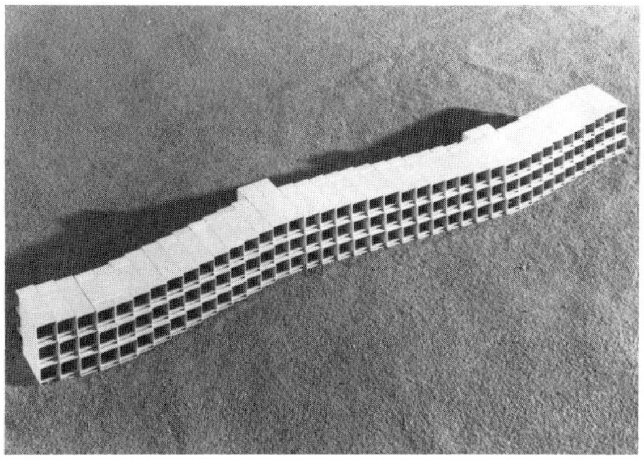

*left*
Low-rise housing development using cellular construction system
1962

*right*
Student dormitory in cellular construction system
Herbert Ohl, Bernd Meurer
1962

tecture pursues within itself and in the world at large is anarchy.

Once architecture had organized the pathways, the dwellings, and the spatial meanings, design rushed in to take charge of the smaller, inaccessible spaces, the signposts, and the tactile realm. A reduction that released design from the totalitarian dimension and allowed it detachment and clear identity. For originally design had (as it still imagines itself to have) a human dimension: it was a service, making its work available and accessible — in contrast to architecture, which inevitably overwhelms, sorts, and pigeonholes its subjective recipients, and which can never be taken in at a glance except from the air.

Nevertheless, design in its turn eventually set out to steal the show and take on the role of a power focus within society. This was no longer design as a service and an intermediary, but as an end in itself. Design took advantage of the real weakness of objects, which had become no more than salable goods, and defined itself as the factor that gave objects back their object quality. If objects wear thin through constant replacement, design provides them with a new *Gestalt* every time, thus stepping out of its marginal

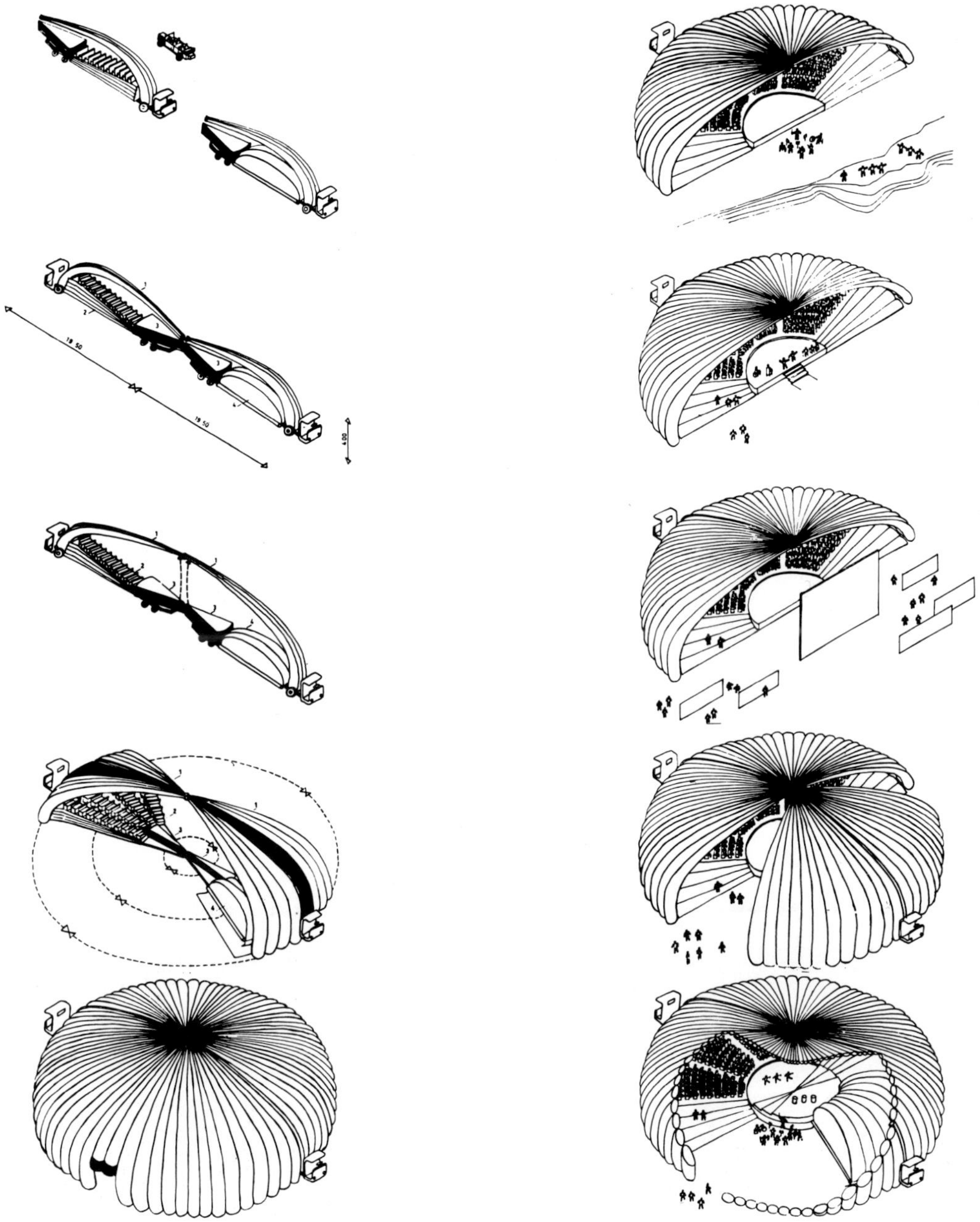

Mobile theater
Willi Ramstein
Third-year student, 1960 – 61
Instructor: Herbert Ohl

A design for a mobile theater with seating for 500 spectators, using an inflatable structure. Two semitrailers transport the whole theater. Once unloaded they provide the necessary power and serve as box offices. Competition entry for UIA Congress, London 1961.

position to acquire a key role in society; at the same time, and conversely, it accelerates the obsolescence of the objects, a process in which its new function is reinforced by that of fashion. Design thus assumes a virtuous and legitimate guise: not superfluous packaging, adding the odd crazy corner or curve, but a social necessity. This is just what the arts and crafts movement was fighting for; so was De Stijl; so were the Russian Constructivists and Suprematists (remember Arvatov's production aesthetic); so was the Bauhaus; and so was Ulm.

Above all, once endowed with this new meaning, design was presented with access to larger spaces, even to the total space of the whole. Engineers became architects, and architects became partly designers. And this widened the scope still more, because all of a sudden everything could be designed: interiors and exteriors, the tactile and the visual, hand manipulation and traffic flow. Max Bill's often-paraphrased motto, "From spoon to city," is a clear statement of this very position — rightly so, because, from the point of view of social relevance, architecture and design must aim at totality (another repression contest) and define themselves with an eye to that totality.

Combinable tubular-frame domes
Prototype in PVC tubing on square plan, 6x6 m (20x20 ft)
Heinz Dobrinski, Horst Schu, Max Thanner
Second- and third-year students, 1962 – 63
Instructor: Rudolf Doernach

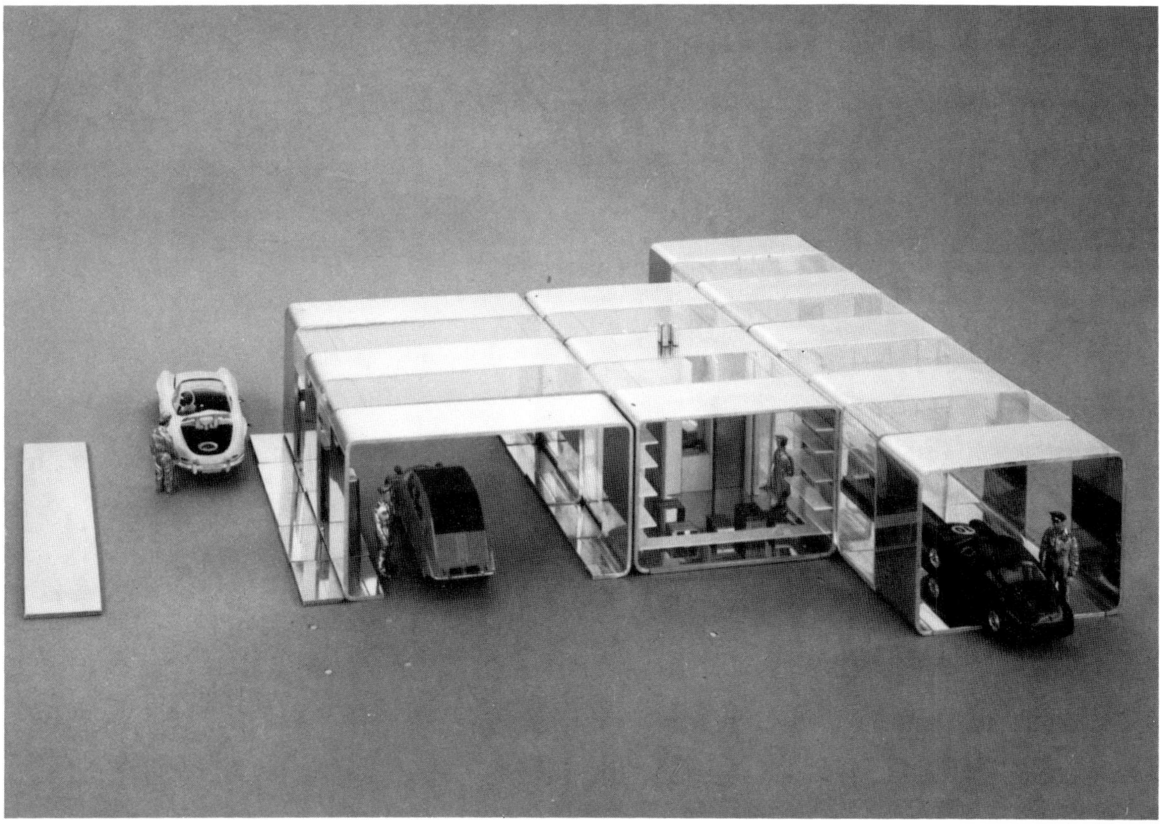

Filling-station system
Ring cellular construction system in steel
Herbert Ohl, Bernd Meurer
1962

Supposing, that is – and here most German architects and designers were covert or overt Hegelians – supposing that the whole was to be equated with the true and not with the untrue.

If this is so, design is simply a self-legitimizing compulsion to plunge into the midst of life and to interpret the creation of form as the creation of social form. For unlike art, which formulates its truths from a marginal position and goes unheeded, design comes across as nothing less than the real truth of appearances.

To be substantiated, this claim must be translated into action; and it was at that time that design began to involve itself in space. Environments were set up, overriding structures imposed, systems formed; and designers went out onto the streets, controlled transportation and highways, created something called urban design, and scaled the ecology mountain (the designer in the friendly guise of a cross between robber baron and Robin Hood, sweeping down from his citadel, widening and then exhausting fields of research, all with the best of intentions).

As an increasingly fragmented society went on developing an enormous demand for unifying standards, design expanded into an abstract

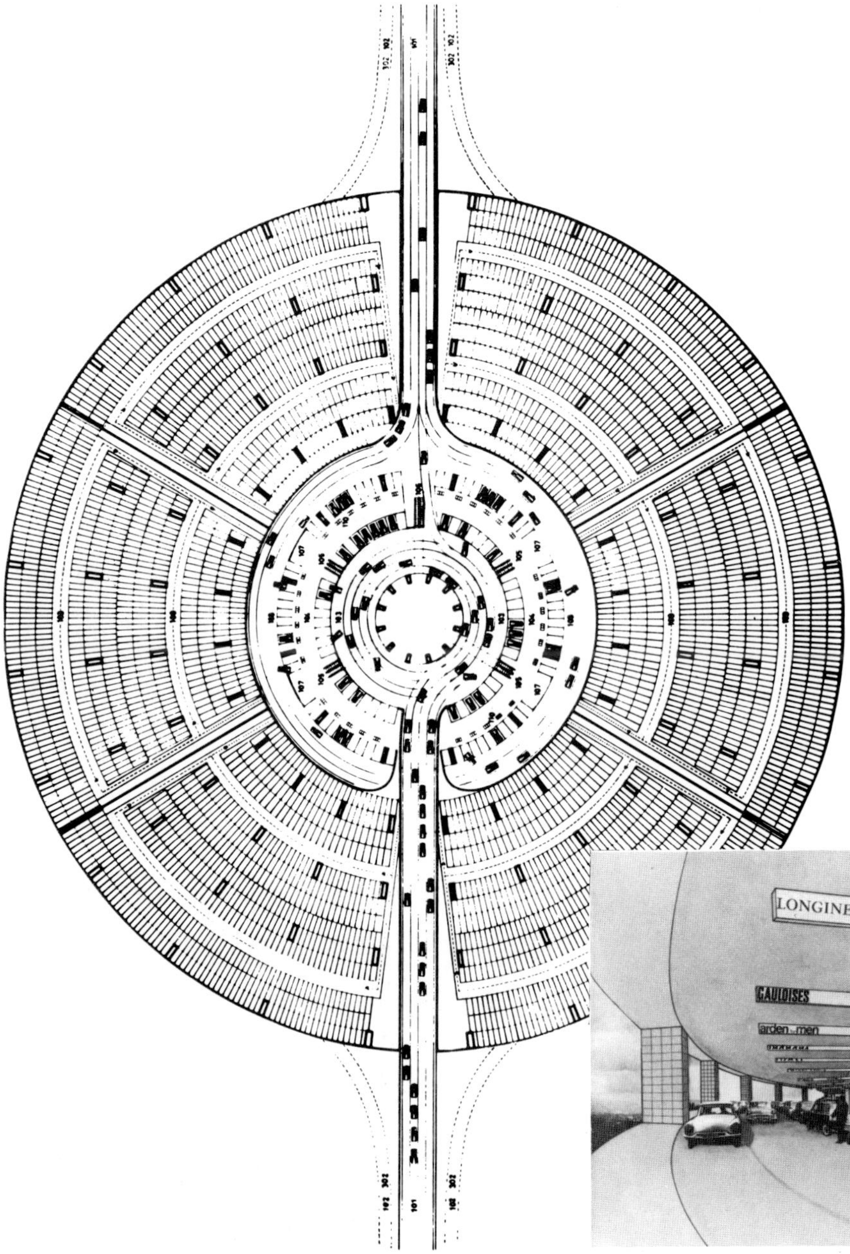

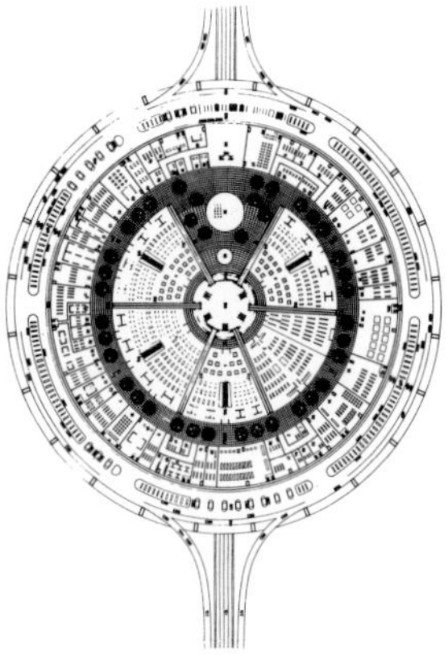

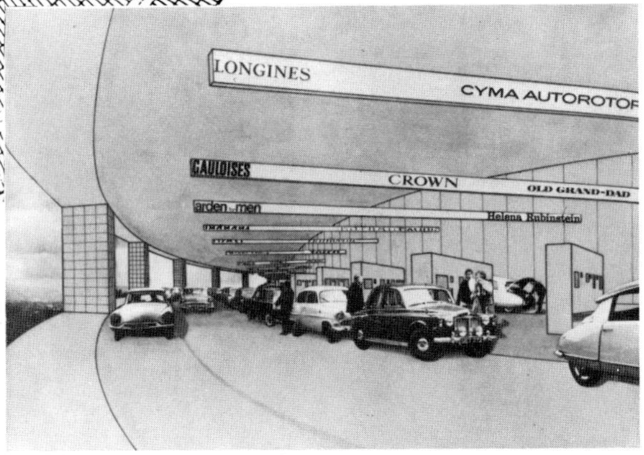


Shopping center
Willi Ramstein
Diploma work, 1962

Plans of both levels

Outer ring road with
drive-in stores

realm as an expression of the imposition of order. That is why so many people are so keen to use the word "design": they talk of "designing research," "designing society," and much else of the same kind. To give form to the immaterial at last: corporate identity — the problem of identity itself — has become the latest, for the time being, of the key functions assumed by design in urban civilization. And however much the HfG may have striven to modify, sublimate, and criticize the Expressionist explicitness of Gropius's Bauhaus credo, it was taking a leading role from the very outset in the expansion of the field of design: system design, urban design, and much more. Not naively so, however; for many of the manifestations and theses of the HfG pursue a coordination of parts that does not depend on the idea of a whole, but exists in the tension between totality and insight, between beautiful untruth and true beauty.

Instructor Frei Otto

*facing page*
Photomontage with highway access

Angled slab construction system for housing
Rudi Dahlmann,
Eberhardt Köster,
Ernest Muchenberger
Third-year students, 1961–62
Instructor: Herbert Ohl

Curtain wall with sun protection for office buildings
Karl-Heinz Allgayer,
Gerhard Curdes
Diploma work, 1963

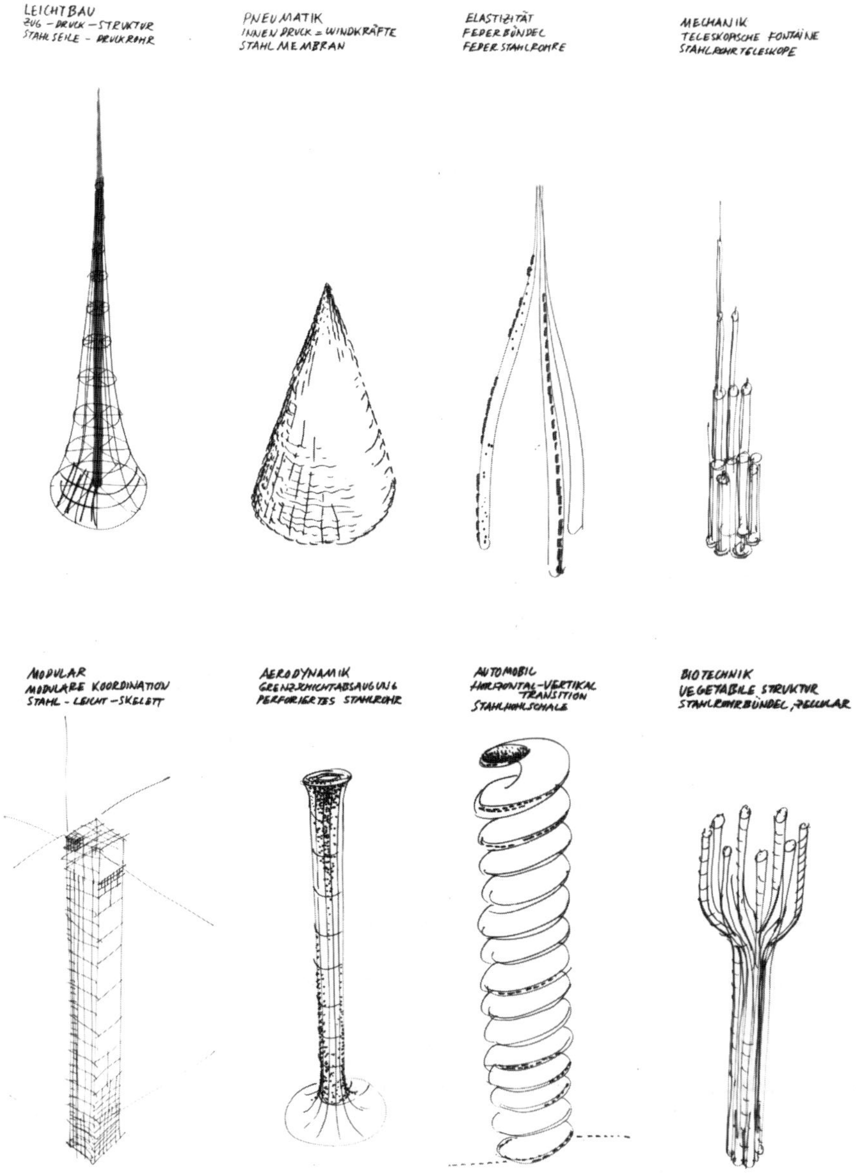


Studies for a tower
H. Ohl, B. Meurer, G. Schmitz
1963

I was only a bird of passage at Ulm; I perched on a flimsy twig for a few weeks, twittered away cheerfully in the winter fog, and left behind me a few blots and a few students.

For me the HfG was unique and fascinating. I was in my element. The residents' territorial disputes did not affect a migrant like me. I would have liked to see the HfG have a

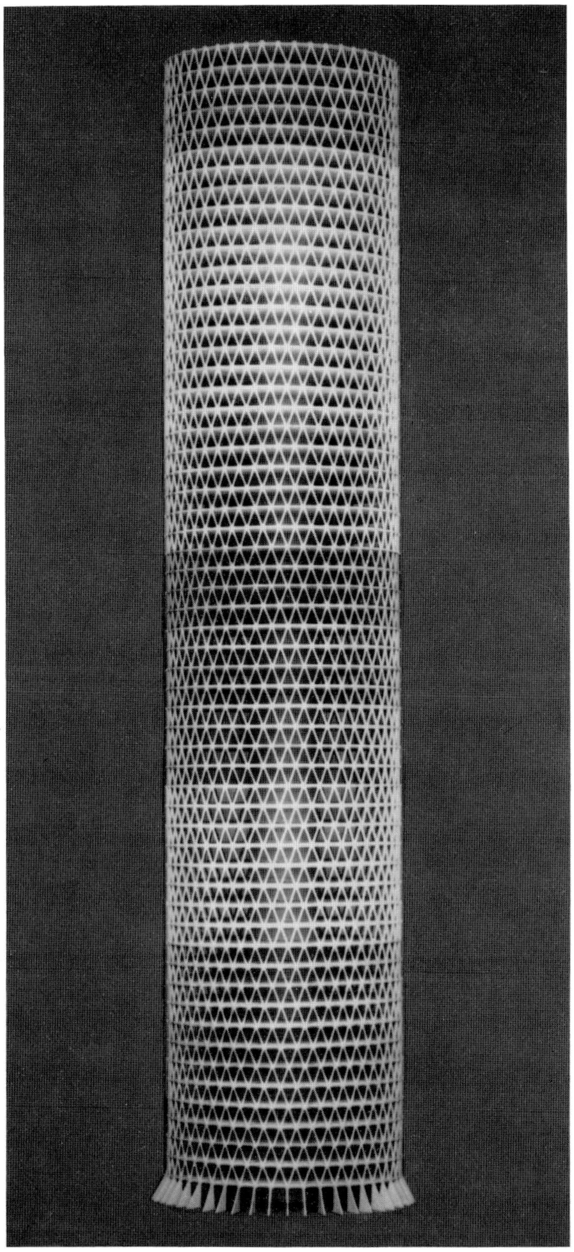

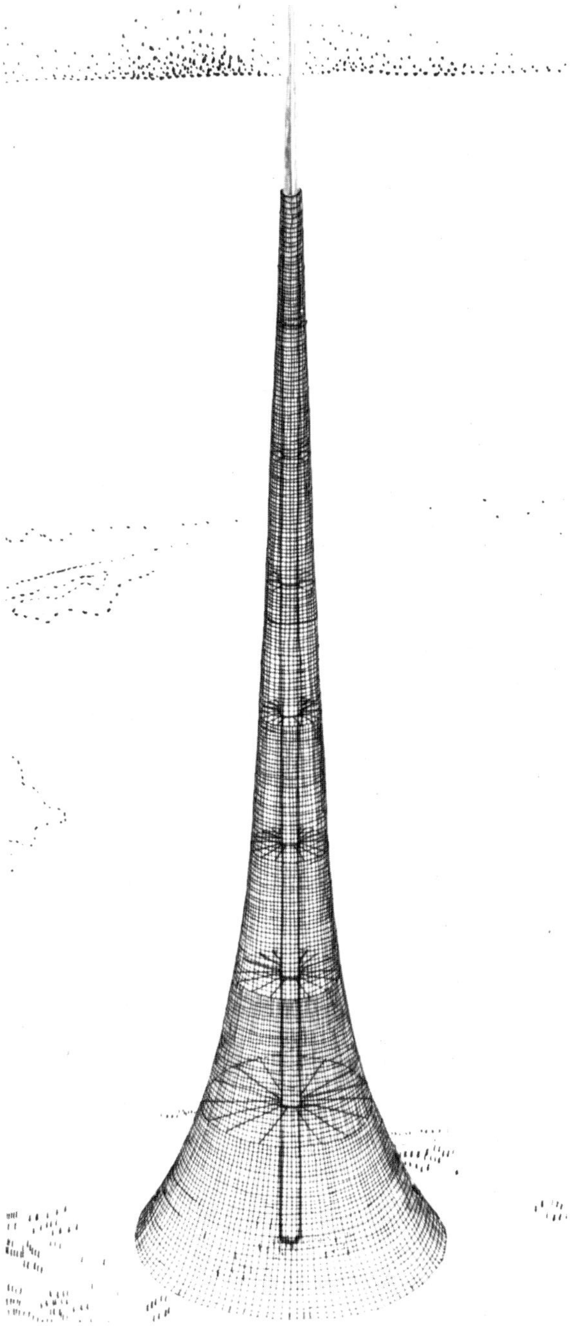

*left*
Concept for an office-building
Student: Forné, 1964
Instructor: Peter Sulzer

long life, or at least a cheerful, clear-cut demise marked by a big party. Its death was demeaning.

Anyone who goes into the old building today is liable to be agreeably surprised. Doctors are researching there in a highly creative way, clearly affected by the genius loci: a tribute to Max Bill.

Frei Otto, 1987

Housing in Ulm
Use of the ring cellular construction system in reinforced concrete
Fundel, Hufenus, Rusconi, Stocker, Stuber
Third-year student, 1965 – 66
Instructor: Herbert Ohl

the objective of the department of building is to train architects who are competent to solve the problems arising from the industrialization of construction. the department of building offers a training that is roughly half-way between a complete and a post-graduate course. the course is intended to bridge the gap between the architect's traditional training and the demands that arise from new requirements, economic change, and technological progress. the aim is to

*right*
Construction and assembly system for large industrial spaces
Werner Wirsing
1967 – 68

*left*
Furnishing system for student dormitories
Herbert Ohl, Bernd Meurer
1964 – 65

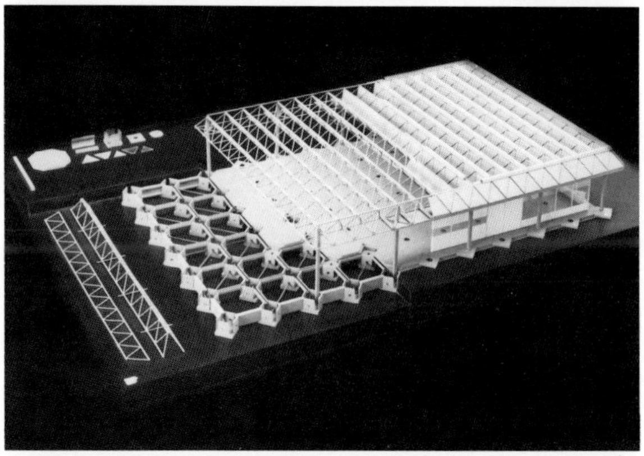

*right*
Werner Wirsing

*left*
Claude Schnaidt

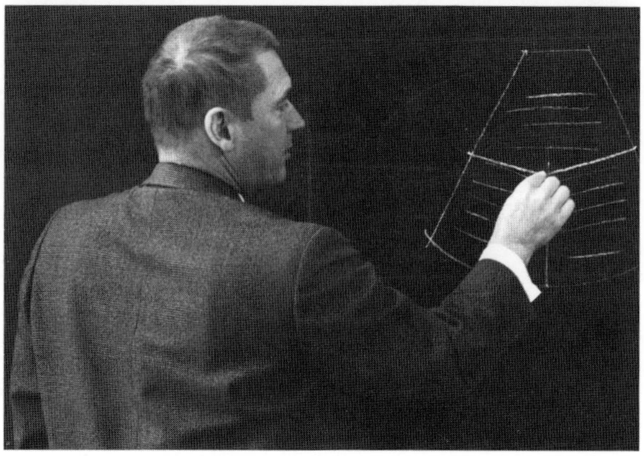

prepare architects for the increased responsibility that results from the industrialization of building. this industrialization of building embraces the growing volume and also the growing anonymity of the demand; the structural changes occurring within the organizations that project and construct buildings; the increasingly continuous nature of design work; the closer nature of the collaboration between all those involved in construction; the greater demands for precision and organization; the mechanization of construction work; the application of new materials and processes; the growth of standardization; and the ever more frequent recourse to scientific research and systematic experimentation. the existence of prior conditions of this kind intensifies the complexity of the architect's task. the department of building therefore sets out to prepare its students to fulfill these important functions.

From the teaching program of the Hfg, 1967 – 68

Satellite city at Beaumarais, near Saarlouis
Housing project using stacked ring cell modules
Herbert Ohl, Claude Schnaidt
1968

Claude Schnaidt

# Architecture and the Scientific and Technological Revolution

I make no claim to provide an overview of all the problems that face architecture in the scientific and technological revolution. I shall restrict myself to those that seem to me to represent a matter of imperative urgency.

The motive forces of the scientific and technological revolution are as follows:

— automation, which has reduced the role of the human being in directly productive activity and displaced it back into the preparatory stages of the production process;

— chemicalization, which creates custom-made materials and thus reduces the mechanical element in the production process;

— the use of nuclear energy, which provides unlimited resources of energy.

The areas in which these motive forces manifest themselves are those at the cutting edge of production, technological innovation, and scientific research. They spread themselves wide and give an evolutionary impetus to the whole economy. Gradually, science pervades the whole of industry, the whole of social life. Between research, planning, production, and organization, new relationships are established. Science becomes the decisive factor in economic growth. It is no longer possible to see it as a "negligible quantity." Previously, growth was promoted by the concentration of all resources on direct production. Those times are past. Now it is necessary to single out what might provide a scientific potential, in order to ensure that better and better solutions are introduced into production, and that this is done in good time, in order to show a return on investment.

The expansion of the economy demands the provision of financial resources, but it also depends to a large degree on the stage of development reached by the individuals concerned. The consequence of the scientific and technological revolution is that the full development of human beings, their capacity for research, invention, and innovation — for creative work — is not only the best way in which the productive forces of society may reproduce themselves: the full development of the human individual becomes a necessary precondition for the progress of civilization. Over the last few centuries civilization has been making great strides forward at the expense of exploited, maimed, alienated human beings. What mattered was the amount of capital and the quantity of labor. Culture was peripheral. Today it has become central to any further social evolution. The scientific and technological revolution is thus equivalent to a cultural revolution. It is giving culture and education a radically new place in society, because it is making material production dependent on the evolution of the human race as a whole.

On the distant horizon, the scientific and technological revolution reveals the first signs of the elimination of work in the traditional sense of the word. This elimination, for Marx and Engels, was the central issue of the social revolution. When the human being ceases to produce things, because things themselves can produce them in his stead, and when the human being therefore devotes himself to the discovery of new solutions and to creative work, i.e., to his own self-realization, then the oppositions between work and pleasure, between work and leisure, will disappear. This change in the character of work can be achieved, however, only if social conditions and the human environment are also changed. I cannot imagine how work could ever become a pleasure in an exploitative society, or how individuals could ever develop themselves fully if they stayed on in the caves inherited from their forefathers.

In the factory, in the office, in the home, in town and country, in all the products and all the locations that pertain to social cooperation, many of our contemporaries feel isolated, confined, out of place. They can see that their environment is out of their control, that it is taking them over and turning against them. If this alienation is to be overcome, then of course the social order must be changed; but there must also be a living environment that acts as very good tools do, and multiplies the strength and the freedom of the human being.

This is the challenge of the scientific and technological revolution. We must make the transition from a human world based on dominance to a human world based on liberation. Putting it differently, and bearing in mind that a large part of our inheritance is now in a critical state: we shall have to reorganize larger and larger areas of housing, work, services, culture, leisure, and transportation into entirely new structures.

From the standpoint of environmental design, the way in which the added leisure is distributed is of great importance. According to which option is selected — a shorter working day, or week, or month, or year, or lifetime — the necessary organization of space will vary. Conversely, overhasty planning decisions on a city level can severely hamper the future evolution of leisure time. Whether the city is pretty to look at or not, whether it is built of steel or of concrete, if it does not match up to the needs of its time it arouses the desire to escape from it. And once you have done all you can to make it easy to get out of the city it is too late, and evidently absurd, to start thinking about other possibilities for the use of leisure time. At present, weekend villages and vacation communities are the necessary counterpart to the urban monsters from whose precincts nature has been banished. These places of recreation, with their attendant crowds, noise, and pollution, no longer have anything to do with their original function, and the human beings whom they attract exhaust themselves in the constant effort to get there and back.

This impasse (which is by no means the only one: there is the relationship between the places where people live and those where they work, and the conflict between collective and individual living) shows us that the traditional town is an outworn model. As the scientific and technological revolution proceeds, it becomes more and more imperative to find new solutions, which can only go in one direction, namely that of removing the opposition between town and country. The adherents of the counterculture, who preach a return to nature, have no more chance of success than the prophets of urban living, or the multitudes of weekend commuters. The scientific and technological revolution makes a fundamental change necessary in the procedures through which humanity makes the world controllable.

Demands for comfort grow, and their satisfaction is increasingly entrusted to energy-consuming mechanical devices. The house, once a passive thing, is becoming active, like a machine. The scientific and technological revolution is accelerating this change. Between the house, the environment, and the inhabitant, new relationships are already starting to appear. Not so long ago, it was the efficiency of the heating boiler that compensated for heat losses. Now the aim is to get control of the whole thermic behavior of the house. The equation now includes the energy that building can derive from outside as well as from inside without additional cost. This new phase of the transition from the passive to the active house means that people no longer seek to defend themselves against the climate but to involve it. Solar energy, for instance, can be correctly used only if the variations in outside temperature, the effect of the wind, the hygrometry, and so on, are studied as a whole. The factual base is thus not solar energy but the climate.

The shell of the bioclimatic house is made in such a way that it functions as a thermal filter, which at the same time insulates and captures energy, stores it and dispenses it. With information technology and telematics, the age of the interactive building has come. Large buildings are already being built that, like servomechanisms, constitute a self-regulating system sensitive to information from outside, in which the equilibrium between command and response is maintained. Data concerning the behavior of the building is stored and allows the optimal use of the equipment. The idea of feeding weather forecasts into the control system no longer belongs to the realm of science fiction.

The penetration of energy, and of electronic and mechanical equipment, into the house, makes it necessary to rethink the nature of building, as a product, from first principles. It makes a scientific approach all the more essential. An interactive architecture means that organization, forethought, and planning are of primary importance. It presupposes a dynamic approach to existing reality. There can no longer be any question of passively incorporating the facts of a situation into one's work. The needs must be grasped dialectically, in their dynamic state, and we must therefore operate on a level that allows us an overview of the whole of the reality that surrounds us.

In the past every generation inherited the conditions under which it operated, and this inheritance predefined, in principle, its whole existence. Now, on the other hand, every one of us passes through a number of far-reaching transformations in the course of a single lifetime. These are not transitory upheavals, after which one can settle down into a state of repose, but a constant, ever-accelerating vortex that involves all the structures of the civilization. The evolution of science and technology has led to a quantum leap in man's relationship to time. The lead time between a scientific discovery and its use in production has shrunk since the beginning of the century from forty years to ten. This means that the human individual, previously sheltered by the durability of the results of change, can now expect to witness three or four transformations, which may well be quite unexpected, of the nature of production and thus of consumption.

In the scientific and technological revolution, mobility becomes the key human quality. This serves to reinforce the hypothesis of an open and flexible architecture — that is, an architecture governed by the increasing impermanence, complexity, and indeterminacy of the programs to which it works. Many will say that there are other ways of overcoming the conflict between the stable, rigid character of the buildings and the constantly changing demands made by their inmates. It is true that those demands can be influenced by imposing stasis upon them through a policy of zero growth. The experience of capitalist countries in crisis shows that this solution, as propounded *sotto voce* by the rich, is not accepted by the masses. Again, it is possible to assume that people are adaptable enough to live in makeshift houses and tolerate the shortcomings of their environment. The life span of buildings can be further reduced, and throwaway houses can be produced, to be replaced when they wear out. In that case the product would have to be supplied at prices that could be amortized in around ten years. This is unthinkable at present. A switch to cheap, ultralightweight housing would need a transformation requiring a shift of social, cultural, and technological levels for which the preconditions do not at present exist. It would also be immensely expensive. For the immediate future, therefore, this is not a realistic option.

A policy that aims to improve the living conditions of the whole people as far and as quickly as possible cannot restrict itself to new building. New production contributes only partially, and slowly at that, to such an improvement. New building can only achieve an annual growth of between 2 and 4 percent in the housing stock. The problem of inadequate housing cannot simply be solved by replacing old buildings with new ones. There is thus no choice: the whole building stock must be included in the policy. Here two complementary possibilities offer themselves: maintenance, which counters the technological aging process; modernization, which enhances use value.

Maintenance and modernization can no longer be regarded as marginal areas of the construction industry. The improvement of housing depends just as much on the efforts that are expended on the existing stock as on the capacity that is available for the production of something new. The question, however, is this: how are the available financial resources, labor, materials, and brain power, to be fairly divided between new building, modernization, and maintenance?

To plan with future maintenance and modernization in mind is going to upset the apple cart. What will happen is that certain universally practiced construction techniques will be called in question; they will be revealed as the last of architecture's metaphysical illusions, and they will vanish. When that happens, we shall not be entirely empty-handed, thank heaven. Adaptable architecture exists already: in factories, in government and educational institutions. It has been tried out in housing, although with the somewhat different purpose of rearranging a dwelling to suit the occupier's requirements and tastes — both of which, at the moment of planning, remained unknown quantities.

These solutions are characterized by more or less large, fixed structures within which the spatial divisions can be varied. Or, more probably, they consist of a cellular structure with or without a load-bearing framework. There will be other possibilities, as soon as research has progressed beyond its infancy. It is unquestionable that flexibility is not compatible with concrete transverse walls and heavy panels in the usual sizes. We must think up something else. If we apply the formula of a skeleton frame with wide spans and light, movable partitions, we shall have to tackle acoustic problems and problems of fire safety to which there is not, as yet, any adequate solution. The compatibility of heating, ventilation, and plumbing systems will raise other difficult questions. Without coordination of dimensions it is impossible to modify internal layouts or to switch and recombine the elements. Standardization will have to go a lot further. Through standardization it will become possible to cover the costs arising from the use of a more efficient technology.

But flexibility, entailed by the need for modernization, is not the only governing factor. Maintenance must be reduced to a minimum, through new materials, replaceable connections, and accessible installations. One day, maintenance and modernization will no longer be separate activities, tasks that are performed too late. They will blend into a single process of constant transformation. The building will in a sense become a permanent building site. Its architecture will never be "finished."

This revolutionary approach to new building will affect existing buildings too. Instead of seeing a historic building, so called, as perfect, finished, monosemic — in short, as a given fact — we shall see it as an open question that allows of more than one answer, and thus as an invitation to add significance. We shall no longer behave as if ancient stones were untouchable. We shall use light, uncamouflaged devices, which can be dismantled when they have outlived their usefulness, in order to awaken the old building to new life. Architecture will be desacralized; it will be emancipated from individual creativity and once more become something for everyone to handle. All this on one condition: that the revolution does not remain a purely technological and scientific one.

Paris, 1987

# The End: A Record

As a private school the HfG enjoyed freedoms that were denied to public institutions, but from the very beginning it came under financial pressure. For even though the HfG employees (instructors, workshop leaders, and others) worked on ill-paid short-term contracts, the Scholl foundation was not able to finance the College on its own. It needed additional funds from public and private sources. And this – in a country where there was no tradition of pride in private institutions – was bound to be a constant source of conflict.

The freer the HfG became, and the more clearly it struck off down a path of its own, the less its paymasters liked it, and the worse its financial crisis became. By 1963, at the latest, it was clear to all concerned that closure of the HfG was a distinct possibility; and by 1967 the position was seen to be hopeless. Misunderstanding piled on misunderstanding. The delicacy of the situation tended to dramatize everything that was done at the College, and even well-meant attempts to resolve the crisis became suspect. When the end came, it came in a rush.

The following abbreviated factual record may serve to present this terminal phase of Ulm in all its hectic drama:

Early 1967
The Federal Government adopts the so-called Tröger report, which states that the HfG is not a shared state and Federal concern, because culture is a matter for the individual states. As a result the Federal subsidy of DM 200,000 is withdrawn. The Baden-Württemberg legislature (the Landtag) refuses to raise its contribution by a corresponding amount. The financial crisis seems hopeless; it is aggravated by an existing burden of debt. This leads to drastic cuts in faculty posts and therefore in classes. The Filmmaking department makes itself independent.

March 1967
The Scholl foundation recommends the closure of one department (Industrialized Building). An HfG committee works on drastic economy measures.

Late 1967
Herbert Ohl becomes Rector of the HfG. The HfG proposes a state takeover as a last resort.

December 7, 1967
The Baden-Württemberg government recommends that the HfG be amalgamated with the Ingenieurschule Ulm.

February 13, 1968
The architecture faculties of the Federal Republic of Germany declare their support for the HfG.

February 29, 1968
The Deutscher Werkbund, the Bund deutscher Architekten (BDA), and the Verband deutscher Industrie-Designer (VDID), vote against a merger between the HfG and the Ingenieurschule.

March 1968
It is decided that in the case of a forcible amalgamation with the Ingenieurschule the HfG will disband. The Scholl foundation decides to close down on September 30, 1968.

# The Constant and Catastrophic End

Michael Erlhoff

In April 1968, issue number 21 of the periodical *Ulm* appeared with a black cover and an editorial announcement signed by Gui Bonsiepe and Renate Kietzmann (and dated February 23): "With this 21st issue the review *Ulm* ceases publication as the official organ of the Hochschule für Gestaltung in Ulm. The editors do not know whether the publication of the review will be continued at a later time under different circumstances, under the same or another title, with the same or different objectives, with the same or different attitudes."

The finale had begun. The financial support from the Scholl Foundation, inadequate in any case, was endangered, and the Baden-Württemberg government was prepared to offer support only if assigned the same powers of intervention and control as in other specialized institutions of higher education.

The finale: "Individuals are not responsible for this sad ending, since a decision of this kind is the result of a large variety of factors. When disintegration reaches a certain point, hard fact takes over; the outsider's objective view takes on a life of its own and comes to appear absolute: the iron law of budgetary necessity. Even so, it is wrong to allow oneself to become bogged down in externals: behind every objectivity there stands a subjectivity." *(Ulm,* 21.)

But however abruptly the harsh reality may have loomed, the end had always been foreseeable. For the HfG had always been an anomaly in its own time, had always gone its

March 6, 1968
In testimony to a committee of the Landtag, Stuttgart University declares itself in favor of a link with the HfG.

March 19, 1968
The Senate resolves to create a body consisting of equal numbers from both institutions that is to work out, by June 15, 1968, the structure and constitution of a future HfG. The existing constitutional and decision-making structure of the HfG is dissolved in favor of a university council.

May 1968
Attacks by students on instructors intensify. The HfG puts forward four closely argued options for the continued existence of the College.

May 26, 1968
Founding of the Gesellschaft zur Förderung der HfG Ulm e.v., a nonprofit company to promote the HfG. The students' union at the HfG adopts the aims and categories of the Sozialistischer deutscher Studentenbund (SDS). There is a total breakdown of relations between students and faculty.

June 19, 1968
The Kleiner Senat decides to send the students of two of the years out to do practical work.

June 27, 1968
The students refuse further cooperation with the Scholl foundation. Strike.

June 28, 1968
The instructors produce a plan for cooperation with Stuttgart University on a postgraduate level. The students send their own separate representatives to Stuttgart.

Summer 1968
The Landtag votes once more to maintain the subsidy of DM 900,000, but ties this to a seven-point ultimatum — among other things the Christian Democrat caucus demands the introduction of equal student participation. The students take up a negotiating posture hostile to the faculty. The HfG is renamed by its students the "Karl-Marx-Schule."

October 1968
In the absence of the personal and financial provision necessary for a resumption of a full pro-

own way, and had thus always been vulnerable both in its internal and external relationships. From the very start, it came under political — and thus also economic — pressure. Bonsiepe formulated this in a summary of the HfG's position in that same farewell issue of *Ulm*: "The continued existence of the HfG was being called in question before the building was completed. Alongside the political hostility inspired by the HfG's ingrained anti-Nazism, there was and is an element of simple provincialism, obtuseness, and cultural conservatism. This is one institution that has never fitted into traditional cultural patterns. There can be no place for environmental design in a view of culture that concentrates on educating economically independent individuals and leaves society to take care of itself."

Through its connection with the Scholl Foundation, the HfG had accorded itself an unusual degree of autonomy that gave it a special experimental status in institutional — as in other — terms, and that was bound to inspire some nervousness in a provincial government conscious of its constitutional responsibilities. By 1963, accordingly, that government, part of a now confident and strong-seeming Federal Republic, intervened to give itself a voice in the constitution of the HfG and thus curbed its autonomy. Then, in 1968, it offered the HfG the bleak alternative of closing down or being absorbed at a departmental level into the existing system of specialist higher education. This was unacceptable and gram of instruction, the instructors refuse to teach. The Rector announces that without the necessary resources and teaching posts the work of the College cannot be properly carried out.

November 13, 1968
Stuttgart University puts to the Grosser Senat a resolution providing for the absorption of the HfG. The HfG regretfully declines, on the grounds that no agreement with the students is any longer possible.

November 28, 1968
One day before the crucial Baden-Württemberg cabinet meeting, students and instructors agree on the last outstanding demands of the Landtag's seven-point ultimatum. The Landtag votes to close the HfG.

December 5, 1968
The Baden-Württemberg premier, Hans Filbinger, declares: "We want to make something new, and for this we need to liquidate the old."

# Bauhaus Weimar Exodus 1

# Hochschule für Gestaltung Ulm Exodus 2

Schon zu Beginn der damaligen »Systemzeit« wurde den Mitgliedern des Bauhauses in Weimar das Leben erschwert und schließlich unmöglich gemacht. Sie zogen eine Konsequenz:
Sie zogen aus.

Sie fanden eine neue Bleibe in Dessau, wo sie unter besseren Umweltbedingungen ihre Arbeit fortsetzen konnten.

Die HfG steht heute vor einer ähnlichen Alternative wie die Bauhäusler. Den an der HfG Arbeitenden wurde am Montag, 19. Februar 1968 unterbreitet, daß eine Kündigung der Arbeitsverhältnisse zum 30. September 1968 stattfinden wird. Sie müssen (und dürfen) sich also nach einem neuen Betätigungsfeld umsehen. Denn als was auch die Operationen zur Weiterführung der HfG deklariert werden – als Sanierung, Rettung, Konsolidierung, Verschmelzung in einen Dachverband mit einer anderen Institution – und was auch die Details solcher Pläne sein mögen, eines steht fest:

Von der HfG bleibt, wenn überhaupt, allenfalls der Name übrig und eine Erinnerung an bessere Zeiten. Sonst nichts.

Deshalb suchen die Mitglieder der Hochschule für Gestaltung Ulm eine neue Möglichkeit zu arbeiten, und zwar in einem Klima,
das ihnen gestattet, nicht nur zu überleben nach fremden Vorstellungen, sondern zu leben nach eigenen Gedanken.

Sie bringen einige Ideen und Erfahrungen mit aus einer Institution, die einmal als eine der besten und fortschrittlichsten Designhochschulen der Welt bezeichnet wurde. Sie möchten Produkte entwerfen, Filme machen, visuelle Kommunikation betreiben, Architektur planen, Theorien entwickeln, eine Designwissenschaft vorbereiten.

Zu fragen ist:
welche Person,
welche Institution,
welches Unternehmen,
welche Stadt
welches Land,
welcher Staat
bietet der HfG einen Aufenthalt, um produktiv tätig sein zu können und fortzusetzen, was so vielversprechend begann?

Bitte wenden Sie sich an die Hochschule für Gestaltung, 79 Ulm, am Hochsträss.

19. Februar 1968

[redacted] Dozenten, Assistenten, Studenten, Mitarbeiter und alle **wirklichen** Freunde der Hochschule für Gestaltung

would have debased the College into just another public educational institution. And so the HfG chose – if you can call it a choice – the end.

As always, it was economic problems that had produced this dependence on the Baden-Württemberg government. In the long run, the Scholl Foundation was not able (and at times it was not willing, either) to support the HfG and its inexorably growing budget. In 1968 it was no longer possible to work within the financial resources that had been just adequate in the 1950s. And so what remained was the purely hypothetical choice between a closer link with industry, which – supposing it to have been feasible – would have turned the HfG into a mere appendage of capitalist interests, and a total submission to state authority, with all that entailed in the shape of control of the form and content of the College's teaching.

Gui Bonsiepe again, in his wise position paper, "ber die Lage der HfG" ("On the Situation of the HfG"), in *Ulm*, 21: "Freedom means first of all economic freedom. And this was in a precarious state from the very start. The assumption that a training institution could and should be financed by accepting commissions to work for industry was an erroneous one. Training cannot be self-financing. The HfG has therefore had to depend on public monies, and relies on the understanding and goodwill of the elected representatives of the people. Neither understanding nor goodwill has been forthcoming without

Protests at the opening of the Bauhaus exhibition in Stuttgart, 1968
Walter Gropius with bullhorn

strings attached. The responsible bodies have often voted to continue grants only in the teeth of determined internal opposition and by narrow majorities. As the international reputation of the HfG has grown, the scope for responding to the demands attendant on that reputation has shrunk: totally inadequate resources have made a mockery of the College's aspirations and ambitions."

It is true that the students and instructors at the HfG might have put up more show of resistance — or even subversion — than they did. But an additional problem was that the HfG had embroiled itself in conflicts and unresolved issues generally — including the central issue of social commitment — to such an extent that it was bound to come to grief. The "wishful dream that a world of commodities might somehow grow rational" (Stanislaus von Moos) turned out to be an illusion, if an indispensable one. For a long time, those at the HfG succeeded in concealing this fact from themselves, by eclectic strategies and methods and by trying to personalize issues that were really structural ones. By 1968, with a widespread movement of political protest going on all around, the HfG's central contradiction could no longer be sustained. Suddenly, everything was open to debate, certainty was a thing of the past, and contradictions were exacerbated to the point where explosion became inevitable. The protest movement — and with it some at least of those at Ulm — took the ostensible

Tomàs Maldonado

# Looking Back at Ulm

The products of the HfG are fairly well known, and so are the texts in which the College's didactic ideas are set out in theoretical terms.

Little or nothing is known, on the other hand, of the cultural influences that explicitly or implicitly formed the basis of that conception. The fact is that this important aspect was in a sense smothered by the importance attached to the theoretical issues attendant on the relationship between Ulm and the Bauhaus. This had a distorting effect: it is simply not true that the theoretical framework for the HfG was exclusively the product of our in-house discussions of the Bauhaus. I am convinced that, quite apart from this debate, there were other disciplines and theoretical trends that in their own way had a decisive influence on our understanding of design and of design education.

What must be remembered is not only the limitless curiosity that we had in those years about anything that was — or seemed — new. That was a feverish, insatiable curiosity directed above all at the new disciplines that were then coming up: cybernetics, information theory, systems theory, semiotics, ergonomics. But our curiosity went further than this: it also extended, in no small measure, to established disciplines such as the philosophy of science and mathematical logic.

The mainspring of all our curiosity, our reading, and our theoretical work was our determination to find a solid methodological basis for the work of design.

This was a highly ambitious undertaking, admittedly: we were seeking to force through, in the field of design, a transformation equivalent to the process by which chemistry emerged from alchemy. Our efforts were, as we now know, historically premature. The bits of methodological knowledge that we were trying to absorb were too "hand-crafted"; and our instrumentation was virtually nonexistent. We did not have what we have today: the personal computer.

We also lacked a full understanding of the notion of "limited rationality," which Herbert Simon was just beginning to develop. And so we remained prisoners of the theoretical generalities of a form of "problem solving" that was nothing more than a Cartesian "discourse upon method."

But in the midst of our limitless faith in method — and we were already dimly aware that it might have a negative aspect in "methodolatry" — there lay some powerful intuitions that the evolution of information technology, especially since 1963, has to a large extent confirmed.

There is no doubt, for instance, that the idea that many of us had in the 1950s, that a kind of symbiosis of calculation and graphic representation could be created within the process of problem solving, is basic to the widespread present-day use of computer graphic techniques. We also deserve some credit for having foreseen, in 1958, the influence that miniaturization would have on the whole range of objects produced by industrial society — as in the enormous changes that are taking place today under the influence of microelectronics.

However, it must be said that our theoretical constructs did not always have such good fortune. Some of our formulations may well seem disarmingly self-evident to the present-day reader. But those were far-off days, and a certain naiveté in expression was the norm.

values of bourgeois society literally, and landed in a morass.

And so — after the death of one of the HfG's major design practitioners, Hans Gugelot, at the end of 1965 — the College's fundamental theoretician, Maldonado, left in June 1967; and by the time the College closed in 1968, Otl Aicher was deeply involved with the commission to provide the visual design for the 1972 Munich Olympics, which would have made it impossible for him to go on working regularly at the HfG. By the time the HfG was closed down, it was already breaking up.

Even so: with new instructors and new students, it might have gone on evolving and finding new forms and new criteria (as in Bonsiepe's idea of an "environmental institute"). There was enough intellectual, political, and practical potential there to create a utopian institution.

Furthermore: "You don't shut down a college" (Tomàs Maldonado), and certainly not an internationally respected and important cultural institution. But the Premier of Baden-Württemberg at that time, Hans Filbinger, who had once been a judge in Fascist Germany, brought the existence of the Hochschule für Gestaltung to an official end on December 5, 1968, with the very German words: "We want to make something new, and for this we need to liquidate the old."

Ideologically, the norm was positive thinking. This goes a long way to explain why in times like the present, in which negativism prevails (or in which people would like it to do so), the things that were said and done at Ulm seem intolerably high-minded at times. The criticism is partly justified, but only partly. Positivism at Ulm was never conformist, but always critical. This is something you cannot say of some of today's negativism.

It must be admitted, however, that many of us were predisposed to assume the mantle and the language of the preacher – to be, in short, more Catholic than the Pope. Perhaps this was just because we believed in our ideas so fervently. It must be said that this sort of attitude is fast disappearing at the present time. It was an attitude that often gave us the conviction that we were, in good faith, the bearers of a gospel of salvation.

But there were fine distinctions among us. I personally, for instance, always rejected the entrenched belief, which is still very widespread, that "design" is a means toward the global redemption of industrial society. I have, in fact, always rejected "design chauvinism" in all its forms.

We lived and worked in relative isolation on a mountain, and it was hard to evade the temptation to be a Zarathustra, dropping admonitions, exhortations, and calls to action on people's heads from a great height. In fact, that is why we at Ulm were so solemn sometimes. Without bitterness, however. And though our ideas may sometimes have been overbold, they were never wild.

Ulm was based on one basic idea, which we all shared in spite of disagreeing on absolutely everything else: the idea that industry is culture, and that there exists the possibility (and also the necessity) of an industrial culture. Some of us, however – including myself – accepted this premise only with reservations.

At that time I was particularly receptive to some of the thinking of the Frankfurt School. Although my own cultural orientation was strongly marked at that time by Neopositivism (I was eagerly reading Carnap, Neurath, Schlick, Morris, Wittgenstein, Reichenbach, etc.), the presence of Adorno in Frankfurt represented for me, as it were, a contradictory intellectual stimulus.

I have to confess that his impressive speculative fertility, his complicated and somewhat cryptic way of writing, his telling and sometimes provocative aphorisms exerted a fascination on me that was anything but rational: "The useless is eroded, aesthetically inadequate. But the merely useful lays waste the world," he once said to me in an attempt to cool my enthusiasm for the industrial culture of usefulness. This was a recurrent theme in his thinking, at that time, and he took it up in similar terms in his memorable lecture on Functionalism to the Werkbund in Berlin.

These and other reflections in the spirit of Adorno, and later also of Habermas, led me to examine the relationship between industrial culture and the culture industry, and to undertake a critical investigation of the role played by "design" in between these two realities.

When people ask me, as they often do, about the contemporary relevance of Ulm, my answer has to be hedged about with qualifications. Times have changed – it would be foolish to ignore the fact. But the problems are ultimately

There have not been many unorthodox, imaginative new foundations in German higher education since the war. The HfG was certainly one of them. It was a place for creative initiatives. That is why so many people denounced it.
Ralf Dahrendorf, 1976

the same, or almost the same. I would say that Ulm remains a convincing model if one thinks of the most highly developed sectors of industry, the sectors of instrumentation, whether for production or communication. On the other hand, Ulm is an outdated model if the reality that is under examination is the one that we at Ulm tended, if not to neglect, at least to underestimate: the reality of those objects that Adorno consigned to the "eroded" realm of the useless. Objects that are needed, all the same. And the attempt to free them from their "eroded" state is a legitimate, even a laudable one. A difficult task, certainly: because the limitless freedom of the useless favors the most elitist extravaganzas. But let us face the truth: it is also difficult to prevent the useful from laying waste the world.

Milan, 1987

# Aftermath

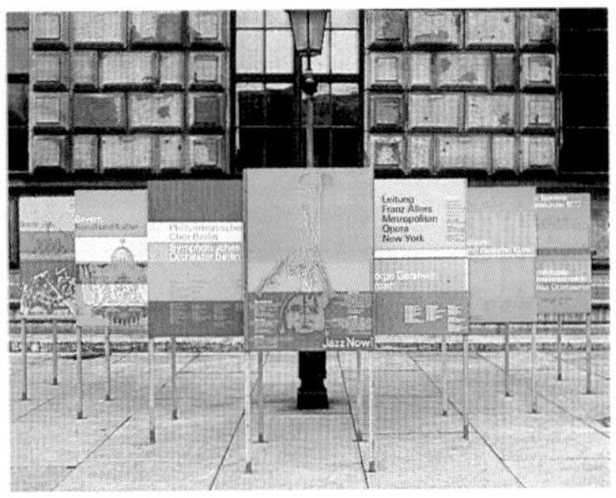

Corporate identity
Olympic Games, Munich
1972
Otl Aicher
Rotis 1972

# Ulm: Not the End

Michael Erlhoff

Only the stupid presumption of a politician could have allowed him to be so sure, in the late 1960s, that he had put an end to the Hochschule für Gestaltung Ulm. As if history were made by men; as if an institution could be casually shifted out of the way.

Self-deception on a megalomaniac scale. Not only does the present Premier of that very same constituent state of the Federal Republic wish he still had the HfG — or else that he could have it back — now that the cultural and economic value of design is being proclaimed on all sides: more significantly, although Ulm as such may have come to an official end, it was not really the end at all. For many of the students, and above all the instructors, at the HfG, the official closure of the College meant a new beginning: a break with monasticism and a plunge into a world teeming with opportunities to spread the Ulm gospel. They scattered like the debris of an explosion; they went out into the world like apostles, to make Ulm universally known. They joined the faculties of universities, design schools, and art academies (in Milan, Paris, Berlin, Hamburg, and elsewhere); they set up their own production facilities and offices, which became new havens of research, trial, design, and teaching; they made Ulm a reality in distant continents; they became consultants for regional, city, and industrial planning, took high office in institutions and professional bodies, wrote weighty books, organized exhibitions and the catalogues thereof,

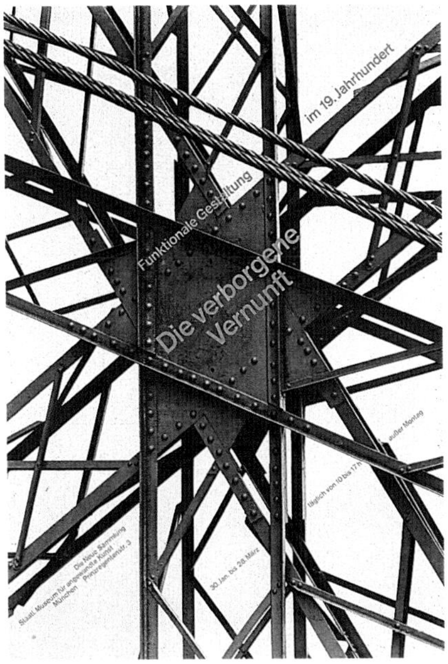


*left*
Computer work station
Hans von Klier
Milan 1983

*right*
Catalogue
Rolf Müller
Munich 1974

Ulm city works department
Hochstrasser, Bleiker
Ulm 1974 – 78

developed designs and design strategies (architecture, theory, the visual *Gestaltung* of everyday life); they gave corporate shape and identity to towns and to corporations. Even if not all its alumni were equally successful, Ulm became a criterion by which everything that came later was judged.

The closure of the College was an issue of principle, and a bitterly fought one; but in fact it was the official end of the HfG that made Ulm official. It was only then, in the process of dispersal and and dissemination, that the quality and the power of Ulm became generally known. So that, at the very least, design and visual communication in the so-called western world, and the training that goes with them, are unthinkable without the influence of Ulmers. These are the teachers and designers, the creators of concepts — and, yes, the custodians of Ulm ideas.

Many Ulmers have become molders of the contemporary image of modernity, thus resolving for themselves the ambivalence — endemic in the HfG itself — between critical detachment and involvement with the status quo. The experiment that was Ulm has become part of normality; it has even become the norm. The rebels have been invested with social responsibilities that they have had no choice but to accept — with more or less sacrifice of Ulm ideas. And so Ulm has come to define the present-day image of good design, *die gute Form*. Meanwhile, largely unobserved by the general public, those

S-Bahn transit system,
West Berlin
Herbert Lindinger
Hanover 1985 – 87

Muscle-powered aircraft
Günter Rochelt
Munich 1984 – 85

who belong to Ulm are still arguing as to what exactly that is. What is at stake, still, is the legitimate succession: a subject that led to controversy after the demise of the Bauhaus, too. There are always plenty of heirs to claim the true inheritance (and the attendant perquisites).

Ulm took root. Without embarking on the futile, purely philosophical or taxonomic controversy over what was directly influenced by Ulm, what was indirectly influenced, and what evolved parallel to it, one can be sure of this: that Ulm was absorbed as a style — how else — or as an obsession; that this became generalized; that it has a conscious or unconscious function as the guiding image of modern design; and that it permeates every department store, every parkland development, every suburb, every utensil. There is no traveler who does not depend on adaptations of Ulm designs to find his way around; no television sport fan who would not lose track without them; no homeowner who does not know the value of prefabricated building components; no artist or art historian who could do without his slide carousel; no medium who spurns the high-tech look; no student of architecture or design who could afford to give Ulm the go-by.

In all this, the HfG inheritance is persistently misinterpreted, finds itself debased into a mere trademark, and most often produces no more than an inferior paraphrase, a second- or third-hand Ulm; but in a sense this is no more than

Chair
Herbert and Jutta Ohl,
for Rosenthal
Darmstadt 1983

*Endless Loop*
Max Bill, for Deutsche Bank,
Frankfurt
Zürich 1987

a normal process. Everything seems to be Ulm these days, and what began as rationality (as manifested at Ulm itself in the founding of the disciplines of industrial design, urban design, design theory, and design planning) has degenerated into mere plausibility. But all this merely reflects the customary plight of those in bourgeois society who are left on stage at the end of the tragedy (an inherently bourgeois art form) and complain that *that* was not what they meant at all; the same applies in many respects to the protest movement of 1968. What is really depressing about this "normal process" is that the true qualities of Ulm products and Ulm ideas threaten to be obscured from view by a mass of standardized pap.

This is why there is still a vague but widespread feeling that we have had altogether too much of Ulm, and that it has deprived us of a cherished – if ill-defined – sense of coziness, beauty, and security. This feeling is the source of constantly renewed attacks on real Ulm design, and on the ideas that go with it. A little Ulm, after all, means no more than "a little peace."

For this reason, over the last few years, the real enthusiasm for Ulm, and the real work on it, have come from what might at first sight seem surprising quarters. For a number of younger people, in particular, Ulm, and the instructors whose names are associated with it, have become such outsize cult objects (or cult figures, as the case may be) that the indebtedness must be worked out

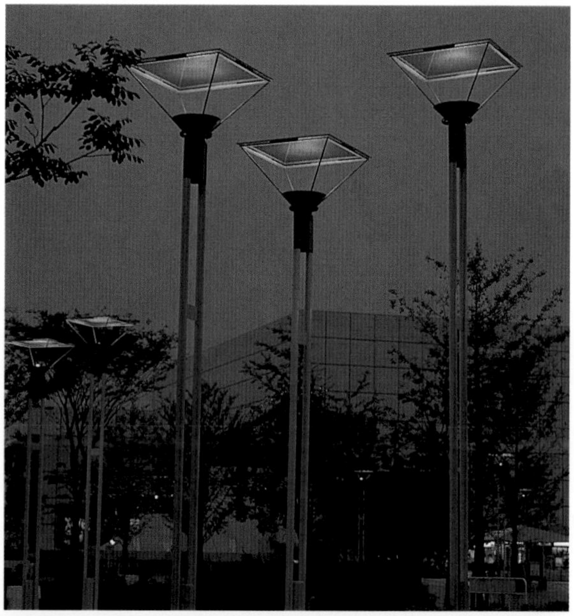

Tsukuba Expo '85:
signs and lighting
Takeshi Nishizawa
Tokyo

Seating system
Klaus Franck, Werner Sauer,
for Wilkhahn
Hanover 1984

and worked through. So intensively has this been done that many of these admirers and disciples have become radicalized in their turn: under the spell of Ulm they have risen in revolt against their own teachers and all that they admire.

These are the people who define — with many a slip — the attempts to create a New German (or another International) Design; and quite often, if you look carefully, they turn out to be wedded to Ulm concepts and Ulm precision. In this way, too, Ulm has taken root — often rather against the better judgment of the Ulmers themselves.

And so the Hochschule für Gestaltung has set its mark even on the opposition to itself. The glib term "post-Ulm" does not (yet) exist; and yet in defiance of its own intentions Ulm has become a myth: today's projective dream of the 1960s. And of course it will come back into fashion — as is only normal, in this world in which time is so quick to devour everything.

Perhaps, however, under conditions that we can only partly define, the rediscovery of Ulm might lead to more: to a serious debate on functionalism, on the object, on modernity, on concepts of training, on architecture; on visuality, perception, and society. If we can be frank about Ulm.

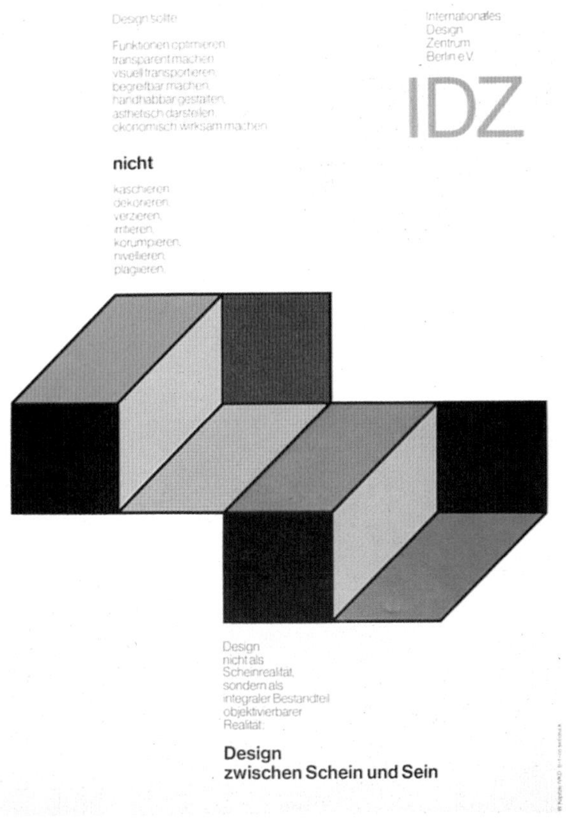

Poster
Herbert Kapitzki, for
International Design Center
Berlin 1976

Personal computer
Ferdinand Porsche,
Dirk Schmauser
Zell, Austria, 1984

Indoor swimming pools
Willi Ramstein
Milan 1976

Folding chair
Andries van Onck
Milan 1983

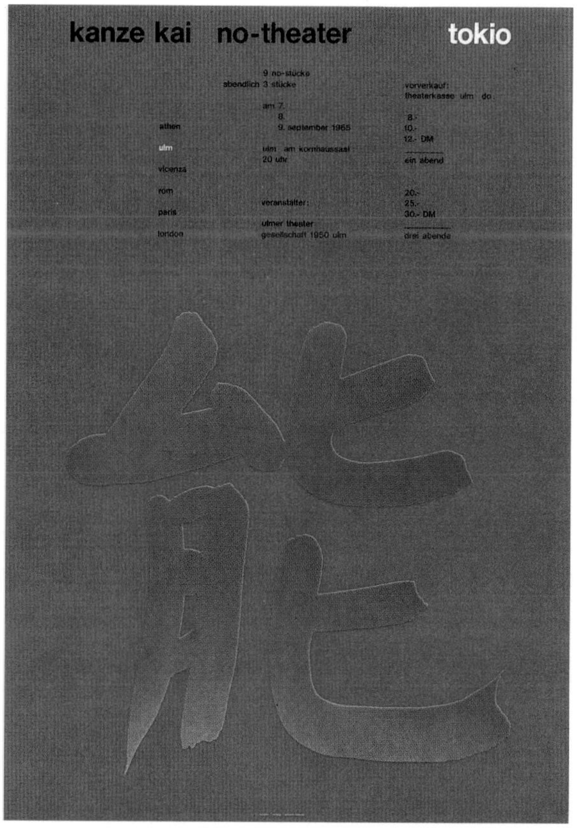


Poster
Almir Mavignier
Hamburg 1965

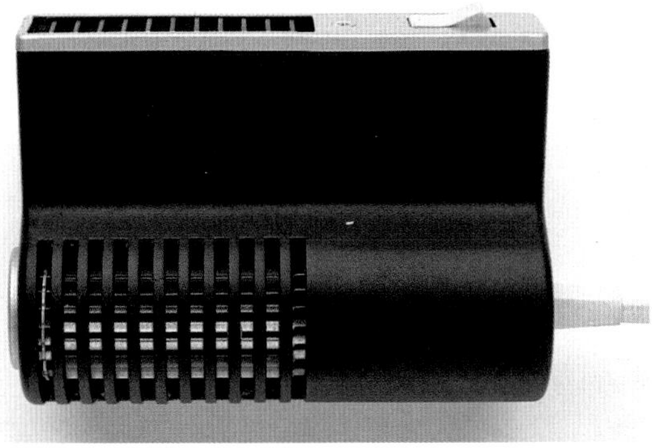

Hair drier
Reinhold Weiss
Braun Product Design
Frankfurt 1963

Fukuwatari Building
Isao Fukuwatari
Tokyo 1984

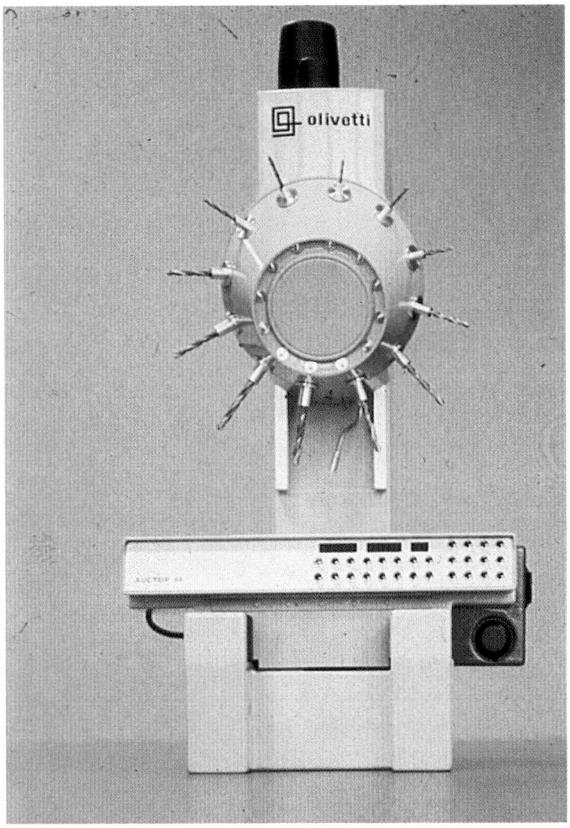


Tool assembly
Rodolfo Bonetto
Milan 1968

# Ulm International

Michael Erlhoff

In principle, the Hochschule für Gestaltung Ulm was a very German institution. It was conceived as a counter to the German political life that had led to a German Fascism; it was rooted in the experience of German history; and it was underpinned by a very German combination of thoroughness, categorical thinking, idealism, utopian imagination, esotericism, and combativeness. It is no surprise, therefore, to find that foreigners, in their reminiscences of Ulm, constantly stress the — to them, often rather exotic — German quality in the HfG.

In principle, the HfG nevertheless also arose from a decidedly internationalist ambition. It was the first significant international institution to appear on the territory of the Federal Republic of Germany. And this was not only because from the outset it made a point of including foreigners as staff instructors, guest instructors, visiting speakers, and students — the people who then spread the Ulm model all over the world — but also through an internationalism that was inherent in the conception of the College.

In practical terms, the HfG represented a continuation of the tradition, rooted in the 1920s, that had become known as the International Style. In addition, Ulm was a part of a seemingly universal pattern of cultural stirrings that, after a long period of chauvinist and nationalist upheaval — which in the 1930s and 1940s was far from restricted to Germany alone — now enthusiastically pursued and praised

Building on Pont des Sauges
Prefabrication technique by
Yokoyame and Gilliard;
architects Calame, Schlaeppi
Lausanne 1968 – 71

Track-laying vehicle
Zlatan Medugorac
Ulm 1981

internationalism. There were many projects that envisaged supranational processes of cultural development.

This internationalism, which harked back to the cultural achievements of the 1920s, culminated in the hope of creating a common language that would transcend all frontiers. Only the destruction of language barriers could create common ground. There were a lot of elements involved in this. On one hand there was the Cold War, and the interests of the two emerging power blocs in consolidating their position as occupying and dominant powers – both laying claim to an internationalism that was in practice a cloak for economic and political power games, an exploitation of the enthusiasm of those who had a genuine interest in breaking down barriers. On the other hand there were all those energetically pursued utopias, such as a United Europe and other new sociopolitical dispensations worldwide: the UN, the Treaties of Rome, international art groups (like Zero and Cobra), the Esperanto revival, international trade, the International Style, the spread of the English (or rather American) language in the western world, "industrial design," abstract and concrete art as international formal languages, concrete poetry, and the extension of the concept of language under the influence of the revived science of linguistics.

If language could throw off its lexical chains and work in terms of phonetics, semiotics, and semantics, then – or at least so it seemed for a while

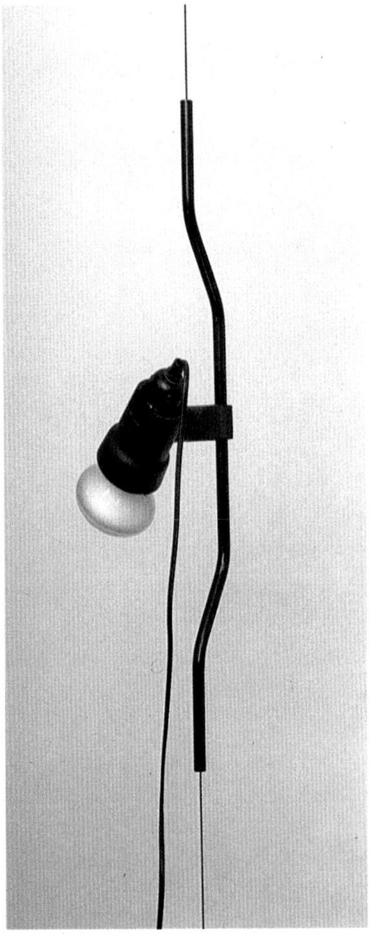

Table fan
Reinhold Weiss
Braun Product Design
Frankfurt 1960

Magazine cover
Tomàs Gonda, for *Casabella*
New York 1981

Video display unit
work station
Horst Diener,
for MBB-Sysscan
Ulm-Gögglingen 1984

Parentesi adjustable
pendant light
Pio Manzú, Achille Castiglioni,
for Flos
1971

— the language barriers, all of them inherently lexical, could speedily be overcome and marginalized. What was more, it was possible (or conceivable) that an international language of signs might be devised in which architecture and design could play their part.

Alongside these new — or newly rediscovered — linguistic devices for overcoming national compartmentalization, it was concrete poetry, above all (with its links to phonetic and visual poetry), that set out to destroy the boundaries of national literatures and at the same time demonstrate the essentially international nature of speech and language. Bense and Gomringer are the first names that come to mind, but there also were the members of the Brazilian group Noigandres; the Dutch poet Paul de Vree; a number of British poets such as Iain Hamilton Finlay and Bob Cobbing, or the Frenchmen Henry Chopin and Bernard Heidsieck, and the American Emmett Williams, who edited the *Anthology of Concrete Poetry*. These were truly international movements, and a network of international communication extended across all frontiers — even across the Iron Curtain to Poland, Hungary, and Czechoslovakia. This phenomenon is most strikingly evident in the magazines of the period: as when *Los huevos de la Plata* in Uruguay juxtaposed texts by concrete poets from Europe and South America, or when a Belgian or English periodical printed contributions from every country under the sun.

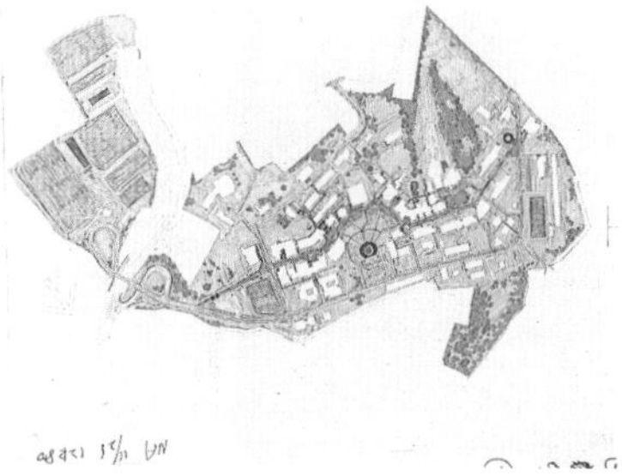

Yokohama University,
with site plan
Shoichi Kawai
Tokyo 1983

The HfG played its part in this widespread campaign to overcome national mental limitations, and promoted it to the limit of its powers. To this day, from an internationalist point of view, it can be seen to have been an exemplary place. And so, in the memory of Ulm, the utopia of a true internationalism lives on.

Magazine design
Karl Heinz Krug, for *form*
Leverkusen, from 1962

Intercity Experimental train
Alexander Neumeister,
for Deutsche Bundesbahn
Munich 1983 – 85

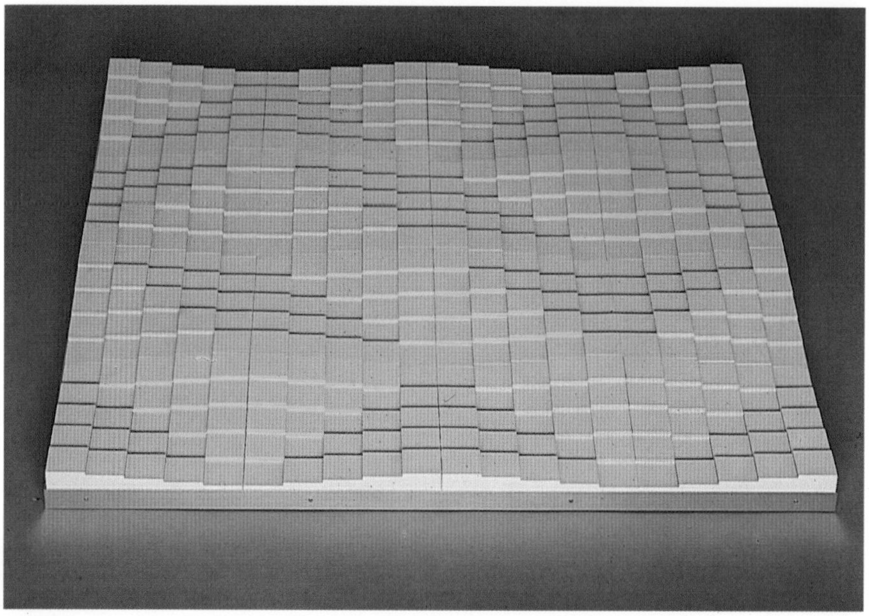

Relief: Color Shadow 62
Shizuko Yoshikawa
Zürich 1978

Sign system for
University Wuppertal
Helmut M. Schmidt-Siegel,
Eckhard Jung, Franco Clivio
Düsseldorf 1975

Audi 80/90
Horst Kretschmar
Ingolstadt 1983 – 86

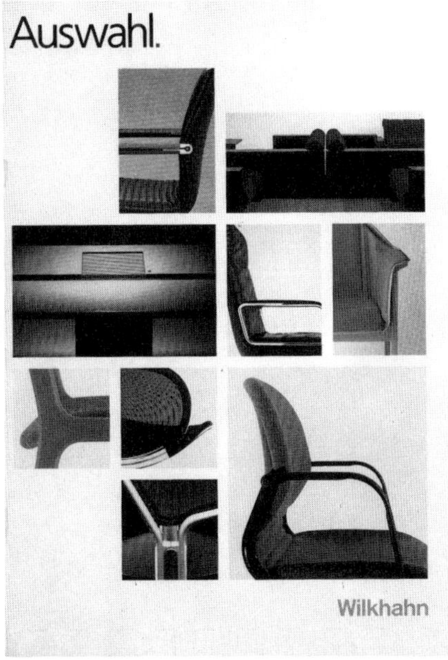


Corporate Image
Herbert Kapitzki,
for Historisches Museum,
Frankfurt
Berlin 1971

Corporate Image
Tomàs Gonda, Bernd Franck,
Rolf Müller, Klaus Franck,
for Wilkhahn
Hanover/Düsseldorf,
from 1964

Swiss National Exhibition
Max Bill
Zürich 1964

Faucet assembly
Michel Millot,
for Solus Usus Design
Paris 1985

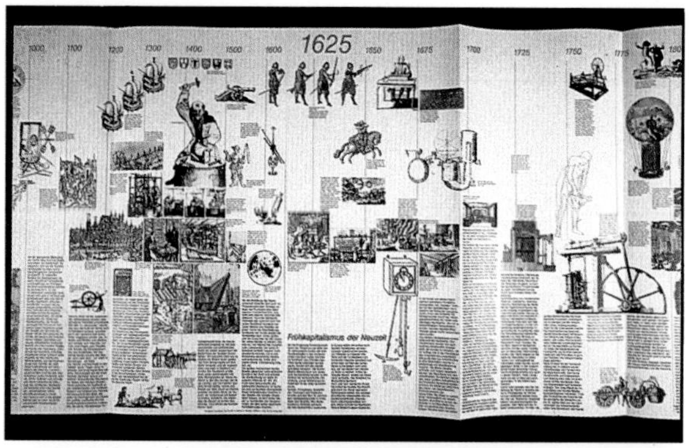


Synoptic chart of industrial history
Michael Klar
Schwäbisch-Gmünd 1982

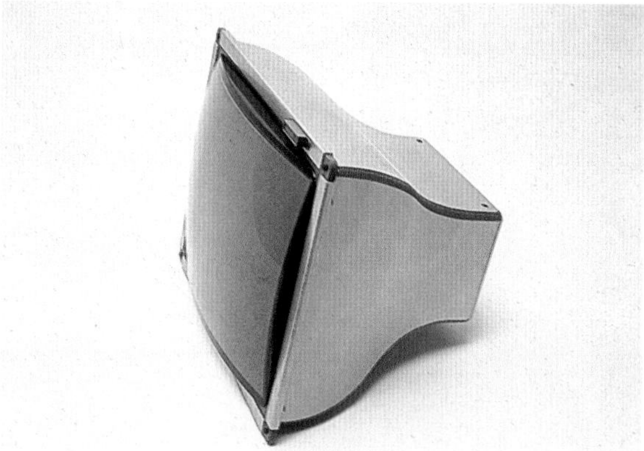

Television set
Bernd Meurer
Darmstadt 1985

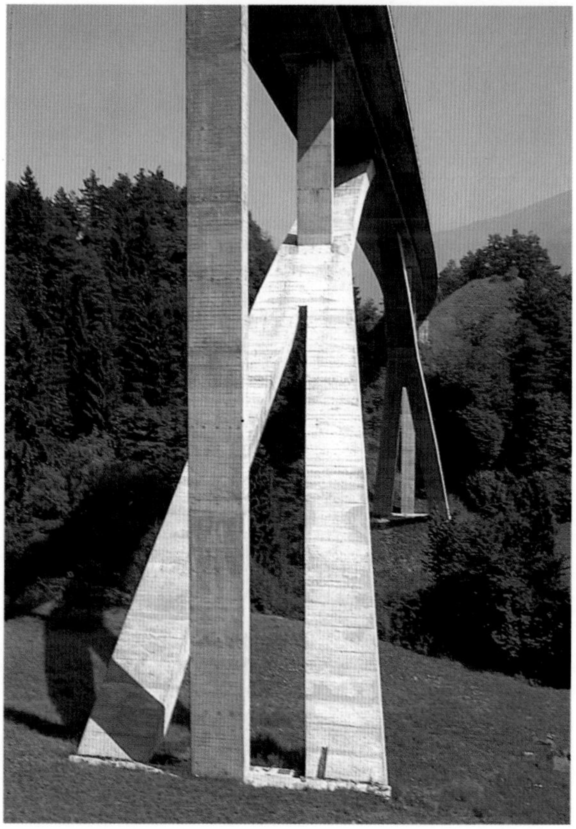

Bridge over Lavina Tobel,
Grisons, Switzerland
Max Bill
Zürich 1966 – 67

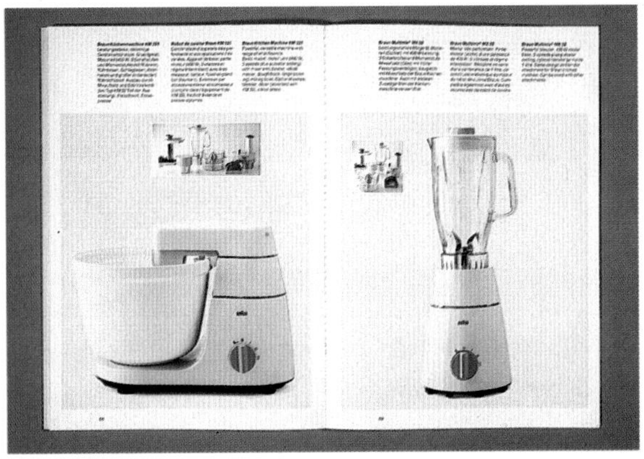

Hose connection system
Dieter Raffler, Franco Clivio,
for Gardena (Kress und
Kastner)
Ulm 1968

Catalogue
Walter Müller, with
Dieter Skerutsch, for Braun
Frankfurt 1984

Kyma toy
Olivio Ferrari, for Naef
Blacksburg 1980

Catalogue
Walter Müller, with
Dieter Skerutsch, for Braun
Frankfurt 1984

Corporate image
Hermann Ay, Josef Breuer,
for the city of Ulm
Augsburg 1986

Mensuration center
Rodolfo Bonetto, for Olivetti
Milan 1965

Typewriter keyboard
Hans von Klier,
Ettore Sottsass
Milan 1965

Distorted Circle
Almir Mavignier
Hamburg 1962

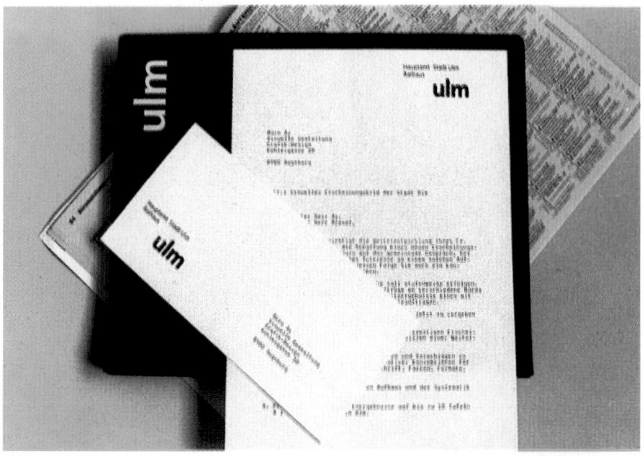

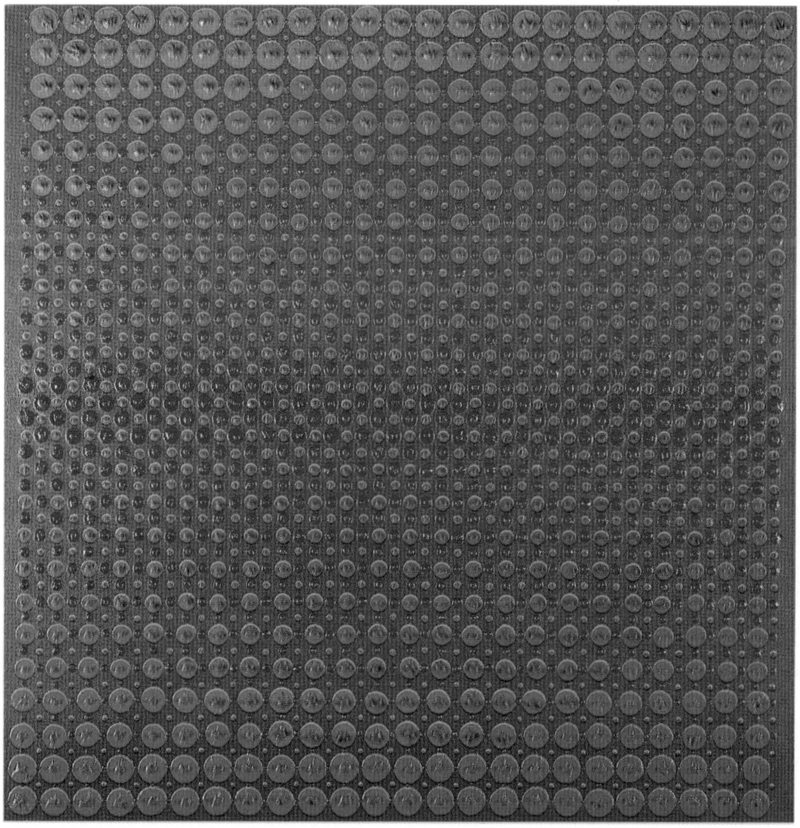

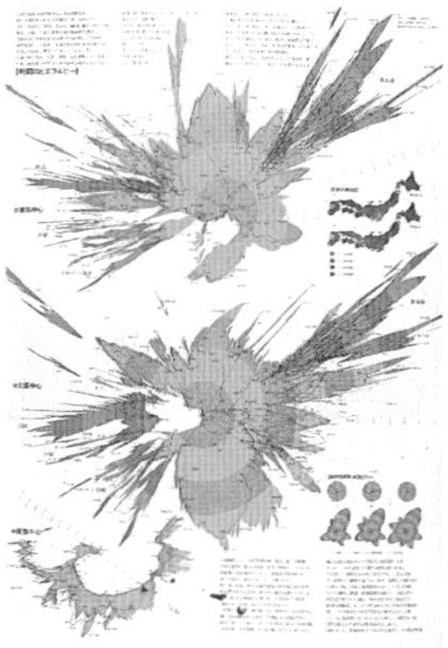

Mass seating
Karl-Heinz Bergmiller
Rio de Janeiro 1982

Time Axis Map of Japan
Kohei Sugiura, with Nobuo Nakagaki, Osamu Takada
Tokyo 1969 – 72

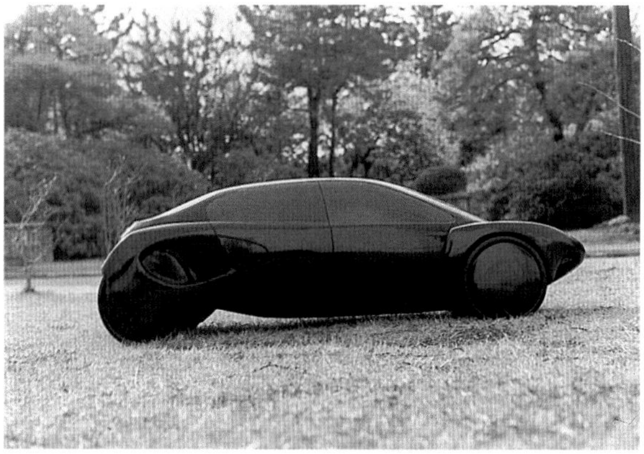

Carbon fiber study
Norihiko Mori
Chibashi 1985

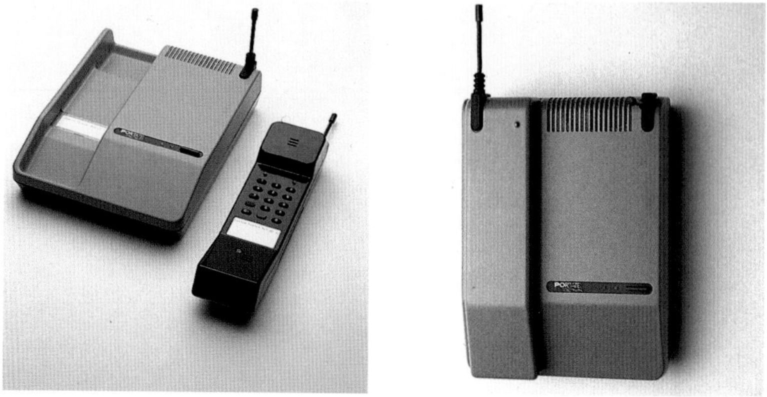

Cordless telephone
Werner Zemp, for Autophon
Zurich 1986

Machine factory
Fred Hochstrasser,
Hans Blieker, Walter Custer,
for Heberlein & Co.,
Wattwiel, Switzerland
Ulm 1964 – 70

Corporate Design for the
University of Bologne
Design: Giovanni Anceschi
1988

Pocket lighter
Frank Hess
Ulm 1975

Home computer system
Dieter Lassmann
Urbigny/Lormes 1984 – 85

Toy
Michel Mellot,
for Solus Usus Design
Paris 1985

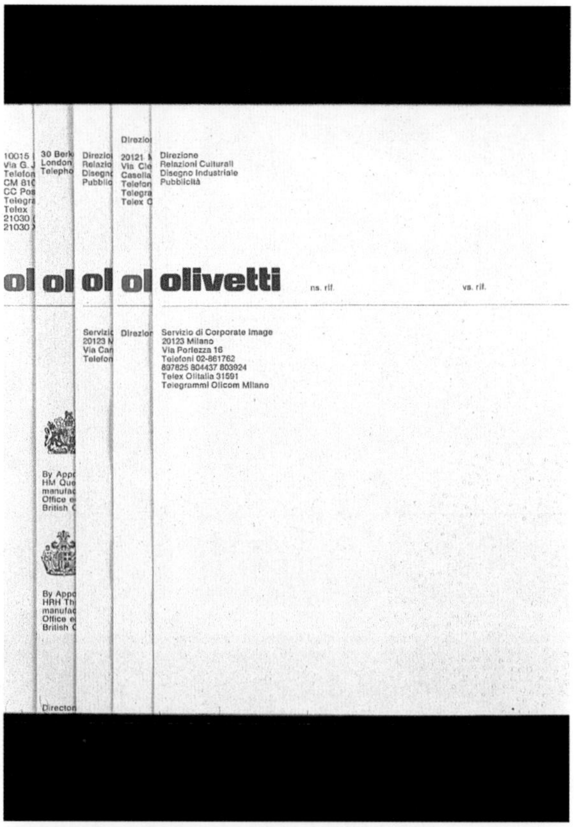

Corporate Image
Hans von Klier, for Olivetti
Milan 1970

Corporate Image
for the Olympic Games
Los Angeles 1984
Sussman-Prejza L.A.

West German standard
city bus
Herbert Lindinger
Hanover, from 1973

Office chair
Klaus Franck, Werner Sauer,
for Wilkhahn
Hanover 1980

Poster
Rolf Müller, for Kiel Festival
Munich 1972

Corporate image
Peter Seitz,
for Minneapolis Zoo
Minneapolis 1979

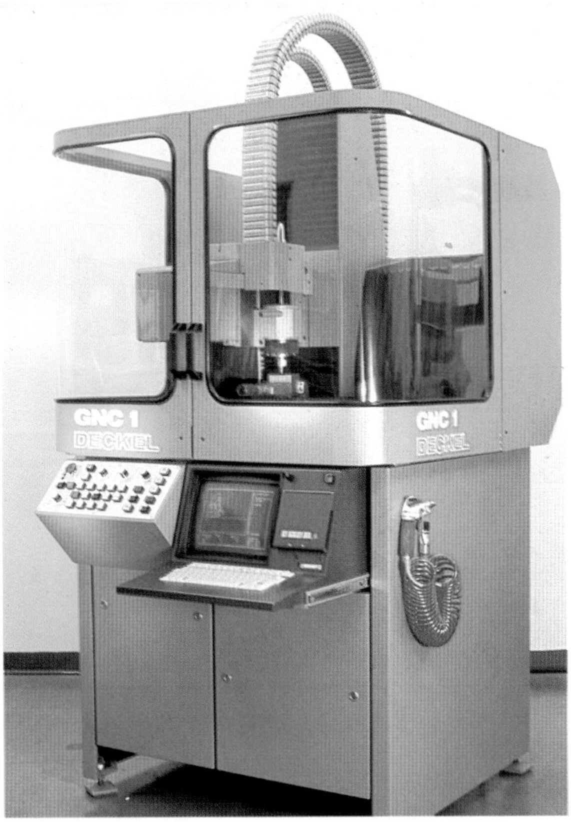

Computer-controlled
engraving machine
Peter Hofmeister, for Deckel
Munich 1984 – 85

Ski binding
Dieter Raffler, for Geze
Neu-Ulm 1982

Portable charcoal brazier
Sudhakar Nadkarni
Bombay 1976

Computer system
Eric Brenzinger, for Unidata
Paris 1972 – 74

School desks
Karl-Heinz Bergmiller
Rio de Janeiro 1970

Memorial Hall
of Kunio Yanagida
Isao Fukuwatari
Tokyo 1975

Luisenplatz, Darmstadt
Herbert Lindinger,
Eric Boss, René Mitscherlich,
Eduardo Vargas
Hanover 1978

Stage set
Frank Hess, Ralf Milde
Ulm 1981

Automobile interior
Rodolfo Bonetto, for Lancia
Milan 1984

Playbill, part of
corporate image
Hermann Ay, Josef Breuer,
for city of Augsburg
Augsburg 1981

Microfiche readers
Dieter Lassmann
Urbigny/Lormes 1976

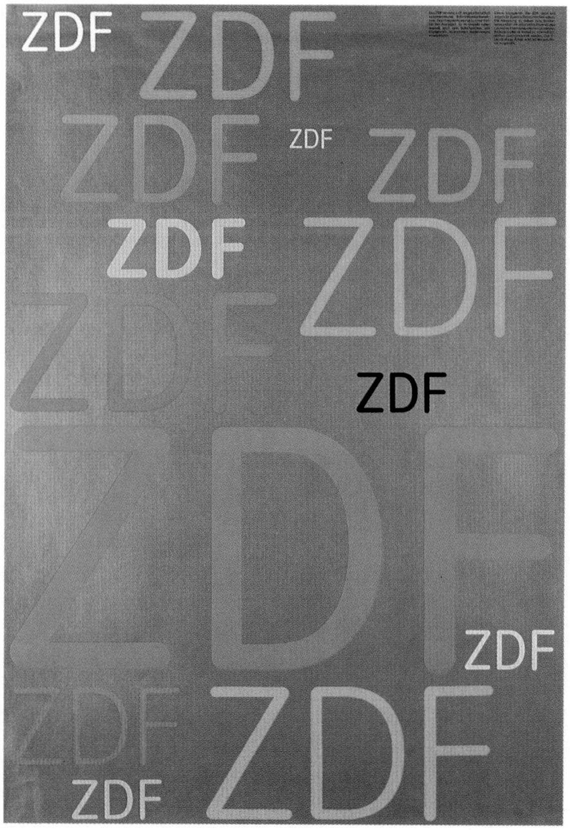


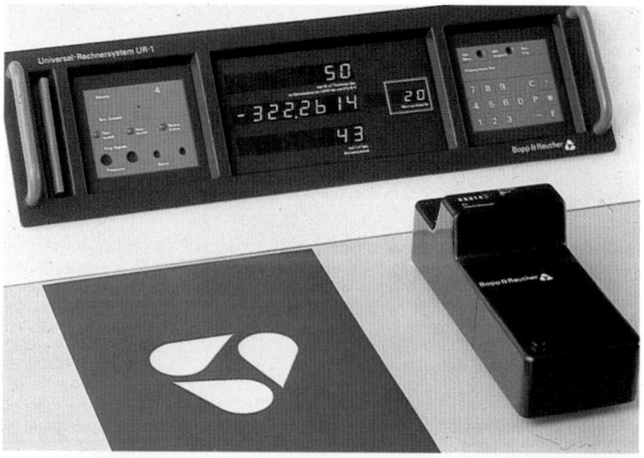


*center right*
Corporate design
Franco Clivio, Rolf Müller,
for Bopp und Reuther
Zürich/Munich 1980

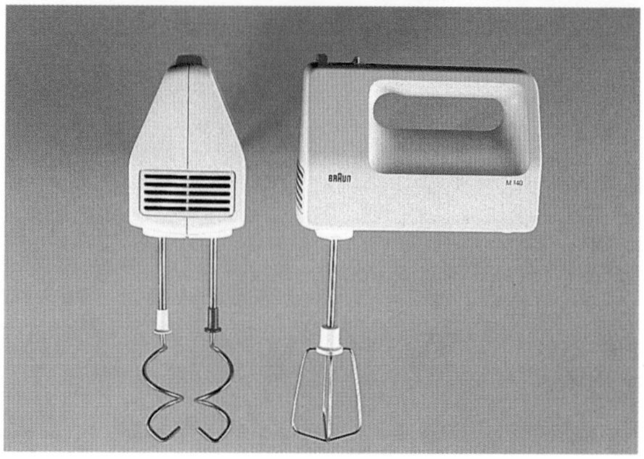


*below right*
Hand mixer
Reinhold Weiss, for Braun
Frankfurt 1965

*left*
Corporate image
Otl Aicher, for ZDF Television
Rotis 1973 – 75

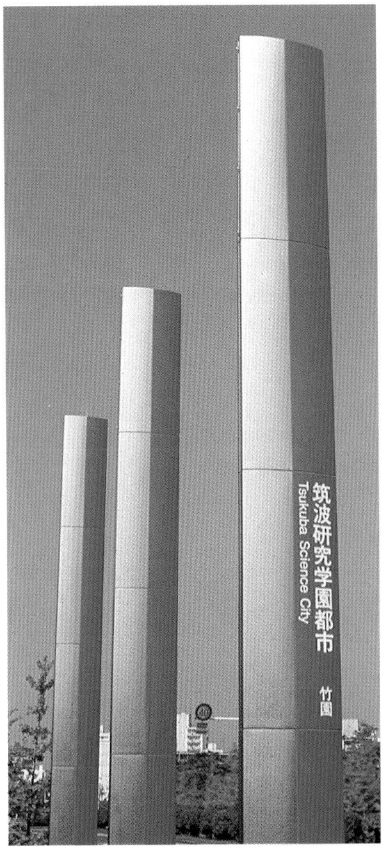


Chair
Herbert Ohl, for Wilkhahn
Darmstadt 1978

Exhibition stand
Hans von Klier, for Olivetti
Milan 1983

Sign system
Takeshi Nishizawa,
for Tsukuba Science City
Tokyo 1985

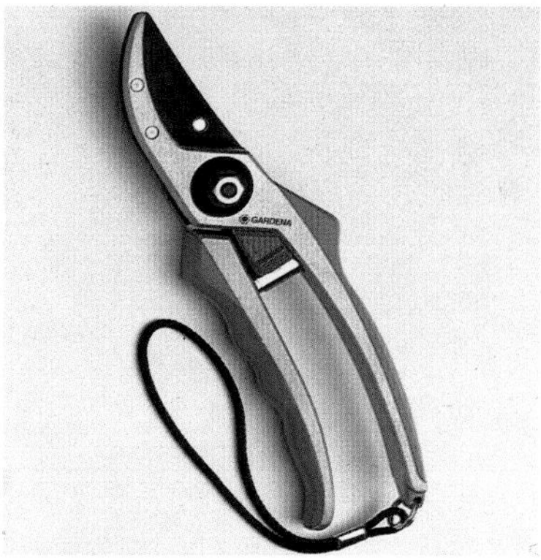


Pruning shears
Franco Clivio
Zurich 1986

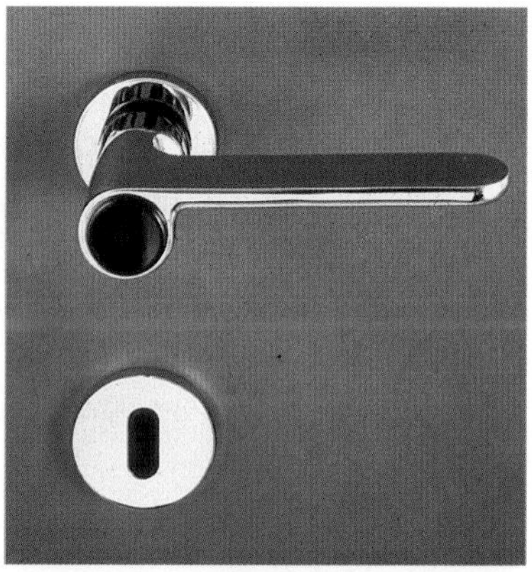

Door handle
Andries and Hiroko van Onck
Milan 1981

Stadtbahn transit system, Stuttgart
Lindinger, Kusserow, Staubach, Weinert
Hanover 1979 – 83

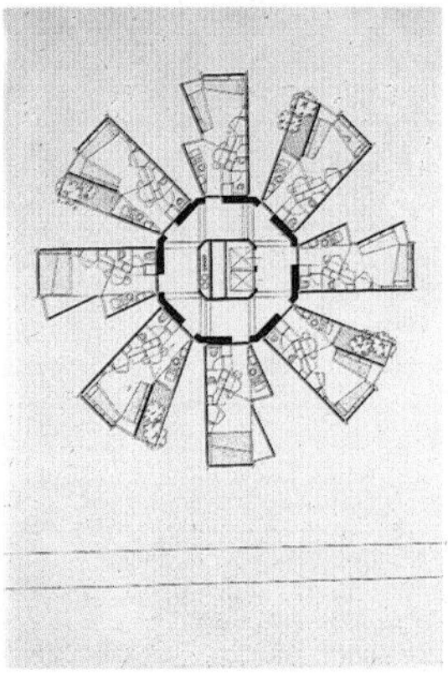

Abitainer accommodation module
Willi Ramstein
Milan 1980

Max Bill in his studio
Zurich 1978

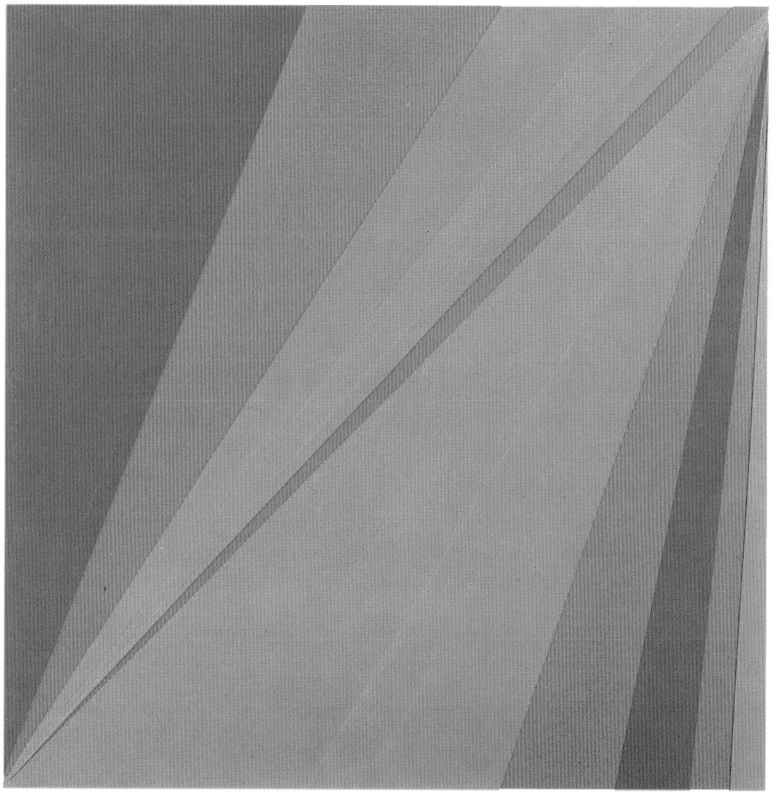

Progression with four
quantitatively equal colors,
ranged parallel
Max Bill
Zurich 1985

*right*
Machine factory
Fred Hochstrasser,
Hans Blieker, Walter Custer,
for Heberlein & Co.,
Wattwiel, Switzerland
Ulm 1964 – 70

*left*
Stadium sign system
Bernd Franck,
Herbert Lindinger,
for Stadion Hannover
Hanover 1974

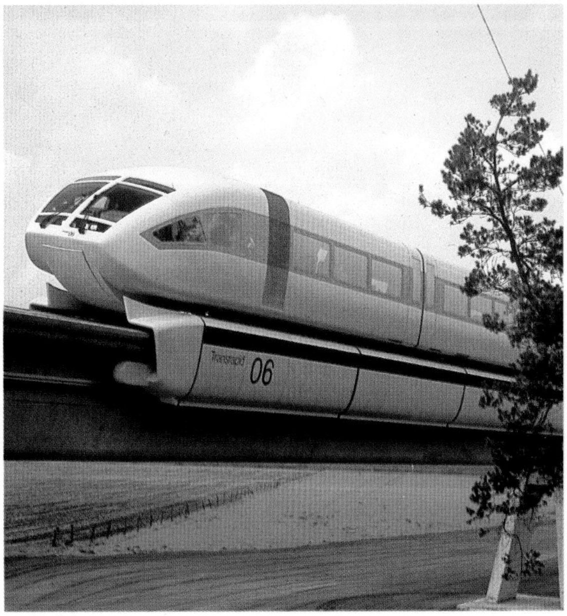


*left*
Transrapid 06 linear-motor
transit system
Alexander Neumeister
Munich 1979 – 82

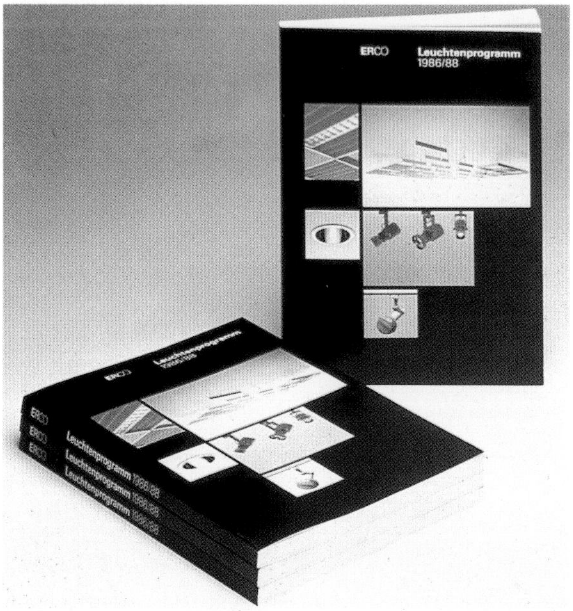

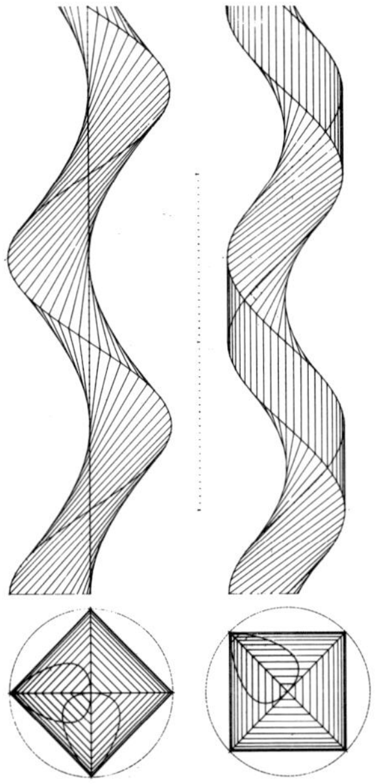

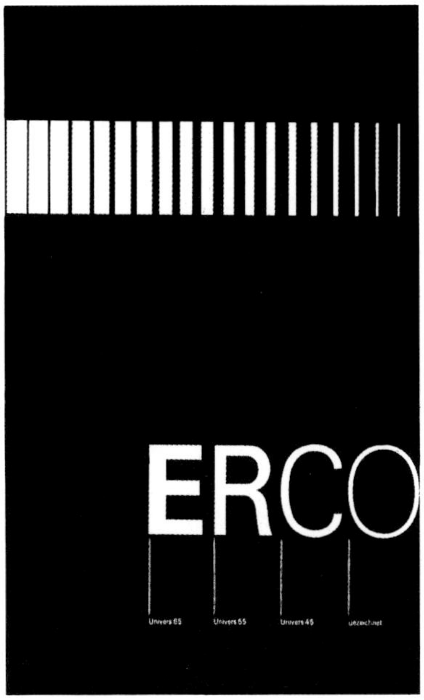

*right, and facing page*
Corporate image
Otl Aicher, for ERCO
Rotis 1977

*Spindle Column*
Walter Zeischegg
Neu-Ulm 1980

Machine tool
Walter Zemp, Zevico Design,
for Baltec Maschinenbau
Zurich 1982

Optometric device
Willy Herold, for Rodenstock
Munich 1961–62

Concrete Poem
Eugen Gomringer
Düsseldorf 1965

**kein fehler im system**
**kein efhler im system**
**kein ehfler im system**
**kein ehlfer im system**
**kein ehlefr im system**
**kein ehlerf im system**
**kein ehleri fm system**
**kein ehleri mf system**
**kein ehleri ms fystem**
**kein ehleri ms yfstem**
**kein ehleri ms ysftem**
**kein ehleri ms ystfem**
**kein ehleri ms ystefm**
**kein ehleri ms ystemf**
**fkei nehler im system**
**kfei nehler im system**
**kefi nehler im system**
**keif nehler im system**
**kein fehler im system**

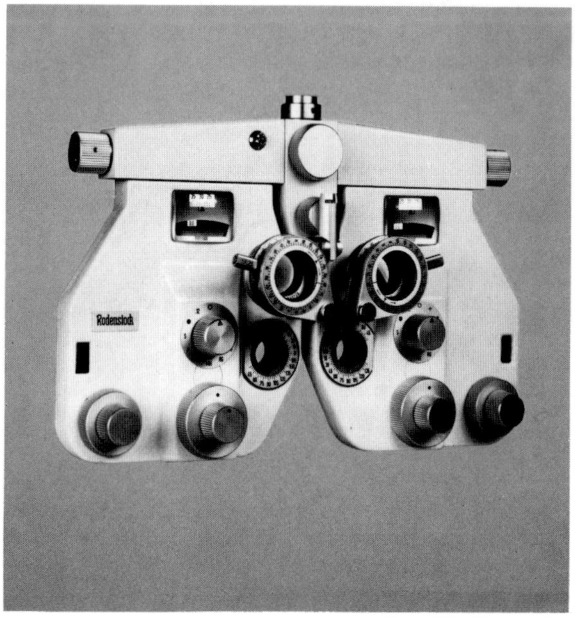


. . . an interaction between one rational aesthetic and another — for which the way was paved by a Max Bill retrospective in São Paulo, in 1950, the influence of which was already visible by 1951 in the work of the painters Mavignier, Serpa, Palatnik, and Maluf, and of the sculptor Mary Vieira. Almir Mavignier, who went to Ulm in 1953, brought a "tropical" element into the Ulm model of visual communication. He not only injected some color into the "Ulm gray," but with his posters he contributed some of the typographical refinement that marked the "text planes" of the concrete poetry of the internationally renowned Noigandres group, founded in São Paulo in 1952 by Haroldo and Augusto de Campos and Décio Pignatari.

The term *Verbovocovisualidade* is Brazilian in origin: a "concrete" text presents a three-dimensional informational space, whole "meaning" is presented by concrete, material means. This ideographic form of poetry has close affinities with advertising copy (instant

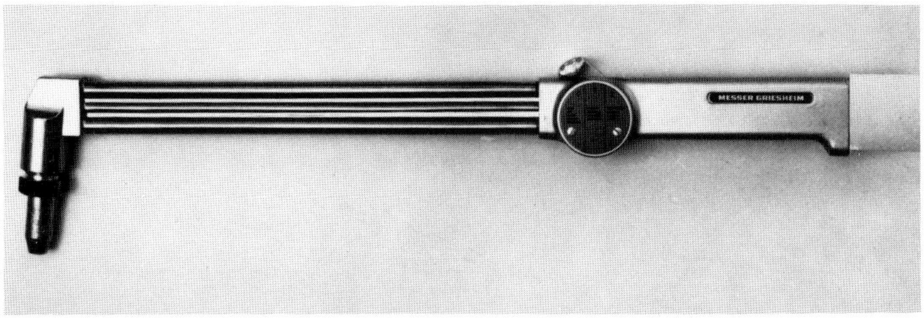


Cutting torch
C. W. Voltz,
for Messer Griesheim
Lindenfels 1970

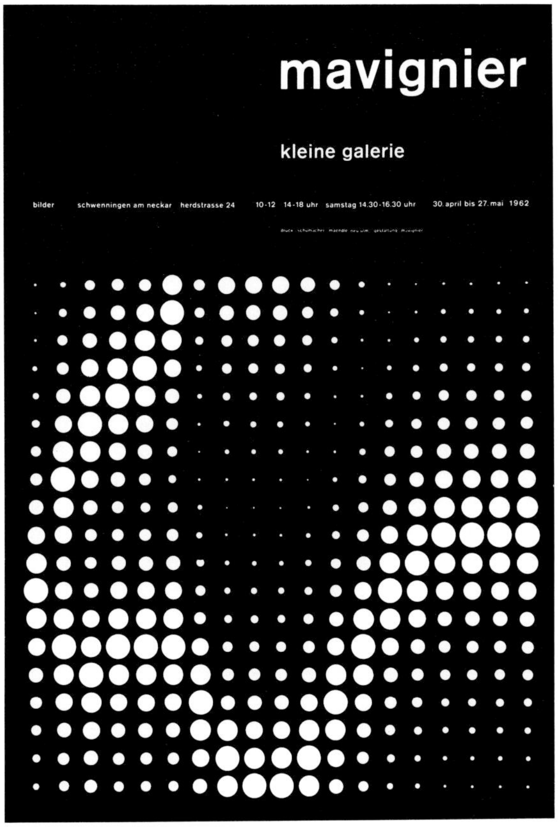


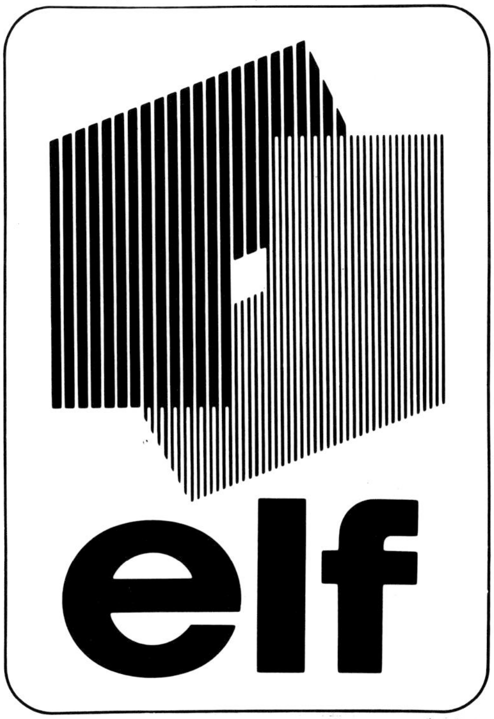


*right*
Corporate image
Eric Brenzinger, with
Chourgnoz, for ELF Oil
Paris 1966 – 67

*left*
Poster
Almir Mavignier
Hamburg 1962

impact, memorability); and in Ulm, with Gomringer present, this led to an increased interest in "concrete" methodology: a product based on a functional aesthetic has a dignity of form that calls for an appropriate, formulaic economy in its advertising.

In Rio in 1953, Bill spoke on architecture to an audience of like-minded people who were working with Lcio Costa on the Brasilia projectú (1956 onward). In 1959 a "neoconcrete" group was formed in Rio to compete with the one in São Paulo: its members included Gullar, Spanidis, and the sculptor Lygia Clark. Also in Rio, in the same year, Maldonado and Aicher addressed audiences of designers whose style shows clear affinities with that of Ulm (Cordeiro, Wollner, de Barros, and others). This serves to underscore the global validity of what Bense, the major theoretical intermediary between Ulm and Brazil, called "the expression of a Cartesianism that has turned into Total Design."

J. P. Mardersteig, 1987

Four logos
Alexandre Wollner
São Paulo 1964 – 68

Wall relief
Shizuko Yoshikawa
Zurich 1972 – 73

Partitioning system
Willy Herold,
for Pohlschröder
Munich 1969 – 71

Espresso Machine
Luca Meda
for Girmi
Milan 1983

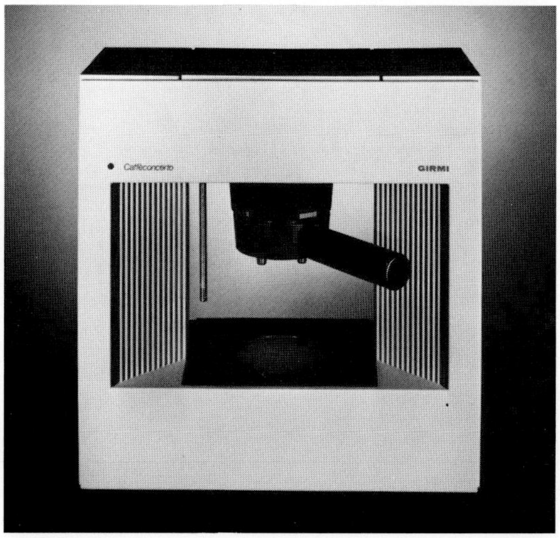

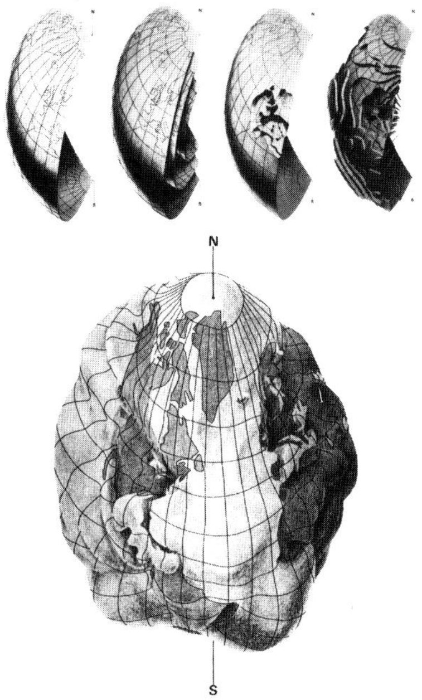


Time Axis Map of Japan
Kohei Sugiura
Tokyo 1969

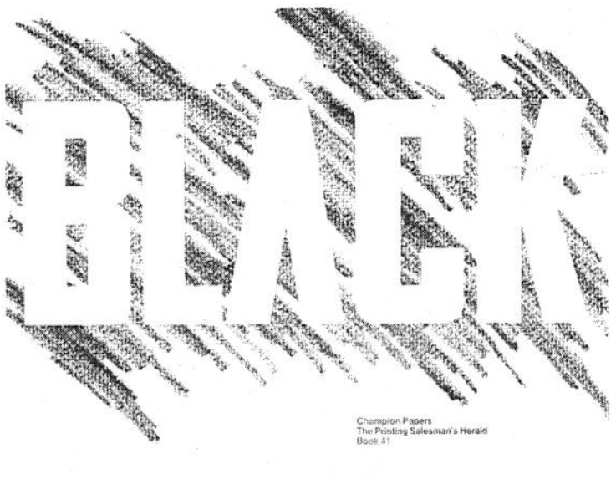

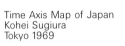

Title page
Tomàs Gonda
New York 1979

Slide projector
Stefan Lengyel,
for Liesegang
Essen 1985

木　　林　　木

林　森　森　林

森　森　森　森

林　　　　　　林

森　森　森　森

林　森　森　林

木　　林　　木

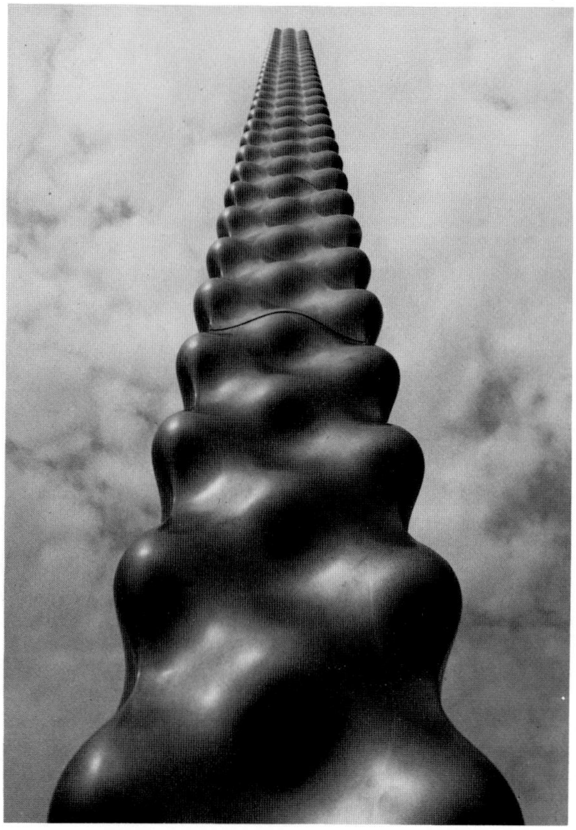

*Undulant Plane Column*
Walter Zeischegg
Neu-Ulm 1978

Poem
Sutaro Mukai
Tokyo 1978

Rolf Schroeter
Zurich 1985

Photographic study
Christian Staub
Seattle 1982

Books by Ulm Authors:
A Selection

Otl Aicher:
*Gehen in der Wüste*
Frankfurt 1982

*Die Küche zum Kochen*
Munich 1982

*Kritik am Auto*
Munich 1984

Aicher and M. Krampen
*Zeichensysteme der visuellen Kommunikation*
Stuttgart 1977

Aicher, G. Greindl, and W. Vossenkühl
*Wilhelm von Ockham: Das Risiko modern zu denken*
Munich 1986

Max Bense:
*Aesthetica*
Stuttgart 1954

*Ästhetik und Zivilisation*
Baden-Baden 1958

*Aesthetica*
Baden-Baden 1965

*Einführung in die informationstheoretische Ästhetik*
Reinbek 1969

*Zeichen und Design*
Baden-Baden 1971

Gui Bonsiepe:
*Designtheorie 1: Design im Übergang zum Sozialismus*
Redaktion Designtheorie 1974

*Teoria e pratica del disegno industriale*
Milan 1975

*A "tecnologia" da tecnologia*
São Paulo 1983

Bonsiepe, P. Kellner, and H. Poessnecker
*Metodologia do desenho industrial*
Brasilia 1984

Bernhard E. Bürdek:
*Designtheorie: Einführung in die Designmethodologie*
Hamburg 1975

Ulrich Burandt:
*Ergonomie für Design und Entwicklung*
Cologne 1978

Peter Croy:
*Grafik – Form und Technik*
Göttingen 1964

*Die Zeichen und ihre Sprache*
Göttingen 1972

Gerhard Curdes:
*Politik und Planung 13: Teilräumliche Planung 2*
Cologne 1980

*Künstlerischer Städtebau um die Jahrhundertwende: Der Beitrag von Karl Henrici*
Cologne 1981

Curdes and K. Stahl
*Umweltplanung in der Industriegesellschaft*
Hamburg 1970

Curdes, G. Piegsa, and M. Schmitz
*Politik und Planung 14*
Cologne 1985

Manfred Eisenbeis:
Eisenbeis, A. Henrich, and M. Marschall
*Programm-Mosaik 2: Handbuch für die Gestaltung von Bildschirmtext*
Nuremberg 1985

Eugen Gomringer:
*Josef Albers*
Starnberg 1971

*Das Stundenbuch*
Starnberg 1980

Ilse Grubrich-Simitis (Ed.):
*Sigmund Freud: Übersicht der Übertragungs-Neurosen*
Frankfurt 1985

Ilse Grubrich-Simitis, Ernst Freud, Lucie Freud:
*Sigmund Freud. Sein Leben in Bildern und Texten*
Frankfurt 1978

Harald Kaas:
*Uhren und Meere*
Munich 1979

Gert Kalow:
*Poesie ist Nachricht: Mündliche Tradition in Vorgeschichte und Gegenwart*
Munich 1975

Herbert Kapitzki:
*Programmiertes Gestalten: Grundlagen für das Visualisieren mit Zeichen*
Karlsruhe 1980

Shoichi Kawai:
*Dialogue in Architectural Design*
Tokyo 1987

Alexander Kluge:
*Neue Geschichten, Heft 1–18: Unheimlichkeit der Zeit*
Frankfurt 1977

Kluge and K. Eder
*Ulmer Dramaturgien/ Reibungsverluste*
Munich 1980

Kluge and O. Negt
*Geschichte und Eigensinn*
Frankfurt 1981

Elke Koch-Weser-Ammassari:
*Stili di vita del tempo libero e categorie socio-professionali*
Rome 1985

Martin Krampen:
*Zeichensystemen der visuellen Kommunikation*
Stuttgart 1977

*Meaning in the Urban Environment*
London 1979

*Die Welt als Zeichen: Klassiker der modernen Semiotik*
Berlin 1981

*Semiotica — Icons of the Road*
Amsterdam 1983

*Umwelt, Gestaltung und Persönlichkeit*
Hildesheim 1986

Krampen, J. Frantz, D. Schempp, T. Seidel, and D. Wolter
*Grüne Archen: In Harmonie mit Pflanzen leben. Das Modell der Gruppe LOG ID*
Frankfurt 1982

Klaus Krippendorf:
*Communication and Control in Society*
New York 1979

*Content Analysis: An Introduction to Its Methodology*
Beverly Hills 1980

*Information Theory: Structural Models for Qualitative Data*
Beverly Hills 1980

Krippendorf, G. Gerbner, O. R. Holsti, W. J. Paisley, and P. J. Stone
*The Analysis of Communication Content*
New York 1969

Tomàs Maldonado:
*La speranza progettuale*
Turin 1970

*Umwelt und Revolte*
Hamburg 1972

*Avanguardia e razionalità*
Turin 1974

*Disegno industriale: Un riesame*
Milan 1976

*Tecnica e cultura*
Milan 1979

*Il futuro della modernit*
Milan 1987

Bernd Meurer:
Meurer and H. Vincon
*Kritik der Alltagskultur*
Berlin 1979

Meurer and H. Vincon
*Industrielle Ästhetik: Zur Geschichte und Theorie der Gestaltung*
Giessen 1983

Shutaro Mukai:
*Origin of Design*
Tokyo 1978

*Semiosis of Form*
Tokyo 1986

Mukai and M. Katsumi
*The Essence of Modern Design Theories*
Tokyo 1966

Mukai, Y. Kawazoe, and H. Chijiwa
*Handbuch für die Farbgestaltung*
Tokyo 1980

Mukai, M. Katsui, and I. Tanaka
*Dictionary of Today's Design*
Tokyo 1986

Takeshi Nishizawa:
*Street Furniture*
Tokyo 1983

Eva Pfeil:
*The Effectiveness of Visual Versus Experimental Teaching Techniques for Aesthetic Color Theory*
Auburn 1979

Alf Pross:
*Zwei Hühner werden geschlachtet*
Frankfurt 1969

Walter Schaer:
*Undergraduate General Education Programs*
Auburn 1976

Günter Schmitz:
*Architectural Fabric Structures*
Washington 1985

Claude Schnaidt:
*Hannes Meyer: Bauten, Projekte und Schriften*
Stuttgart 1965

*Umweltbürger, Umweltmacher*
Dresden 1982

Margit Staber:
*Sophie Taeuber-Arp*
Geneva 1970

*Max Bill*
St. Gallen 1971

*Fritz Glarner*
Zurich 1976

Carlo Testa:
*Industrialisierung des Bauens*
Zurich 1972

Testa and T. Schmid
*Bauen mit System*
Zurich 1969

Gui Bonsiepe

# The "Ulm Model" in the Periphery

The question of the relevance of the "Ulm model" to the developing world presupposes an outline definition of this model itself. It was certainly no coincidence that the faculty and student body of the HfG were markedly international in composition. In addition, the program of the HfG displayed characteristics that were relevant far beyond the local circumstances of the Federal Republic. This does not mean that the HfG laid any claim to universal relevance. It was conceived within the context of the industrialized countries and was primarily adapted to their needs and their resources. But its influence was not restricted to the comparatively small number of industrialized countries — in global terms, the so-called center or "metropolis" — but also reached those countries of the "periphery" that see industrialization as a tool with which to reduce their technological dependency, to create economic wealth, and eventually to create an autonomous material culture of their own.

The Ulm model of teaching was based on the realization that the modern environment is decisively conditioned by industry — especially the processing industries, the communications industry, and the construction industry — and that it is not possible to deal with these problems within the framework of a traditional university-level education. The HfG filled a niche that could never have been filled by the traditional university.

It was no longer a matter of adding art on to industry as a civilizing element from outside — the basic fallacy underlying the "decorative arts" — but of developing the possibilities of formal creation inherent in industry itself. This openness to industry as a cultural manifestation did not in the least entail an uncritical attitude to industry as such. It was precisely the functional, aesthetic, and social shortcomings of industrial production that the HfG's program was intended to help correct.

Industrialization is now in progress on a planetary scale, in spite of the occasional signs of disillusion that appear in the metropolitan countries. It therefore comes as no surprise to find that the Ulm model has an attraction for those peripheral countries that possess the basis, however minimal and disorganized, for the growth of manufacturing industry in any of its forms — from the tiny businesses of the "informal" sectors, with their often rudimentary production techniques, by way of medium-sized and larger concerns, to large-scale businesses.

This attraction to the Ulm model is explained not only by a positive attitude to the word "industrialization" — in spite of all the justified criticism there has been from ecological quarters — but as a response to an educational system that stands in need of drastic reform, especially the universities. These often operate only as degree factories, rather than as dynamic institutions prepared to concern themselves with the real needs and potentials of the societies that support them. It is part of the colonial legacy that training for technological careers, particularly in the engineering sciences, is designed to enable people to manage imported technology rather than to create technological innovations of their own. Neither in their definition of their objectives, their organization, their curricula, nor their teaching methods do these universities train the kind of technical and cultural intelligentsia that is essential to the creation of a material culture on a modern, which means an industrial, basis.

The first objective set by the planning bureaucracies was import substitution; but it was not long before the inadequacy of this as an industrial development policy became evident. When taken seriously, industrialization does not consist of turning out products without enquiring into their origins: it must include the dimension of invention and innovation. Without this to create an industrial dynamic, the countries of the periphery will never get any further than a merely passive, reflex industrialization.

By the early 1960s the Ulm model had reached some parts of the periphery, as in the founding of the Escola Superior de Desenho Industrial (ESDI) in Rio de Janeiro, where a number of HfG alumni came to work as instructors (Karl-Heinz Bergmiller, Alexandre Wollner). Again, HfG influence had a part in the founding of the National Institute of Design (NID) at Ahmadabad in India, where HfG faculty members gave guest courses (Hans Gugelot, Herbert Lindinger, Wolfgang Siol, Christian Staub, and others). These institutions based themselves, in policy, design, curriculum, and teaching methods (problem-based learning in design courses), on the experience of the HfG. This experience was brought to them through contacts with HfG faculty members, through Ulm alumni who came there to teach, and also through the publications of the HfG, especially the magazine *Ulm*.

In these institutions the autonomy of product design and visual communication, as disciplines distinct from that of architecture, had been established. As is well known, there is no lack of endeavors to interpret design as a continuation of architecture on a smaller scale, and to suggest that an architectural training constitutes a professional qualification for solving design problems in general.

The Ulm model further disposes of the assumption that a training in machine construction, supplemented by a crash course in applied aesthetics, enables an engineer to solve design problems more or less in passing. Ulm teaching reveals the inadequacy of the criterion of mechanical efficiency, as used in technological training. Both in their approach to design and in their design methodology, engineering theory and industrial design are diametrically opposed; and attempts to treat design as an appendage to a training in engineering have had no very satisfactory results. The classical conception of machine construction breaks down when it comes to aesthetics. It is incapable of recognizing aesthetics as a constituent part of design work.

The ESDI in Brazil and the NID in India were new foundations modeled on the HfG, i.e., institutions that lay outside the framework of traditional university-style institutes of technology. The Industrial Design Centre (IDC) in Bombay, on the other hand, was set up as a postgraduate school within a technological training institution, albeit with a high degree of autonomy. Its first director was an alumnus of the HfG, Sudhakar Nadkarni.

Both the NID and the IDC combine their teaching activity with consultant work for industry and public bodies, as was done as a matter of policy at the HfG. This detail is especially important for the countries of the periphery, in which, as a rule, a social and cultural gulf separates the insular world of the universities, on the one hand, from the world of society and industry, on the other. In addition, companies are often forced by their own financial weakness to turn to public institutions for what often amounts to subsidized design work. In the periphery, state support for the design field plays a role that must not be underrated. In addition to the traditional role of national design centers in the metropolitan countries, which are usually restricted to the organization of exhibitions and the provision of documentation, design centers in the periphery operate as centers of practical design work.

India thus has two exemplary educational centers for product design and visual communication, in which the experience gained at the HfG has been assimilated.

In Latin America, too, design has been taken up as an instrument of industrialization. Governmental institutions play a major part in industrial development generally, and industrial design has been included in multilateral and bilateral programs of technical cooperation — as, for example, in Chile in the late 1960s. The government of Salvador Allende, in particular, set out to liberate design from its associations with oversophisticated luxury products for the privileged 5 percent of the population, and to extend it to the field of capital goods (Gui Bonsiepe was involved here). The resulting experience, which bore the clear impress of the Ulm model, formed the basis of a working document issued in 1973 by UNIDO (the United Nations Industrial Development Organization): perhaps the first occasion on which any international organization had addressed the problem of defining the role of industrial design in the periphery.

All this would have been well-nigh impossible in the absence of the Ulm idea of design as an activity with a technological — and thus discursive and rational — base. What had happened previously was that the failure to present any such rationale had excluded design from any access to industry, and to the governmental institutions that were in a position to decide whether design had any role to play in the process of industrialization.

It is in this process that the political dimension of design reveals itself: for design is an instrument by which the ruinous technological dependency of the Third World can be reduced. Present-day adherents of postmodernism or of "radical design" — so radical that it leaves everything just the way it was before — regard the HfG's design rationalism as an abomination, or else dismiss it with a languid gesture as outmoded. But it is precisely this rationalism that has shown itself to be a workable option in the Third World. It acts as a powerful antidote to those frivolous, "ludic" attitudes that make a deep semantic issue out of stylizing a door handle or a table lamp, deploying the needs of the psyche as a cloak for a retreat from technological and economic parameters. The rationalism of Ulm shuts the door firmly against any romanticization of poverty or idealization of "appropriate technology," and dispels the paternalistic attitudes of "aidism." Ultimately, the design problems of the periphery can be solved only in the periphery. Design "for" the Third World is no more and no less than ideology.

A glance at the offerings of design programs in the metropolitan countries serves to underline the enduring relevance of the Ulm model; for, in spite of the efforts of those who have tried so hard to bury Ulm rationalism, no viable alternative has yet been created that could serve the countries of the periphery as a point of reference — even a critical one — for the development of a product culture of their own. Even less are metropolitan training institutions prepared to investigate the specific design problems of the periphery, and to vary their instruction accordingly. Such suggestions as that design in the periphery should be "tropicalized" are more an expression of a taste for tourist clichés than of a serious option. The reduction of design to stylistic issues is another way in which the problem is obfuscated; but, in view of the mass of short-lived formal innovations that are documented in the numerous specialist publications of the metropolitan world, the Ulm morphology itself is a legacy of the HfG whose importance is certainly not to be dismissed as trivial.

The HfG curriculum, with its wide range of theoretical disciplines that opened the way to critical reflection on the students' part, has frequently served as a guide to curricula in the periphery, as for example in the first postgraduate course in product design at the Universidad Autnoma Metropolitana, Azcapotzalco, Mexico.

In Brazil, industrial design was enshrined in a program issued by the National Council for Scientific and Technological Development (Gui Bonsiepe, 1981). Similar developments may be observed in Cuba, where in 1984, in the context of a UNIDO project, the national Office of Industrial Design (Oficina nacional de diseo industrial) embarked on an exploration of the role of design in all its many manifestations (initial design, product testing, standardization, training, promotion).

There is still no documentary record of the Ulm "diaspora." But it is already evident that the presence of the Ulm model, and the influence of the Ulm experience, will form a chapter in any future history of design in the periphery.

Florianópolis, 1987

# Appendixes

## The Faculty

The faculty of the HfG was composed of the permanent staff instructors (or *Festdozenten*), the guest instructors (*Gastdozenten*), the workshop leaders of the various workshops, and the assistants.

### Staff Instructors

#### Inge Aicher

Née Scholl, born Ingersheim (Württemberg), 1917. After graduating from girls' high school in Ulm, where her father was in private practice as an accountant, she trained in her father's office as an assistant auditor. In her spare time she pursued an interest in music, art, literature, and — especially — philosophy and music, in the company of a group of friends and her brother and sister, Hans and Sophie. An initial enthusiasm for the Hitler Youth turned to a strong dislike.

The decisive event in her life was the fate of Hans and Sophie, who joined the White Rose student group in Munich in active resistance against the Nazi regime and were sentenced to death and executed in February 1943. With the rest of her family, Inge was held by the Gestapo for several months, and her youngest brother, who had been drafted to the Eastern Front, was shortly afterward posted missing.

In Ulm, after the end of the Nazi regime, she founded the Volkshochschule, for adult education, in the conviction that democracy had no strong basis unless the citizens were basically well informed. She ran the Volkshochschule from 1946 through 1974, and in 1968, with the opening of the Einstein-Haus, she achieved her aim of housing the school in a modern building in the center of Ulm.

In the late 1940s, building on her experience of working in adult education in a devastated city, Inge Scholl began to work with Otl Aicher and a number of friends, including Max Bill and Hans Werner Richter, toward the foundation of a Hochschule für Gestaltung. As the person responsible for this independent, privately run college, she set up the Scholl Foundation in memory of her brother and sister. The United States High Commissioner, John J. McCloy, was impressed enough by her energy and sense of purpose to promise her a million marks on condition that she could succeed in collecting an equal sum from German sources. She did succeed, and McCloy handed over his check in mid-June 1952.

Inge Scholl married Otl Aicher in 1952. They have had five children. In 1972, four years after the closure of the HfG, they moved to Rotis, in the Allgäu, where she has since worked on the archive devoted to her brother and sister, and has also been active in the Peace Movement. Inge Scholl has written an account of Hans and Sophie Scholl and their resistance to Nazism, which has been published as a paperback and also as a documentary volume (*Die weiße Rose*) by S. Fischer.

#### otl aicher

born ulm, 1922. in 1946 studied in the sculpture class at the munich academy. started his own graphic design studio in ulm in 1947, in munich in 1967, and at rotis in the allgäu in 1972.

an initiator and founder member of the hochschule für gestaltung in ulm. In 1954–66, instructor in the department of visual communication. in 1956–59, co-rector. in 1962–64, rector. guest lecturer at yale university and in rio de janeiro.

developed corporate identity for firms such as braun electric, deutsche lufthansa, zdf, erco lighting, frankfurt airport, westdeutsche landesbank, dresdner bank, severin und siedler publishers. in charge of design 1967–72 for the olympic games in munich, and developed an international system of pictograms. in 1984 founded the institut für analoge studien, rotis.

author of flugbild deutschland, 1968, with rudolf sass; im flug über europa, 1980, with rudolf sass; zeichensysteme, 1980, with martin krampen; die küche zum kochen, 1982; gehen in der wüste, 1982; kritik am auto, 1984; innenseiten des kriegs, 1985; wilhelm von ockham, 1986, with wilhelm vossenkuhl and gabriele greindl; greifen und griffe, 1987, with robert kühn.

lives at rotis.

#### Max Bense

Born Strasbourg, 1910. Studied physics, mathematics, and philosophy at the universities of Bonn and Cologne. Science doctorate (Dr. phil. nat.).

After World War II he became Registrar, then in 1949 Professor of Philosophy and Scientific Theory, at Jena University. From 1949, Professor of Philosophy and Scientific Theory at the Technological University, Stuttgart.

In 1954–58 and in 1966 he was an instructor at the HfG in philosophy and semiotics, responsible for devising and teaching a syllabus for the department of Information.

Publications: *Quantenmechanik und die Daseinsrelativität,* 1938; *Geistesgeschichte der Mathematik,* 2 vol., 1946, 1949; *Philosophie als Forschung,* 1946; *Hegel und Kierkegaard,* 1948; *Technische Existenz,* 1949; *Literaturmetaphysik,* 1950; *Philosophie zwischen den Kriegen,* 1951; *Naturphilosophie,* 1958; *Aesthetica,* 1–4, 1954, 1956, 1958, 1960; *Theorie der Text,* 1962; *Aesthetica,* 1965; *Ungehorsam der Ideen,* 1966; *Semiotik,* 1967.

Died in 1990.

Max Bill
Born Winterthur, 1908. In 1927–29 studied at the Bauhaus (Hochschule für Gestaltung), Dessau. Professionally active from 1930 as an architect and painter, from 1932 as a sculptor, from 1936 as a writer. Extensive theoretical writings in all fields. Sculpture *Endless Loop,* 1935; sixteen lithographs, *Fifteen Variations on a Single Theme,* a first attempt at serial and systematic thinking in painting. Since the early 1940s, his painting has concentrated on the creation and control of color energies in a plane through geometrical structures. Buildings: Hochschule für Gestaltung, Ulm; Cinevox, Neuhausen; "bilden und gestalten" section of Swiss National Exhibition, Lausanne; administration building, Imbau, Leverkusen; radio studios, Zurich.

1950: cofounder of the HfG, and first Rector. 1951: Grand Prix for sculpture, Biennale, São Paulo, and Biennale, Milan. 1968: Cultural Prize of the city of Zurich. 1979: Cultural Prize of the city of Winterthur. Grand Cross of Merit, Federal Republic of Germany. Honorary Doctor of Engineering, Stuttgart University. 1982: Kaiserring of the city of Goslar and Order of the Crown, Belgium. 1985: Commander of the Order of Arts and Letters, France, Honorary Fellow of the American Institute of Architects, member of the Academy of Arts, Berlin, external member of the Royal Academy of Science, Literature, and Fine Arts of Belgium, corresponding member of the Academy of Architecture, Paris, honorary member of the Academy of Art, Düsseldorf, corresponding member of the Academia Nacional de Bellas Artes, Argentina, member of International Academy of the Philosophy of Art.

For further information on biography, bibliography, exhibitions, publications, see Eduard Hüttinger, *Max Bill,* ABC-Verlag, Zurich.

Lives at Zumikon, near Zurich.

Gui Bonsiepe
Born Augsburg, 1934. High school diploma, Stuttgart. Studied at the HfG, 1955–59; took diploma in Information department. Taught at HfG 1963–68. Has worked since 1968 in Latin America (Chile, Argentina, Brazil). In public development institutions 1971–73 (Chile) and since 1981 (Brazil, National Council for Scientific and Technological Development). Director of Design Institute, Florianpolis, since 1984. Currently on a scholarship at Berkeley in the field of software development.

Design activity in fields of product design and visual communications. Worked as adviser for UNIDO and ILO in India, Portugal, and Cuba. Books: *Teoria e pratica del disegno industriale,* 1975; *Diseño industrial, tecnologa y dependencia,* 1978; *A "tecnologia" da tecnologia,* Brasilia 1983, and around 80 articles.

Around 60 publications in specialized journals on industrial design, packaging design, the politics of technology, design training, and methodology (Argentina, Brazil, Canada, Chile, Costa Rica, Cuba, Ecuador, France, West and East Germany, India, Japan, Mexico, Spain, Switzerland, UK, USA, USSR, Yugoslavia). Vice-president of ICSID in 1973–75.

Lives at Florianópolis.

Anthony Fröshaug
Born London, 1920. Studied graphic design at Central School of Arts and Crafts, London, 1938–39. Studied science at the University of London, 1940–43. Worked from 1940 as freelance typographic designer and exhibition designer.

Taught at Central School, 1948–49; head of typography department, 1952–53. Instructor in the department of Visual Communication at the HfG, Ulm, 1957–60. Was also much involved in the Basic Course. In charge of design for the periodical *Ulm,* numbers 1–5. Taught typography at Royal College of Art, London, 1957–61. Taught typography at Watford College of Art, 1964–67. Studied architecture at Architectural Association, London, 1967–69. Taught graphic design at Central School of Arts and Crafts, London, 1970–84.

Took part in the design of the exhibitions *Britain Can Make It,* 1947; *Festival of Britain,* 1951; *Purpose and Pleasure,* 1952. Books: *Typographic Norms,* 1964; *Art without Boundaries,* 1972.

Died in 1984.

Hans Gugelot
Born Macassar, Celebes, in 1920. Studied architecture in Lausanne and at the Eidgenössische Technische Hochschule (ETH), Zurich, 1940–46. Traveled and played jazz guitar. Worked in a number of architectural practices including Max Bill's, 1948–50. Worked independently from 1950. Designed furniture for Wohnbedarf, Zurich. Experiments with furniture systems.

Instructor at HfG, Ulm, 1954–65; co-Rector, 1960–61. Was the decisive influence on the department of Product Design. Devised a pioneering and influential range of radio receivers for the firm of Braun, Frankfurt, 1954–57. He set the design parameters of the firm and was largely responsible, together with Otl Aicher, for Braun's new corporate identity. He intensively promoted the idea of system construction. A particularly influential project in this respect was his M 125 furnishing system, which was followed onto the market by many others. Gugelot always sought to incorporate technical or operational advances in his new designs — as in his Carousel slide projector for Kodak.

Founded his own development office outside the HfG in 1962. In 1961 and 1965 was visiting professor at the National Institute of Design, Ahmedabad, India, where he advised on the planning of the school.

Distinctions: Grand Prix of the Triennale, Milan, 1957, and Compasso d'Oro, Milan, 1962.

One-man exhibitions: Ulm 1960; posthumous touring exhibition, Neue Sammlung, Munich 1984; Zurich 1985; Stuttgart 1986.

Died in 1965.

Gert Kalow
Born Cottbus, 1921. Graduated from high school, 1939, and enrolled at Jena University. Late 1939, drafted into army. Prisoner of war until 1947. From 1947, studied at Hamburg and Heidelberg universities: philosophy, sociology, music, and literature, under Jaspers, Weber, and Georgiades. From 1950, an increasing level of journalistic activity (as theater and art critic and author of magazine and radio features; also radio plays).

Appointed head of the Information department at the HfG, 1956. Co-Rector and chairman of the rectorial college, 1960–62. Developed the Film department together with Alexander Kluge.

Head of arts/feature (Abendstudio) department of Hessian Radio, 1963–86. Honorary professor at Hochschule für Gestaltung, Offenbach, from 1974.

Principal books: *Zwischen Christentum und Ideologie,* 1956; *Hitler — Das gesamtdeutsche Trauma,* 1967 (British and American editions, 1968 and 1970); *erdgaleere,* poems, 1970; *Poesie ist Nachricht,* 1975. Many contributions to anthologies in English, French, Italian, Japanese, etc. Editor of numerous compilation volumes, some with Alexander Kluge.

Rockefeller Scholarship, 1962; Villa Massimo, Rome 1971–72. Vice President of West German PEN Club, 1976–80.

Lives in Heidelberg.

Herbert W. Kapitzki
Born Danzig, 1925. Studied fine arts in Danzig, 1942–43, and in Hamburg and Stuttgart under Professor Willi Baumeister, 1946–52. Freelance designer in the field of visual communication from 1953. Member of Alliance Graphique Internationale (AGI).

Instructor at HfG, 1964–68; head of department of Visual Communication from 1965. Vice President of Federation of German Graphic Designers (BDG), 1969–71. Cofounder of Institut für Umweltgestaltung, Frankfurt, 1969. Professor of Visual Communication at Staatliche Hochschule für bildende Künste (HfbK), later Hochschule der Künste (HdK), West Berlin; deputy director of HfbK, 1974; head of Area 4, HdK, 1975.

Jury member for the Federal Design Prize awarded by Rat für Formgebung, 1976–77. Work includes: corporate images for Design Center, Stuttgart; IDZ, Berlin; Schering AG; Historisches Museum, Frankfurt; information systems in city government buildings, West Berlin.

Exhibition venues include: Stuttgart, Washington, New York, Kassel (*documenta 3*), Hamburg, Offenbach, Helsinki, Warsaw, Berlin.

BDG prize for poster of the year, 1955, 1964.

Lives in West Berlin and Wildberg/Calw.

Hanno Kesting
Born Gelsenkirchen, 1925. Studied philosophy, modern history, and sociology at Heidelberg University, 1947–52. Research associate at social studies research unit, Münster University: research into industrial sociology.

Instructor in sociology and political science at HfG, 1957–60. Co-Rector, 1958–60. Founding editor-in-chief of the magazine *Ulm.* Left the HfG in 1960.

Professor of industrial sociology at Bochum University, 1967–75.

Books: *Lenin und das Wesen des Kommunismus,* with H. Popitz, H. P. Bahrdt, and E. A. Jüres; *Technik und Industriearbeit,* 1957; *Das Gesellschaftsbild des Arbeiters,* 1957; *Geschichtsphilosophie und Weltbürgerkrieg,* 1959.

Died in 1975.

Alexander Kluge
Born Halberstadt, 1932. Studied law and took doctorate, Marburg University, 1949–53. Studied history, Frankfurt University, 1956–58. Worked as attorney and from 1959 as filmmaker. From 1962, head of Institut für Filmgestaltung, initially as part of the HfG.

Publications: *Die Universitätsverwaltung* (with Hellmut Becker), 1958; *Lebensläufe: Erzählungen,* 1962; *Schlachtbeschreibung,* 1964. In collaboration with Oskar Negt: *Öffentlichkeit und Erfahrung,* 1972; *Geschichte und Eigensinn,* 1981; *Lernprozesse mit tödlichem Ausgang,* 1975; *Neuere Geschichten,* 1–18; *Unheimlichkeit der Zeit,* 1979.

Films: *Abschied von gestern,* 1966; *Artisten in der Zirkuskuppel: ratlos,* 1968; *Gelegenheitsarbeit einer Sklavin,* 1972; *In Gefahr und grösster Not bringt der Mittelweg den Tod,* 1974; *Der starke Ferdinand,* 1976; *Die Patriotin,* 1979; *Der Kardinal,* 1981; *Der Angriff der Gegenwart auf die übrige Zeit,* 1985; *Vermischte Nachrichten,* 1986.

Awards: Silver Lion, Venice, 1966; Golden Lion, Venice, 1968; International Film Critics' Prize, 1976; Federal German Film Prizes; Fontane Prize, 1979; Bremen Prize for Literature, 1979; Kleist Prize, 1985.

Lives in Munich.

Georg Leowald
Born Düsseldorf, 1908. Studied architecture at Düsseldorf Academy. Worked as an architect in industry until 1941, then in private practice in Berlin. Built industrial installations.
Professor of architecture, HfBK, West Berlin, 1947–51. Then took up product design as well, including office furniture (Pohlschröder), office chairs (Wilkhahn), and office machines. Taught industrial design at WKS, Wuppertal, 1957–59. First Prize in the international competition for industrially manufactured furniture at the Museum of Modern Art, New York, 1947.
Died in 1969.

Herbert Lindinger
Born Wels, Austria, 1933. Studied graphic design in Linz. Exhibition designs. Studied visual communication and product design at the HfG, Ulm. Associate to Gugelot until 1961, and assistant at HfG; instructor 1962–68. Worked on Hamburg elevated railroad, Braun radios, and HfG exhibitions of 1958 and 1963. Visiting professor, Ohio State University (1965) and National Institute of Design, Ahmedabad (1970). Worked at Institut für Umweltgestaltung, Frankfurt, 1969–71.
Since 1971, full professor and director of Institut für Industrial Design, Hanover University; dean 1976, member of senate 1981–83. President of Verband deutscher Industriedesigner (VDID), 1970–76. Board member of Rat für Formgebung und gute Industrieform, Hanover; from 1971 representative of VDID at European Design Federation; juror for numerous prizes in Germany and elsewhere.
Design for urban transit systems: Hamburg, Hanover, Stuttgart, West Berlin. Helicopters, standard city buses, graphic design (Olivetti, EBV). Design of city squares in Hanover, Darmstadt, Heidelberg. "Junge Generation" art prize, Berlin; international design and architecture prizes. Gold medal of the 10th Design Biennale. Member of the Deutscher Werkbund and the Akademie der freien Künste, Mannheim.
Lives in Hanover.

Tomàs Maldonado
Born Buenos Aires, 1922. Professor of environmental design in the faculty of architecture, Technological University, Milan. From 1954 through 1967 he was an instructor at the HfG, Ulm. Lethaby Lecturer, Royal College of Art, London, 1965. Fellow of the Council of Humanities, Princeton University, 1966; Class of 1931 Professor at the School of Architecture, Princeton, 1968–70. Design Medal of the Society of Industrial Artists and Designers (SIAD, Great Britain), 1968. Chairman of the Presidium of the ICSID (International Council of Societies of Industrial Design), 1967–69. Professor at Bologna University, 1971–83. Editor of the magazine *Casabella,* Milan, 1976–81. Research Fellow at Graduate School of Design, Harvard University, 1979.
Published books include: *La speranza progettuale,* 1970; *Avanguardia e razionalit,* 1974; *Disegno industriale: un riesame,* 1976; *Il futuro della modernit,* 1987; he was the editor of the volume *Tecnica e cultura,* 1979.
Has designed industrial products and acted as adviser in the planning of shopping centers in France and Italy. Since 1983 he has been project coordinator for the building of a new city neighborhood in Florence.
Lives in Milan.

Abraham A. Moles
Born Paris, 1920. Studied physics and philosophy in Grenoble, Lyons, Zurich, Paris, New York, and Strasbourg; doctorates in science and philosophy. Worked at Centre national de la recherche scientifique, 1945–54. Rockefeller Scholarship, 1956; scientific director of Edition Kister, Geneva, 1957– 58; at Columbia University, New York, 1957; head of electronic music laboratory, Scherchen, Switzerland, 1959; Centre d'études méthodologiques, 1958–64.
Professor at HfG, Ulm, and Strasbourg University, 1961–66. Director of Institute for Social Psychology, Strasbourg University, since 1966. Visiting professor at Laval University, Quebec (1973), and University of California, San Diego (1976).
Numerous publications include: *Physique du bruit,* 1952; *Structure physique du signal musical,* 1952; *Création scientifique,* 1956; *Théorie de l'information et perception esthétique,* 1958; *Musiques expérimentales,* 1960; *Sociodynamique de la culture,* 1963; *L'Affiche dans la société urbaine,* 1969; *Créativité et méthodes d'innovation,* 1970; *Art et ordinateur,* 1970; *Théorie des objets,* 1971; *Psychologie de l'espace* (with E. Rohmer), 1971; *Kitsch, l'art du bonheur,* 1972; *Micropsychologie* (with E. Rohmer), 1976.
Member of Physical Society of America, Société de psychologie, Société des gens de lettres, International Association of Applied Psychology.
Lives in Strasbourg.

Herbert Ohl
Born Mannheim, 1926. Studied graphics and painting at Academy of Fine Arts, and architecture at Technische Hochschule, Karlsruhe, and Politecnico, Milan (doctorate in architecture). Design work on university and hospital buildings.

Instructor at HfG, Ulm, 1958–68; Rector, 1965–68. Visiting Professor at Columbia University; Harvard University; Carnegie Institute of Technology, Pittsburgh; Princeton University; University of California, Los Angeles; Texas A&M University, College Station; Instituto Nacional de Tecnologa Industrial, Buenos Aires; National Institute of Design, Ahmedabad.

Institut für Umweltgestaltung, Frankfurt, 1969. Technical director of Institut für Formgebung, Darmstadt, 1974–84. Member of board of Rat für Formgebung, and of ICSID (1978–80). Professor of Automobile Design at FHG, Pforzheim, since 1984.

Research and development work on industrial building, automobile construction, and furniture design. Awards from the Museum of Modern Art, New York; G-Mark design prize, Tokyo; Gute Form prize, Darmstadt; Design Center, Stuttgart; Cooper Hewitt Museum, New York.

Lives in Darmstadt and Milan.

Edgar Reitz
Born Morbach, Hunsrück, in 1932. Studied theater, journalism, and art history at Munich University, leaving in 1957. Founded and ran a studio theater. Worked in theater, trained as an actor; practical and theoretical work in motion pictures. First short films as director, from 1958.

Participation in festivals; visits to USA, Latin America, and the Near East. Member of the Oberhausen Group in 1962. Founded and directed the Institut für Filmgestaltung at the HfG, together with Alexander Kluge. Numerous shorts, experimental and documentary films, publications on film theory; founded his own film production firm. Works as an independent film producer and director, based in Munich.

Principal films: *Mahlzeiten,* 1966; *Cardillac,* 1968; *Das Goldene Ding,* 1971; *Die Reise nach Wien,* 1973; *Stunde null,* 1976; *Der Schneider von Ulm,* 1978; *Heimat,* 1980–84.

Distinctions and awards: Adolf Grimme Prize, in gold, 1985, 1986, in silver, 1977; Golden Camera, 1985; Filmband, in gold, 1974; Critics' Prize, 1985; Spiresci Prize, Venice, 1984; German Film Prize, 1963, 1974, 1975, 1976, 1978.

Lives in Munich.

Horst W. J. Rittel
Born Berlin, 1930. Studied mathematics and theoretical physics at Göttingen University, 1949–54, and mathematics and sociology at Münster University, 1954–56. Worked for the firm of Hoesch as a mathematician and physicist. Joined social research unit, Münster University, 1958.

Instructor in methodology, theory of science, and operational analysis at the HfG, 1958–63. From 1963, Professor of the Science of Design, department of Architecture, University of California, Berkeley; Gruppe für Systemforschung, Heidelberg, 1968–72. From 1973 professor and director of the Institut für Grundlagen der Planung (IGP), Stuttgart University. Dean of the faculty of architecture and urban planning, Heidelberg University, 1976–81.

Research and development work: systems analysis and development, especially for national and international bodies (Bundestag, West Germany; several departments of Federal and state governments, United Nations Environment Program, UNESCO, EEC, etc.), and also for research projects and industry.

Writings: *Die Informationswissenschaften* (with W. Kunz), 1972; *UMPLIS – Entwicklung eines Umwelt-Planungs-Informationssystems* (with W. Kunz and W. Reuter), 1980.

Died in 1990.

Claude Schnaidt
Born Geneva, 1931. Studied architecture at Technikum and Geneva University.

Worked in Institut für industrialisiertes Bauen, Ulm, 1958–60. Official of the European Economic Commission of the UN, Geneva. Architect and designer at the Typisierungsbüro für städtische Bauten, Waschau, 1961–62; instructor in the Building department of the HfG, Ulm, 1962–68. Cofounder, chairman, and research director of the Institut de l'environnement, Paris, 1968–71. Consultant to UNESCO for the setting up of the Ecole d'Architecture et d'Urbanisme, Dakar, 1971–73. Professor at the Ecole d'architecture Paris-Villemin and Ecole nationale supérieure des arts décoratifs. Honorary doctorate from Hochschule für Architektur und Bauwesen, Weimar, 1983.

Research on history of modern architecture, Bauhaus, Functionalism, dwelling types, design methods. Numerous papers on technological, pedagogic, and cultural issues of environmental design. Author of *Hannes Meyer – Projekte, Bauten und Schriften,* 1965; *L'Age de la pierre,* 1980; *Umweltbürger und Umweltmacher,* 1982; *L'Imitation et l'invention,* 1983.

Lives in Paris.

Christian Staub
Born Menzingen, Switzerland, in 1918. Studied painting in Paris, 1938–40. Studied photography at Kunstgewerbeschule, Zürich, 1941–44, with H. Finsler; worked as press photographer in Switzerland, 1944–46, and in Vienna, 1946–48. Photography of Swiss watch industry for Foot, Cone & Belding agency, 1948–58.

Instructor in photography and documentary filmmaking at the HfG, Ulm, 1958–63; visiting professor at National Institute of Design, Ahmedabad, 1963–66, and University of California, Berkeley, 1966. Since 1967, Professor of Photography, University of Washington.

Represented in the following museums: Museum of Modern Art, New York; Institute of Art, Detroit; Museo de fotografa, Mexico City.

Photography for numerous architecture journals and for books: *Circus,* 1955; *Bienne — noir sur blanc,* 1957; *Fritz Wotruba,* 1947. Touring exhibition, *Nehru's India,* 1965. Work commissioned by Braun, Stuttgarter Gardinen, and Rosenthal. One-man exhibitions in Seattle, Zug, Lausanne, London, and elsewhere.

Lives in Seattle.

Friedrich Vordemberge-Gildewart
Born Osnabrück, 1899. Studied architecture and painting from 1919 at Kunstgewerbeschule and at Technische Hochschule, Hanover. Worked as painter, typographer, and journalist, from 1924 onward. Member of De Stijl and Der Sturm, 1925. Studio at the Kestner-Gesellschaft, Hanover. Contributor to avant-garde periodicals. Worked with Kurt Schwitters and was a founder member of the "abstraction — création" group, Paris, 1932. Took part in numerous international exhibitions.

Left Germany for Amsterdam, 1938. Taught at art academy, Rotterdam, from 1952. Instructor in Visual Communication department of HfG, Ulm, 1954–62.

Works in many major museums worldwide. Awards and distinctions include the Justus Moser Medal of the city of Osnabrück. Second Prize at São Paulo Biennale, 1953.

Died in Ulm in 1962.

Walter Zeischegg
Born Vienna, 1917. Studied at Baufachschule and Kunstgewerbeschule, Graz, and entered Akademie der bildenden Künste, Vienna, in 1936, to study sculpture. His studies were interrupted when he was drafted into the German army, and he resumed them in 1946 in Fritz Wotruba's master class. Worked as a sculptor and freelance product designer. Research into design morphology. Ergonomic studies. Through Carola Giedion-Welcker, who much admired his sculptures, he came into contact with Max Bill.

Worked on the planning stage of the HfG, 1951–54; instructor in department of Product Design, 1954–68. Designed entire product range for Helit. Extensive research into the applications of magnets; numerous patents, for live track for electric spotlights, one-way mixer faucets, and other items. Constant research and development work in the area of spherical geometry; many sculptures. A by-product of all this research was the ubiquitous Helit ashtray.

Awards and distinctions: Museum of Modern Art, New York; Design Center, Stuttgart; Corporate Design Award of the Hanover Trade Fair.

Died in 1983.

Assistants with Teaching Duties

Bernd Meurer 1963–68
Hans Roericht 1965–66
Günter Schmitz 1963–67

Workshop Leaders and Technical Instructors

Roland Fürst
Born 1939. Served apprenticeship as photographer. Workshop leader at HfG, 1964–67; subsequently freelance, with some teaching. Has taught photography at the Fachhochschule, Mannheim, since 1986.

Ernst Hahn
Born 1921. Leader of the photographic workshop, HfG, 1954–56. Then leader of photo-Lab Siemens Berlin.

Conrad Hildebrandt
Born 1920. Werkkunstschule Kiel. Leader of plastic workshop, HfG, 1959–68.

Paul Hildinger
Born Ulm, 1921. Served apprenticeship as cabinetmaker. Leader of wood workshop at HfG, 1953–67. Self-employed in Ulm since 1967.

Herbert Maeser
Born 1915. Served apprenticeship as compositor. Leader of the typographic workshop at HfG, 1960–68. At Institut für Umweltplanung, 1968–69. Taught at trade school, Memmingen, 1969–79. Teaching and freelance work.

Peter Matthes
Born 1937. Porcelain modeler. Studied at Akademie der bildenden Künste, Nuremberg. Leader of plaster workshop at HfG, 1964–68.

Otto Schild
Born Berlin, 1914. Employed from 1948 onward under Professor Wagenfeld at the Technische Hochschule, Stuttgart. Leader of plaster workshop at HfG, 1953–61. Professor at Hochschule für bildende Kunst, Hamburg, 1969–79.

Josef Schlecker
Born Ehingen, 1926. Served apprenticeship as toolmaker. Institute für Griff-Forschung, 1946–52. Leader of metal workshop, HfG, 1954–68. In the studio of Walter Zeischegg, 1968–85; at Prodesign, Neu-Ulm, since 1985.

Wolfgang Siol
Born Berlin, 1929. Served apprenticeship as photographer. Built up the photographic studio at the Heinrich Hertz Institut, Berlin. Leader of the photographic workshop, HfG, 1956–63. Guest instructor in photography at the HfG, 1963–64; then freelance. Instructor in photography at Fachhochschule, Hamburg, since 1980.

Cornelius Uittenhout
Born Leiden, Netherlands, 1921. Served apprenticeship as goldsmith. Leader of metal workshop, HfG, 1954–61. At WKS, Kassel, 1961–63; from 1963 as instructor. Since 1968, instructor in Product Form at the Academia, Eindhoven, Netherlands.

| Guest instructor | Subject | Period |
|---|---|---|
| Acosta, W. de | Building Construction | 59 – 60 |
| Albers, Josef | Basic Course | 53 – 55 |
| | | 58 – 59 |
| Alsleben, Kurd | Structural Theory | 65 – 68 |
| Altvater, Elmar | Political Economy | 67 – 68 |
| Ammende, Hans P. | Applied Physiology | 61 – 62 |
| Archer, L. Bruce | Product Design | 60 – 62 |
| Arndt, Hans Werner | Linguistics | 60 – 66 |
| Auer, Felix | House Technology | 66 – 68 |
| Authenried, Bernd | House Technology | 61 – 62 |
| Bahrick, Henry P. | Human Engineering | 59 – 60 |
| Baravalle, H. von | Basic Course | 54 – 59 |
| Bartels, Heinz | Ergonomics | 60 – 65 |
| Bauer, Konrad | Technical Construction | 58 – 59 |
| Baumann, Horst | Photography | 63 – 64 |
| Bechtle, Günter | Economics | 66 – 68 |
| Bellmann, Hans | Product Design | 53 – 54 |
| Bense, Max | Information | 54 – 58 |
| Berg, Arnljot | Film | 66 – 67 |
| Berghan, Wilfried | Film | 60 – 62 |
| Berns, Harald | Technical Construction | 60 – 61 |
| Birkel, Eberhard | Technical Physics | 67 – 68 |
| Blaser, Werner | Product Design | 55 – 57 |
| Bonetto, Rodolfo | Product Design | 61 – 65 |
| Bremer, Claus | Literature Seminar | 61 – 68 |
| Brückner, Peter | Social Psychology | 67 – 68 |
| Burandt, Ulrich | Ergonomics | 61 – 67 |
| Burckhardt, Lucius | Cultural Sociology | 58 – 59 |
| Burtin, Will | Visual Communication | 65 – 66 |
| Cincers, Raduz | Film | 66 – 67 |
| Ciribini, Giuseppe | Building | 58 – 59 |
| Cornelius, Peter | Visual Communication | 63 – 68 |
| Curjel, Hans | Cultural History | 53 – 57 |
| | | 61 – 62 |
| Czajka, Wladyslaw | Economics of Building | 63 – 68 |
| Dahrendorf, Ralf | Political Science | 61 – 63 |
| Dietz, A. | Building | 61 – 62 |
| Doermer, Christian | Film | 61 – 63 |
| Doernach, Rudolf | Building | 60 – 63 |
| Dörries, Bernhard | Film | 61 – 65 |
| | | 66 – 67 |
| Dressel, Gerhard | Building | 59 – 61 |
| Dressen, Peter | | 57 – 68 |
| Durst, Hermann | Technical Physics | 61 – 62 |
| Edelstein, Wolfgang | Film | 66 – 67 |
| Eichhorn, Gerhard | Building Technology | 57 – 59 |
| Emde, Helmut | Constructional Geometry | 61 – 68 |
| Enzensberger, H. M. | Information | 56 – 57 |
| Fabri, Albrecht | Information | 56 – 57 |
| Fetscher, Iring | Sociology | 60 – 61 |
| Fieger, Erwin Color | Photography | 63 – 64 |
| Fische, Kurt J. | Film | 60 – 62 |
| Fischer, Ludwig | Rhetoric | 67 – 68 |
| Frank, Helmar | Structure Theory | 63 – 65 |
| Franke, Wolf D. | Geometrical Rendering | 61 – 62 |
| Franzen, Erich | Sociology | 56 – 58 |
| Friedeburg, L. von | Sociology | 62 |
| Frei, Otto | Building | 58 – 60 |
| Frateili, Enzo | Building | 63 |
| Gerstner, Karl | Visual Communication | 61 – 63 |
| Gomringer, Eugen | Information | 56 – 57 |
| Gonda, Tomàs | Visual Communication | 63 – 66 |
| Gotterbarm, Paul | Technical Mechanics | 66 – 67 |
| Grandjean, Etienne | Physiology of Work | 57 – 59 |
| Gredinger, Paul | Visual Communication | 61 – 63 |

| Guest instructor | Subject | Period |
|---|---|---|
| Gregor, Ulrich | Film History | 61 – 63 |
| Gugelot, Hans | Product Design | 62 – 65 |
| Gunzenhäuser | Communications | |
| | Technology | 67 – 68 |
| Haan, Hermann | Building | 57 – 58 |
| Haenle, Siegfried | Technical Drafting | 57 – 68 |
| Hall, Chadwick | Photography | 65 – 66 |
| Hamburger, Käthe | Information | 56 – 58 |
| | | 61 – 62 |
| Hartmann, Erwin | Physiology | 61 – 67 |
| Hartmann, Horst | Product Design | 61 – 63 |
| Hauser, Heinrich | Technical Physics | 63 – 64 |
| Heck, Ludwig | Radio Technology | 61 – 62 |
| Heeger, Fritz | Social Psychology | 63 – 65 |
| Henne, Hermann | Physics of Building | 60 – 65 |
| Hennecke, Hans | Literary History | 60 – 62 |
| Hiestand, Ernst | Packaging | 67 – 68 |
| Hoffmann, Hans P. | Economics | 63 – 66 |
| Horisberger, Bruno | Applied Physiology | 57 – 59 |
| Huff, William S. | Basic Course | 63 – 68 |
| Irle, Martin | Psychology | 61 – 62 |
| Itten, Johannes | Basic Course | 54 – 55 |
| Jens, Walter | Literary History | 63 – 64 |
| Johnsch, Peter | Economics of Building | 67 – 68 |
| Kaiser, Joachim | Literary History | 58 – 61 |
| Kammerer, Guido | Economics | 66 – 68 |
| Kandel, L. | Architectural Criticism | 66 – 67 |
| Kapal, Ewald | Applied Physiology | 58 – 62 |
| Khittl, Ferdinand | Film | 61 – 62 |
| Knoll, Rudolf | Production Theory | 58 – 60 |
| Kotulla, Theodor | Film | 61 – 63 |
| Krammer, Gisela | Cultural History | 61 – 62 |
| Krampen, Martin | Social Psychology | 61 – 63 |
| | | 66 – 68 |
| Kückelmann, N. | General Law | 65 – 67 |
| Künzel, Helmut | Techical Physics, | |
| | Building | 63 – 68 |
| Küsgen, Horst | Economics of Building | 67 – 68 |
| Kutter, Markus | Visual Communication | 61 – 63 |
| Lachenmann, H. | Cultural History | 63 – 64 |
| | | 66 – 67 |
| Ladiges, Peter | Cultural History, | |
| | Literature | 61 – 66 |
| Lakatos, Bertalan | Technical Physics, | |
| | Building | 63 – 67 |
| Lehr, Albert | Economics | 61 – 66 |
| Leonhard, Michael | Building | 59 – 60 |
| Limberg, Klaus | Technical Drafting | 63 – 67 |
| Lindemann, Helmut | Political Science | 63 – 64 |
| Loeper, Hans | Film | 61 – 62 |
| Lussen, Eberhard | Physics of Building | 59 – 60 |
| Lutz, Burkart | Economics | 61 – 68 |
| Mackensen, Rainer | Sociology/Ecology | 60 – 61 |
| Mai, Wolfgang | | 67 – 68 |
| Mainka-Jellinghaus, B. | Film | 66 – 67 |
| Makowski, Z. S. | Building Construction | 63 – 64 |
| Martin, Bruce | Building | 57 – 60 |
| Mauch, Thomas | Literary History | 66 – 67 |
| McBride, Will | Photo Reportage | 65 – 66 |
| Minke, Gernot | Aerofoil Systems | 67 – 68 |
| Mitchell, Neal | Building | 65 – 66 |
| Morschel, Jürgen | Art History | 66 – 67 |
| Müller, Aemilius | Color Theory | 54 – 55 |
| Müller, Hans D. | Film | 63 – 68 |
| Müller-Brockmann, J. | Visual Communication | 61 – 63 |

| Guest instructor | Subject | Period |
|---|---|---|
| Nestle, Fritz | Physics | 63 – 64 |
| Neuburg, Hans | Visual Communication | 61 – 63 |
| Neumann, E. | Visual Communication | 65 – 67 |
| Neusel-Helvaci, A. | Aerofoil Theory | 67 – 68 |
| Nonné-Schmidt, H. | Basic Course | 53 – 58 |
| Norberg-Schulz, C. | Building | 57 – 58 |
| Nussbaum, Walter | | 67 – 68 |
| Oestreich, Dieter | Product Design | 57 – 58 |
| Oestreich, Herbert | Product Design | 60 |
| Passow, Wilfried | Literary History | 61 – 62 |
| Patalas, Enno Film | History | 62 – 63 |
| Patterson, James | Building | 65 – 66 |
| Pee, Herbert | Cultural History | 63 – 64 |
| Pelau, Stas | Domestic Technology | 63 – 65 |
| Perrine, Mervyn W. | Perception Theory | 58 – 61 |
| Pirker, Theo | Industrial Sociology | 62 – 64 |
| Pizzetti, Giulio | Physics of Building | 57 – 60 |
| Platschek, Hans | History of Painting | 62 – 63 |
| Podach, E. F. | Anthropology | 56 – 57 |
| Pörtner, Paul Radio | Production | 63 – 64 |
| Pross, Harry | Sociology | 61 – 63 |
| Pross, Helga | Sociology | 61 |
| Querngässer, Fritz | Visual Communication | 57 – 58 |
| | | 63 – 68 |
| Raacke, Peter | Product Design | 62 – 67 |
| Rago, Thomas | Photography | 57 – 58 |
| Rapp, Alfons | Domestic Technology | 61 – 66 |
| Rauch, Thomas | Film | 66 – 67 |
| Rautenstrauss, W. | Theory of Construction | 61 – 62 |
| Reichl, Ernst | Technical Construction | 66 – 68 |
| Reiher, Hermann | Physics of Building | 59 – 63 |
| Reinke, Wilfried | Film | 66 – 67 |
| Röll, Eduard | Film | 66 – 67 |
| Roelli, Werner | Visual Communication | 64 – 65 |
| Rohrberg, Klaus | Technical Physics | 67 – 68 |
| Roos, Hans Dieter | Film History | 65 – 66 |
| Rübenach, Bernhard | Radio | 57 – 61 |
| Ruehl, Raimond | Film | 61 – 62 |
| Ruge, Gerd | Political Science | 62 – 63 |
| Rykwert, Joseph | Building | 57 – 58 |
| Sader, Manfred | Psychology | 61 – 64 |
| Schatz, Paul | Geometry | 54 – 55 |
| Scheidegger, Ernst | Photography | 55 – 57 |
| Schiller, Wolfgang | Film | 62 – 63 |
| Schleiermacher, D. | Film | 62 – 66 |
| Schloemp, Petrus | Film Practice | 62 – 63 |
| Schober, Herbert | Physiology | 57 – 66 |
| Schön, Johannes | Technical Drawing | 57 – 58 |
| | | 59 – 60 |
| Schönfeld, Arnold | Packaging Technology | 60 – 61 |
| Schrenk, Martin | Economics | 62 – 63 |
| Schröter, Rolf | Applied Photography | 67 – 68 |
| Schütte, Wolfgang | Sociology of Building | 61 – 63 |
| Schütze, Christian | Communications Media | 59 – 60 |
| | | 62 – 63 |
| Schumacher, Karl | Photography | 63 – 64 |

| Guest instructor | Subject | Period |
|---|---|---|
| Schwennicke, Fritz | Film | 61 – 62 |
| Seitz, Fritz | Color Theory | 67 – 68 |
| Senft, Harro | Film | 61 – 63 |
| | | 64 – 66 |
| Siol, Wolfgang | Theory of Photography | 63 – 64 |
| Sörgel, Werner | Work Methodology | 65 – 68 |
| Sombart, Nicolaus | Philosophy | 61 – 62 |
| Spata, Jan | Film | 66 – 67 |
| Speidel, M. | Architectural Criticism | 65 – 66 |
| Sperlich, Hans G. | Cultural History | 57 – 61 |
| Spitzen, Franz J. | Film | 61 – 62 |
| Stachowiak, H. | Cybernetics | 64 – 65 |
| Stankowsky, Anton | Visual Communication | 62 – 63 |
| Stolper, Hans | Domestic Technology | 62 – 63 |
| Stritzinger, W. | Photography | 66 – 67 |
| Strobel, Hans Rolf | Film Practice | 62 – 63 |
| | | 64 – 65 |
| Sugiura, Kohei | Visual Communication | 64 – 67 |
| Sukopp, Hans | Film Technology | 60 – 61 |
| | | 67 – 68 |
| Sulzer, Peter | Building | 62 – 65 |
| Teller, H. | Production Technology | 61 – 62 |
| Thiele, Rolf | Radio Technology | 61 – 62 |
| Thornley, D. G. | Building | 60 – 61 |
| Tmin, Alfred | Building | 62 – 63 |
| Tonne, Friedrich | Technical Physics, | |
| | Building | 59 – 68 |
| Treinen, Heiner | Sociology | 63 – 65 |
| Tucny, Petr | Ergonomics | 66 – 68 |
| Uhde, J. | History of Music | 62 – 63 |
| Urchs, Wolfgang | Film Practice | 62 – 63 |
| Vesely, Herbert | Film | 58 – 59 |
| Vivarelli, Carlo | Visual Communication | 62 – 63 |
| Voss, Herbert von | Patent Law | 58 – 63 |
| Wachsmann, Konrad | Building | 54 – 57 |
| Wallis, Matthew | Building Methodology | 54 – 60 |
| Walser, Martin | Information | 57 – 58 |
| Walther, Elisabeth | Information | 56 – 58 |
| Wasowski, Z. | Domestic Technology | 59 – 62 |
| Wegner, Klaus | Psychology | 60 – 61 |
| Weller, Konrad | Building Construction | 60 – 68 |
| Wicha, Hansjörg | Filmmaking | 65 – 67 |
| Wilde, Monica | Film | 66 – 67 |
| Wirth, Wolf | Film Practice | 62 – 63 |
| Wormbs, Rudolf | Basic Town Planning | 67 – 68 |
| Zillmann, Adolf | Theories of Advertising | 62 – 66 |
| Zimmermann, G. | Sociology | 65 – 68 |

Visiting lecturers included: Reyner Banham, Gillo Dorfles, Charles Eames, Buckminster Fuller, Walter Gropius, Hugo Häring, S. I. Hayakawa, F. H. K. Henrion, Ludwig Mies van der Rohe, Alexander Mitscherlich, Hans Werner Richter, Norbert Wiener.

Students
The HfG was planned to take a maximum of 150 students in all departments over four years.

| | |
|---|---|
| Total enrollment 1953–68 | 640 |
| divided as follows: | |
| Product Design | 220 |
| Visual Communication | 135 |
| Building | 127 |
| Film | 23 |
| Information | 14 |
| Basic Course only | 121 |

Female students made up approximately 15 percent of the total.

Nearly half of the students came from outside Germany: Algeria, Argentina, Austria, Belgium, Brazil, Canada, Chile, Colombia, Finland, France, Great Britain, Greece, Hungary, India, Indonesia, Israel, Japan, Liberia, Mexico, Netherlands, New Zealand, Norway, Peru, Poland, South Africa, South Korea, Spain, Sweden, Switzerland, Thailand, Trinidad, Turkey, USA, Venezuela, Vietnam, Yugoslavia.

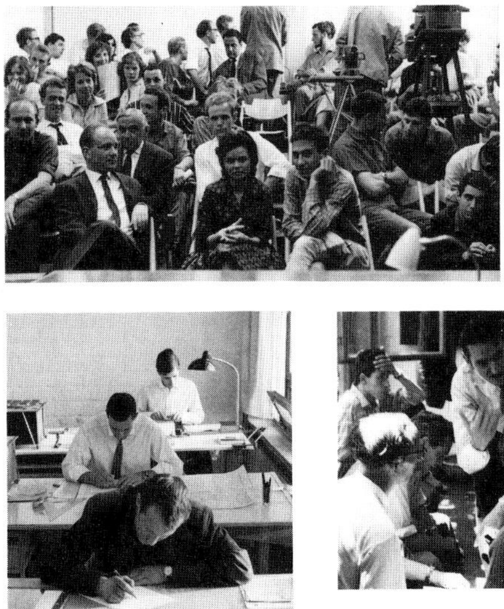
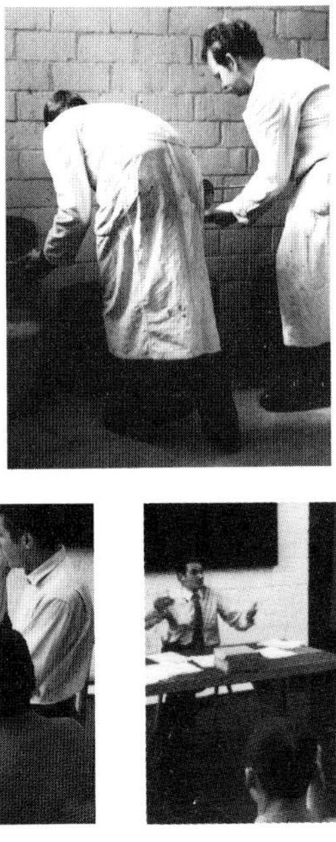

Statistics of Teaching Hours for 1958–59

| | Subject | Hours in each year 1 | 2 | 3 | 4 | Instructor |
|---|---|---|---|---|---|---|
| Shared | Methods of Rendering | 70 | | | | Fröshaug |
| | 20th-Century Cultural History | 140 | | | | Sperlich |
| | Math, Physics, Chemistry | 70 | | | | Eichhorn |
| | Mathematical Operational Analysis | | 35 | 35 | 12 | Rittel |
| | Methodology | 70 | | | | Rittel |
| | Sociology | 70 | 35 | | | Kesting |
| | Patent Law and Miscellaneous | | 22 | | 22 | |
| | Visual Method Exercises | 420 | | | | Fröshaug, Maldonado |
| | Workshop Work | 280 | | | | Hildinger, Schild, Schlecker, Siol, Uittenhout |
| | Theory of Science | 35 | | | | Rittel |
| Product Design | Departmental (D sign) | | 525 | 525 | 525 | Gugelot, Leowald, Zeischegg |
| | Applied Physiology | | 48 | 48 | | Grandjean, Horisberger |
| | Subject History Seminar | | 70 | | | |
| | Production Theory | | 105 | 105 | 36 | Knoll |
| | Mechanics | | 70 | | | Eichhorn |
| | Sociology | | 35 | | | Kesting |
| | Materials and Basic Techniques | | 105 | 105 | | Haenle |

| | Subject | Hours in each year 1 | 2 | 3 | 4 | Instructor |
|---|---|---|---|---|---|---|
| | Building Departmental (Design) | | 525 | 525 | 525 | Ohl |
| | Applied Physiology | | | 48 | 48 | Grandjean, Horisberger |
| | Statics of Construction | | 88 | 88 | 88 | Pizzetti |
| | Subject History Seminar | | 70 | 70 | | |
| | Production Theory | | 72 | 72 | 36 | Knoll |
| | Industrialized Building (Seminars) | | | | 20 | Ciribini, Martin, Frei Otto, Wallis |
| | Sociology | | | 35 | | Kesting |
| | Science of Materials | | 72 | 72 | 36 | Haenle |
| Visual Communication | Departmental (Design) | | 525 | 525 | 525 | Aicher, Fröshaug, Staub, Vordemberge-Gildewart |
| | Subject History Seminar | | 70 | 70 | | |
| | Semiotics | | 70 | 70 | | Maldonado |
| | Sociology | | | 35 | | Kesting |
| | Technology | | 105 | | | |
| Information | Departmental (Design) | | 525 | 525 | 525 | Kalow, Rübenach |
| | Photography, Film, Sound | | 105 | 105 | | Staub |
| | History, Organization Theory, Media | | 48 | 24 | | |
| | History of Modern Literature | | 22 | 46 | | Kalow |
| | Information Theory | | 35 | 35 | | |
| | Linguistics | | 22 | 22 | | |
| | Semiotics | | 48 | 48 | | Maldonado |
| | Sociology | | 35 | 70 | | Kesting |
| | Typography | | 24 | | | Aicher |

Statistics of Teaching Hours for 1966–67

| | Subject | Hours in each year 1 | 2 | 3 | 4 | Instructor |
|---|---|---|---|---|---|---|
| Shared | Ergonomics | 30 | | | | Burandt |
| | 20th-Century Cultural History | 70 | | | | Dörries, Kapitzki, Lachenmann, Lindinger, Morschel, Schnaidt |
| | Cybernetics | | 6 | 6 | | Moles |
| | Methodology | 60 | 60 | | | Alsleben, Moles |
| | Economics | 20 | | | | Lutz |
| | Politics | | 18 | | | |
| | Psychology | 30 | | | | Krampen |
| | Sociodynamics of Culture | | 30 | | | Moles |
| | Sociology | 10 | 72 | | | Lutz, Zimmermann |
| Product Design | Design Theory | | 20 | | | Bonsiepe, Maldonado |
| | Ergonomics | | 60 | 240 | | Burandt |
| | Design Work | 400 | 716 | 632 | | Bonsiepe, Lindinger, Raacke, Zeischegg |
| | Constructional Geometry | 120 | 56 | 60 | | Emde |
| | Modeling and Rendering | 256 | | | | Aicher, Fürst, Hildebrandt, Hildinger, Kapitzki, Schlecker, Stritzinger, Sugiura |
| | Sociology (Counseling) | | | 72 | | Lutz |
| | Technical Physics | 60 | 30 | | | Limberg |
| | Technical Design (Science of Materials, Theory of Construction) | 90 | 112 | 112 | | Haenle, Limberg, Reichl |
| Building | Scheduling | | | 16 | | Schmidt |
| | History of Building and Architectural Criticism | | 40 | | | Kandel |
| | Design Work | 390 | 568 | 854 | 408 | Ohl, Schmitz, Schnaidt, Wirsing |
| | Basic Town Planning | | | 32 | | Ginelli |
| | Domestic Technology | | 30 | | | Auer, Hartmann |
| | Industrialized Building | | | 120 | 40 | Yokohama |
| | Constructional Geometry | 60 | | | | Emde |
| | Modeling and Rendering | 268 | | | | Fürst, Hildebrandt, Hildinger, Maeser, Matthes, Schlecker, Schmitz, Stritzinger |
| | Modular Coordination | 24 | 24 | | | Schmitz |
| | Organization | | | 24 | | Niewerth |
| | Economics of Building | | | 40 | | Jokusch |
| | Sociology (Counseling) | | | 72 | | Lutz |
| | Technical Physics | | 46 | | | Künzel, Tonne, Rohrberg |
| | Technical Design: Production Theory, Construction, Science of Materials, Statics | 142 | 120 | | | Gotterbarm, Haenle, Weller |
| | Aerofoil Theory | | 40 | | | Minke |

| | Subject | Hours in each year 1 | 2 | 3 | 4 | Instructor |
|---|---|---|---|---|---|---|
| Visual Communication | Conceptualization | 30 | | | | Moles |
| | Design Analysis and Criticism | | 30 | 30 | | |
| | History of Communication | | 32 | 32 | | Neumann |
| | Design Work | | 632 | 752 | 222 | Aicher, Cornelius, Kapitzki, Krampen, Sugiura |
| | Communication Theory (Photo, Typo) | 160 | | | | Fürst, Maeser |
| | Communication and Rendering Techniques | 136 | | | | Aicher, Fürst, Kapitzki, Maeser, Stritzinger, Sugiura |
| | Psychology | | 60 | | | Krampen |
| | Sociology (Counseling) | | | 72 | | Lutz |
| | Communication Theory | 30 | 120 | 120 | | Krampen, Moles |
| | Visual Methodology (Exercises) | 672 | | | | Bonsiepe, Huff, Lindinger, Querngässer, Sugiura |
| Filmmaking | Educational Research; Education and Sociological Conceptualization | 35 | | | 35 | Edelstein |
| | German Film Agencies | 7 | | | 7 | Loewenthal |
| | Documentary Filmmaking | 28 | | | 35 | Cincera |
| | Scenario Writing for 2,500 Meter film | by arrangement | | | | |
| | Film History and Analysis | 84 | | | 84 | Mai |
| | Film Production Methods, Production Models | | | | 28 | Kluge, Reitz |
| | Film Exhibition | 7 | | | 7 | Backheuer |
| | Motion Picture Law, Copyright Law, Public Subsidies | 7 | | | 7 | Kückelmann |
| | Filmmaking and Public Responsibility | 7 | | | 7 | Krüger |
| | Filmmaking and Journalism | 7 | | | 7 | Gaus |
| | Film Distribution | 7 | | | 7 | Braun |
| | Camera Instruction | 70 | | | | Reitz, Spata |
| | Print Technique | 14 | | | 21 | Röll |
| | Practical Work as Assistant Cameraman | by agreement | | | | |
| | Production Technique with Blimp and Synchronized Sound | | | | 28 | Reitz |
| | Realization of Filmic Microstructures | | | | 42 | Berg, Kluge, Reitz |
| | Practical Editing Technique | 35 | | | | Berg |
| | Sociological and Political Themes | 70 | | | 91 | Kroher, Sörgel |
| | Scripts | | | | 35 | Müller |
| | Sound Studio, Sound Technique | 112 | | | 35 | Reitz, Wicha |

## Literature on the HfG Ulm


Arbeitsgruppe HfG-Synopse: *HfG-Synopse — eine synchro-noptische Darstellung der Hochschule für Gestaltung Ulm.* Edition HfG-Synopse N. H. Roericht, Ulm 1986.

Petra Kellner, Holger Poessnecker: *Produktgestaltung an der HfG Ulm. Designtheorie,* No. 3, Hanau 1978.

Alexander Kluge, Klaus Eder: *Ulmer Dramaturgien/Reibungs-verluste.* Carl Hanser Verlag, Munich/Vienna 1980.

Norbert Korrek: "Versuch einer Biographie, Die Hochschule für Gestaltung Ulm." Dissertation, Weimar University, 1984.

Martin Krampen, Horst Kächele, eds.: *Umwelt, Gestaltung und Persönlichkeit — Reflexionen 30 Jahre nach Gründung der Hochschule für Gestaltung Ulm.* Georg Olms Verlag, Hildesheim 1986.

Bernhard Rübenach: *Der rechte Winkel von Ulm,* ed. Bernd Meurer. Verlag der Georg Büchner Buchhandlung, Darmstadt 1987.

Eva von Seckendorf: "Die Hochschule für Gestaltung Ulm — Gründung und Ära Max Bill." Dissertation, Hamburg University, 1986.

Hartmut Seeling: "Geschichte der Hochschule für Gestaltung Ulm — Ein Beitrag zur Entwicklung ihres Programms und die Arbeiten im Bereich der visuellen Kommunikation." Dissertation, Cologne University, 1985.

Numerous accounts of the HfG in works on the history of architecture and design.


## Publications of the HfG


Catalogue of the HfG traveling exhibition. Design Center, Stuttgart 1963; Neue Sammlung, Munich 1964; Stedelijk Museum Amsterdam 1965.

*Ulm.* Magazine of the HfG, Nos. 1–21, Ulm 1958–68.

*Output.* Magazine of the students of the HfG, Nos. 1–22, Ulm 1961–69.


## Major Accounts in Specialist Periodicals


"La scuola di Ulm: organizzazione e metodo di lavoro"
*Comunit,* No. 72, Milan 1959.

"Hochschule für Gestaltung Ulm"
*Stile industria,* No. 21, Milan 1959.

"HfG Ulm. Ein Rückblick"
*Archithese,* No. 15, Niederteufen, Switzerland, 1975.

"Tre scuole: Bauhaus, Vchutemas, Ulm"
*Casabella,* No. 435, Milan 1984.

"Il contributo della scuola di Ulm"
*Rassegna,* No. 19, Milan 1984.

"Design aus Ulm — Gestaltung vor 30 Jahren"
*Westermanns Monatshefte,* No. 7, Brunswick 1985.


*who was who.* Address list of Ulm instructors and alumni, held by the "club of ulm."

Approximately 30 radio and television broadcasts.

This documentary record owes its existence very largely to the committed support of many Ulmers and friends in Ulm.

Most of the photographs taken during the life of the HfG were supplied by the then heads of the HfG's Photographic Workshop:
Ernst Hahn
Wolfgang Siol
Roland Fürst

For many photographs of Ulm life, we have to thank the devoted archivists:
Hans G. Conrad
Sisi von Schweinitz

Great assistance with dating was afforded by the *HfG-Synopse* compiled by:
Hans Roericht
Petra Kellner
Marcela Quijano
Sibille Riemann

Also the address list in the Club of Ulm's *who was who*

The following helped us with photographs, documents, negatives, manuscripts, and notes:
Inge and Otl Aicher
Mario Carrieri
Ello Delugan
Horst Eifert
ERCO
Roland Fürst
Ilse Grubrich-Simitis
Günther Hörmann
Peter Hofmeister
William S. Huff
Herbert Kapitzki
Walter Kiehlneker
Joe Klatt
Herbert Lindinger
Lufthansa
Tomàs Maldonado
Almir Mavignier
Bernd Meurer
Christoph Naske
Herbert Ohl
Andries van Onck
Michael Penck
Edgar Reitz
Foto Selinger
C. J. Uittenhout
Ursula Wenzel
Hans Werner
Erdmann Wingert

Almost without exception, the illustrations in the "Aftermath" section were supplied by the designers concerned.

Original works and models were lent by:
Reinhard Butter
Michael Conrad
Thomas Dawo
Günther Elstner
Susanne Eppinger-Curdes
Klaus Erler
Bernd Franck
Klaus Franck
Rolf Garnich
Ilse Grubrich-Simitis
Hans von Klier
Herbert Lindinger
Ernst Moeckl
Josef Müller-Brockmann
Bernhard Rübenach
Estate of Walter Zeischegg

We are also grateful for loans from the following institutions:
Stiftung Hochschule für Gestaltung Ulm
HfG Stadtarchiv Ulm
Bauhaus Archiv Berlin
Neue Sammlung München
The Museum of Modern Art, New York

Thanks due also to other Sponsors:
BASF
Arthur Braun
Max Braun-Stiftung
Erco
Helit
Lufthansa
Pfaff
Rosenthal
Wilkhahn

The following gave us the benefit of their advice:
Richard Bachinger
Max Bill
Bernhard E. Bürdek
Francois Burkhardt
Nicholas Chaparos
Michael Conrad
Kai Ehlert
Klaus Erler
Bernd Franck
Klaus Franck
Fred Hochstrasser
Günther Hörmann
Gert Kalow
Herbert Kapitzki
Hans von Klier
Karl-Heinz Krug
Stefan Lengyel
Tomàs Maldonado
Almir Mavignier
Bernd Meurer
Shutaro Mukai
Gerda Müller-Krauspe
Sudhakar Nadkarni
Herbert Ohl
Sisi von Schweinitz
Cornelia Vargas
Dr. Paolo Viti
C. W. Voltz
Dr. Hans Wichmann
Dr. Renzo Zorzi

Professor Herbert Lindinger
Born 1933, instructor at HfG Ulm, 1962–68
Since 1971, Professor at the Institut für Industrial Design (IID), Hanover University

Professor Egon Chemaitis
Born 1944, studied at the Hochschule der Kunst (HdK), West Berlin
Research associate at IID, Hanover, 1983–86
Since 1986, Professor at HdK, Berlin

Dr. Michael Erlhoff
Born 1946, studied German literature, sociology, art history
Author of several books, radio scripts, and texts, organizer of many exhibitions including the design section of *documenta 8*
Technical director of the Rat für Formgebung (Design Council)

Sibille Riemann
Born 1956, studied at HdK, Berlin
Member of HfG-Synopse working group
Since 1985, associate at IID, Hanover

Helmut Staubach
Born 1949, studied design in Schwäbisch Gmünd and at HbK, Kassel
Research associate at IID, Hanover, 1983–86
Now a freelance designer

First MIT Press edition 1991


This work originally appeared in German under the title *Hochschule für Gestaltung Ulm: Die Moral der Gegenstände.*


"The Ideology of a Curriculum" and "The Development of a Critical Theory" are reprinted from Kenneth Frampton's essay "Apropos Ulm: The Ideology of a Curriculum," *Oppositions,* no. 3, May 1974 (The MIT Press), by permission of the author.



Printed and bound in the Federal Republic of Germany.

Library of Congress Cataloging-in-Publication Data

Hochschule für Gestaltung Ulm. English
Ulm design: the morality of objects / edited by
Herbert Lindinger; translated by David Britt.
p. cm.
Translation of: *Hochschule für Gestaltung Ulm.*
Includes bibliographical references.
ISBN 0-262-12147-6
1. Design – Study and teaching – Germany – Ulm
I. Lindinger, Herbert, 1933 –
II. Hochschule für Gestaltung (Ulm, Germany)
III. Title
NK 430.G4H57813 1990
745.2'0943'473 -- dc20